D0930590

An Introduction to LOGIC and SCIENTIFIC METHOD

BY

MORRIS R. COHEN

Department of Philosophy, College of the City of New York

AND

ERNEST NAGEL

Department of Philosophy, Columbia University

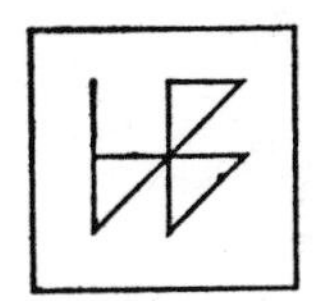

HARCOURT, BRACE & WORLD, INC.

NEW YORK AND BURLINGAME

PRINTED IN THE UNITED STATES OF AMERICA

PREFACE

Though formal logic has in recent times been the object of radical and spirited attacks from many and diverse quarters, it continues, and will probably long continue, to be one of the most frequently given courses in colleges and universities here and abroad. Nor need this be surprising when we reflect that the most serious of the charges against formal logic, those against the syllogism, are as old as Aristotle, who seems to have been fully aware of them. But while the realm of logic seems perfectly safe against the attacks from without, there is a good deal of unhappy confusion within. Though the content of almost all logic books follows (even in many of the illustrations) the standard set by Aristotle's *Organon*—terms, propositions, syllogisms and allied forms of inference, scientific method, probability and fallacies—there is a bewildering Babel of tongues as to what logic is about. The different schools, the traditional, the linguistic, the psychological, the epistemological, and the mathematical, speak different languages, and each regards the other as not really dealing with logic at all.

No task is perhaps so thankless, or invites so much abuse from all quarters, as that of the mediator between hostile points of view. Nor is the traditional distrust of the peacemaker in the intellectual realm difficult to appreciate, since he so often substitutes an unclear and inconsistent amalgam for points of view which at least have the merit of a certain clarity. And yet no task is so essential, especially for the beginner, when it is undertaken with the objective of adjusting and supplementing the claims of the contending parties, and when it is accompanied by a refusal to sacrifice clarity and rigor in thought.

In so far as an elementary text permits such a thing, the present text seeks to bring some order into the confusion of tongues concerning the subject matter of logic. But the resolution of the conflicts between various schools which it effects appears in the selection and presentation of material, and not in extensive polemics against any school. The book has been written with the conviction that logic is the autonomous science of the objective though formal conditions of valid inference. At the same time, its authors believe that the aridity which is (not always unjustly) attributed to the study of logic testifies to the unimaginative way logical principles have been taught and misused. The present text aims to combine sound logical doctrine with sound pedagogy, and to provide illustrative material suggestive of the rôle of logic in every department of thought. A text that would find a place for the realistic formalism of Aristotle, the scientific penetration of Peirce, the pedagogical soundness of Dewey, and the mathematical rigor of Russell—this was the ideal constantly present to the authors of this book.

However inadequately this ideal is embodied in the present text, the embodiment is not devoid of positive doctrine, so presented that at least partial justice is done to supplementary approaches to logic.

1. The traditional view of logic as the science of valid inference has been consistently maintained, against all attempts to confuse logic with psychology, where by the latter is meant the systematic study of how the mind works. Logic, as the science of the weight of evidence in all fields, cannot be identified with the special science of psychology. For such a special science can establish its results only by using criteria of validity employed in other fields as well. And it is clear that questions of validity are not questions of how we happen to think, but of whether that which is asserted is or is not in conformity with certain objective states of fact.

2. On the other hand, the pedagogical applications of psychological logics have not been ignored. We have aimed to present the subject in such a manner that discussion of doctrines new to the student is made continuous with his presumed knowledge at the outset. We have therefore avoided as far as possible the synthetic method of exposition: the method which begins with highly abstract elements and constructs a science out of them. Instead, we have followed what seems to us psychologically a more appropriate method. Illustrations with which a college student may reasonably be supposed to be familiar are usually taken as the text for discussion, and abstract, formal elements are gradually revealed as abstract phases of the subject matter. In this way, we trust, we have removed many of the difficulties which face the young student, and at the same time have indicated to him the important rôle played by logic in all of man's activities.

3. Again, while we have tried to present the significant results of symbolic or mathematical logic to those who have no previous

knowledge of the subject, we have not tried to develop the technique of symbolic manipulation for its own sake. In our opinion, such a technique, while very valuable, belongs properly to mathematics developed as an organon of science, and not to an elementary book on logic. Nor do we share the rather hostile attitude towards the Aristotelian logic expressed by some of the more zealous workers in the newer fields. We have not been sparing in indicating the limitations of the traditional presentation of our subject. But we think that the newer achievements in exact logic have served to extend as well as to correct the Aristotelian logic. We have thus given a great deal of attention to traditional views that might well be left out in a systematic presentation of our present knowledge. For we think that the discussion and correction of the limitations of the traditional views has many pedagogical advantages in making our final ideas clear.

4. We do not believe that there is any non-Aristotelian logic in the sense in which there is a non-Euclidean geometry, that is, a system of logic in which the contraries of the Aristotelian principles of contradiction and excluded middle are assumed to be true, and valid inferences are drawn from them. What have recently been claimed to be alternative systems of logic are different systems of notation or symbolization for the same logical facts. We have drawn freely on the natural sciences for illustrations of logical principles, precisely because the logical structure of these sciences is clearly more than linguistic. We have therefore frankly indicated the metaphysical significance of logical principles, and have not failed to note that the structure of language is itself often a clue to something other than linguistic fact. While maintaining that logic as an autonomous science must be formal, we have insisted that its principles are not therefore without significant content; on the contrary, we have taken the position that they are inherently applicable because they are concerned with ontological traits of utmost generality. We think that the category of objective possibility is essential to logical discussion.

In the main, therefore, we view the history of logic as that of a series of contributions of diverse value by the various schools. If our point of view is consequently somewhat eclectic, seeking to give the student a liberal rather than a narrow view of the subject, we have nevertheless striven hard to maintain clear distinctions as to fundamentals. Florence Nightingale transformed modern hospital practise by the motto: Whatever hospitals do, they should not spread disease. Similarly, logic should not infect students with fallacies and confusions as to the fundamental nature of valid or scientific reasoning.

Different instructors will naturally attach more value to different parts of the book. Not all of it can be presented in a one-semester course, and enough material has been included to occupy the student's attention for a full year. In a one-semester course, the authors have found that the substance of Book II, with the inclusion of Chapters III, IV, and VIII of Book I, gives the most satisfactory results. Those not interested in mathematics may omit Chapter VII. Books are tools which wise men use to suit their own ends. One of the authors, who has given courses in elementary logic for over twenty years, has generally treated the contents of Book II (Applied Logic and Scientific Method) before the formal logic of Book I. There are, to be sure, some topics in Book II which presuppose the solutions of Book I. But experience shows that such difficulties are readily surmountable. It is especially the hope of the authors that general readers as well as students of the natural and social sciences will find this book helpful towards an understanding of scientific method.

M. R. C.
E. N.

The continued demand for this book, which has exhausted three printings of it, has given us a chance to correct certain errors and to revise some statements in the interest of greater clarity.

M. R. C.
E. N.

January 7, 1936

CONTENTS

VII. THE NATURE OF A LOGICAL OR MATHEMATICAL SYSTEM

VIII. PROBABLE INFERENCE

IX. SOME PROBLEMS OF LOGIC

BOOK II: APPLIED LOGIC AND SCIENTIFIC METHOD

XI. HYPOTHESES AND SCIENTIFIC METHOD

XII. CLASSIFICATION AND DEFINITION

XIII. THE METHODS OF EXPERIMENTAL INQUIRY

XIV. PROBABILITY AND INDUCTION

XV. MEASUREMENT

XVI. STATISTICAL METHODS

XVII. PROBABLE INFERENCE IN HISTORY AND ALLIED INQUIRIES

XVIII. LOGIC AND CRITICAL EVALUATION

XIX. FALLACIES

XX. CONCLUSION

AN INTRODUCTION
TO LOGIC AND SCIENTIFIC METHOD

CHAPTER I

THE SUBJECT MATTER OF LOGIC

§ 1. LOGIC AND THE WEIGHT OF EVIDENCE

Most of our daily activities are carried on without reflection, and it seldom occurs to us to question that which generally passes as true. We cannot, however, always remain in a state of unquestioned belief. For our habitual attitudes are frequently challenged by unexpected changes in our environment, if they are not challenged by our own curiosity or by the inquisitiveness of others.

Let us suppose the reader to be seated at his table some late afternoon. The gathering darkness is making his reading difficult. Ordinarily he would turn on the electric light near him and continue with his work. But on this occasion, we suppose, the Shade of Socrates suddenly appears to the busy reader, just as his hand is on the switch, and asks him to please tell what he is doing. The reader has stout nerves, and quickly recovering from his surprise, explains: "I wish to put on the light, and this is the switch. Since your day . . ." "Yes, yes," we can imagine the Shade to interrupt, "I know all about your modern methods and theories of lighting. You needn't take time to tell me about *that.* But I do wish you would tell me how you know that it is the electric switch you were just pointing to." The reader's temper may by this time have been thoroughly ruffled, and after an embarrassed silence, he may reply with pained surprise and some asperity: "Can't you see, Socrates?"—and turn on the light.

What is of interest to us in this imaginary dialogue is that a doubt, however slight, might be raised in the reader's mind about a proposition *This is the electric switch,* which had previously been accepted without question; and that the doubt might be resolved by claiming that any evidence besides *seeing* was superfluous. There are other propositions for which it would be difficult to find any evidence other than a direct seeing, hearing, touching, or smelling. *It is half-past eleven on my watch; My forehead is hot to the touch; This rose I am smelling has a fine fragrance; The shoes I have on are uncomfortable; That is a loud noise.* These are examples of propositions on account of which most of us would lose our tempers if we were pressed to give reasons why we believed them to be true.

Not all propositions, however, are regarded as so obvious. If the Shade should accost the reader entering the office of a life insurance company, and ask him what he is about, the reader might perhaps say: "I am going to buy a life insurance policy." Should the reader be pressed for his motives, a possible answer might be: "I shall die some day, and I wish to provide for my dependents." If Socrates should now demand why the reader believes in the truth of the proposition *I shall die some day,* the answer will no longer be, "Can't you see?" For we cannot literally see our own future death. But a little reflection may suggest the following reply: "All living creatures, O Socrates, must perish some day, and since I too am a living creature, I too shall die some day."

There are propositions, therefore, which we believe to be true because we can find some *other* propositions of whose truth we have no doubt and which we think will serve as *evidence* for the disputed proposition. *The sun is approximately ninety-three million miles away; Caesar crossed the Rubicon; There will be an eclipse of the sun next year in North America; The sum of the angles of a triangle is equal to two right angles.* These are a few propositions in whose truth we may believe because we think others, if not we ourselves, can find supporting propositions for them.

The distinction between propositions which are believed without grounds other than direct observation or apprehension and propositions which are believed because other propositions can be found to serve as evidence for them, cannot always be drawn very sharply. We sometimes believe a proposition to be true partly because we can make direct observations and partly because we can find supporting propositions. If we drop two rocks of unequal weight from the same height at the same time, we believe the proposition, *The two rocks strike the ground at the same time,* not only because we *see* that they do, but because we know a *reason* why they should do so. Moreover, many propositions whose truth seems very clear to us are in fact false. For we often see what we expect to see rather than that which actually happens. Many remarkable advances in knowledge have resulted from our questioning the truth of propositions which we previously regarded as "self-evident." And a critical study of human beliefs reveals how much "interpretation" is present in what at first sight seems like "immediate knowledge." But it is not necessary for our present purpose to settle the question as to what propositions, if any, can be known to be true "immediately."

All that we now require is the recognition of the general need of evidence for what we or others believe or question. In scientific or historic research, in courts of law, and in making up our minds as to all sorts of practical issues, we are constantly called upon to pass on diverse considerations offered in support of various propositions at issue. Sometimes we find such considerations to be irrelevant and to constitute no evidence at all, even though we have no doubt as to their truth, while other propositions we regard as conclusive or demonstrative proof of a point at issue. Between these two extremes we have situations in which there is some testimony or circumstance that points to a given conclusion but is not sufficient to exclude some alternative possibility. For most occasions we are satisfied with a preponderance of evidence, that is, if there is more evidence in favor of a proposition than against it; but in some cases, for example, when as jurymen we pass on the guilt of one accused of a crime, we are required to act affirmatively only if there is no reasonable doubt left, that is, no doubt which a "reasonably" prudent man would act on in the course of his affairs.

Logic may be said to be concerned with the question of the adequacy or probative value of different kinds of evidence. Traditionally, however, it has devoted itself in the main to the study of what constitutes proof, that is, complete or conclusive evidence. For, as we shall see, the latter is necessarily involved in determining the weight of partial evidence and in arriving at conclusions that are said to be more or less probable.

§ 2. CONCLUSIVE EVIDENCE OR PROOF

Let us consider the proposition *There are at least two persons in New York City who have the same number of hairs on their heads,* and let us symbolize it by q. How could its truth be established? An obvious way would be to find two individuals who actually do have the same number of hairs. But this would require an extremely laborious process of examining the scaips of perhaps six million people. It is not a feasible method practically. We may be able to show, however, that the proposition q follows from or is necessitated by other propositions whose truth can be established more easily. In that event, we could *argue* for the truth of the proposition q, in virtue of its being *implied* by the others, and in virtue of the established truth of the propositions offered as evidence. Let us try this method.

Suppose it were known by an actual count that there are five thousand barber shops in New York City. Would the proposition *There are five thousand barber shops in New York City* be satisfactory evidence for q? The reader will doubtless reply, "Nonsense! What has the number of barber shops to do with there being two persons with an identical number of scalp hairs?" In this way the reader expresses the judgment (based on previous knowledge) that the number of barber shops is no evidence at all for the equality in the number of hairs. Not all propositions are *relevant,* even if true, to the truth of a proposition in question.

Let us now consider the proposition *The number of inhabitants in New York City is greater than the number of hairs that any one of its inhabitants has on his head.* We shall denote this proposition by p. Is the truth of p sufficient to establish the truth of q? The reader might be inclined to dismiss p, just as he dismissed the information about the number of barber shops, as irrelevant. But this would be a mistake. We can show that if p is true, q must be true also. Thus suppose, taking small numbers for purposes of illustration, that the greatest number of hairs that any inhabitant of New York City has is fifty, and that there are fifty-one people living in New York City, no one of whom is completely bald. Let us assign a number to each inhabitant corresponding to the number of hairs that he has. Then the first person will have one hair, the second person two hairs, and so on, until we reach the fiftieth person, who will have, at most, fifty hairs. There is one inhabitant left and, since we have assumed that no person has more than fifty hairs, he will necessarily have a number of hairs that is the same as that possessed by one of the other fifty persons. The argument is perfectly general, as a little reflection shows, and does not depend on the number fifty we have selected as the maximum number of hairs. We may, therefore, conclude that our proposition p, *The number of inhabitants in New York City is greater than the number of hairs that any one of its inhabitants has on his head,* implies proposition q, *There are at least two persons in New York who have the same number of hairs on their heads.* The two propositions have been shown to be so related that it is impossible for the first (called the *evidence* or *premise*) to be true, and the second (called the *conclusion* or *that which is to be proved*) to be false.

Other instances of conclusive evidence can be multiplied indefinitely. Thus we can prove that a missing individual is dead by showing that he sailed on a boat destroyed at sea by an explosion that prevented anyone from being saved. So we can prove that our neighbor, Mr. Brown, has no right to vote by showing that he is not yet twenty-one years of age and that the law prohibits such individuals from voting.

Mathematics is, of course, a field in which proof is essential. A distinction, however, must be noted in this respect between applied and pure mathematics. In the former, as in the examples already mentioned, we assume that certain propositions, for example, the laws of mechanics, are *true;* and we prove the *truth* of other propositions by showing that they necessarily follow or are mathematically deducible from those assumed. In pure mathematics, on the other hand, we restrict ourselves to demonstrating that our primary assumptions necessarily imply or entail the theorems which are deduced from them, and ignore the question whether our conclusions as well as our axioms or postulates are in fact true.

It might be of some advantage to use the word "proof" for the former procedure (by which we conclude a proposition to be *true*), and to designate by "deduction" or "demonstration" the procedure which only establishes an *implication* or *necessary connection* between a premise and its conclusion irrespective of the truth or falsity of either. Such a terminology would permit us to say that a proposition is *proved* when, and only when, a premise *implies* that proposition and that premise is itself *true*. But so habitual is the usage which speaks of "proving" theorems in pure mathematics that it would be vain to try to abolish it. It is therefore safer to continue to speak of "proof" in pure mathematics, but to recognize that what we prove there are always implications, that is, that *if* certain propositions are true, certain others must be true. And this, after all, is the phase of all proof in which logic is primarily interested.

In all cases, then, of complete evidence or proof the conclusion is implied by the premises, and the reasoning or inference from the latter to the former is called *deductive*. We *infer* one proposition from another *validly* only if there is an objective relation of *implication* between the first proposition and the second. Hence, it is essential to distinguish *inference,* which is a temporal process, from *implication,* which is an objective relation between propositions. An implication may hold even if we do not know how to infer one proposition from another. Thus an inference to be valid requires that there be an implication between propositions. On the other hand, the being of an implication does not depend upon the occurrence of the psychological process of inferring.

§ 3. THE NATURE OF LOGICAL IMPLICATION

In every attempt at a complete proof of propositions of practical importance we thus find two questions involved:

1. Are the propositions offered as evidence true?
2. Are the conclusions so related to the evidence or premises that the former necessarily follow from and may thus be properly deduced from the latter?

The first question raises what is called a factual or material issue; and the answer to it cannot be assigned entirely to logic without making the latter include all the sciences and all common knowledge. Logic as a distinctive science is concerned only with the second question—with the relation of *implication* between propositions. Thus the specific task of logic is the study of the conditions under which one proposition necessarily follows and may therefore be deduced from one or more others, regardless of whether the latter are in fact true.

As any number of propositions can be combined into one, every instance of implication or logical sequence can be said to hold between two propositions, which might be most accurately designated as the *implicating* and the *implied,*[1] but are generally called *antecedent* and *consequent,* as well as *premise* and *conclusion.* We must, however, note that in using the terms "antecedent" and "consequent," or the expression, "It logically follows," we are referring to an abstract relation which, like that between whole and part, does not directly refer to any temporal succession. The logical consequences of a proposition are not phenomena which follow it in time, but are rather parts of its meaning. While our apprehension of premises sometimes precedes that of their conclusion, it is also true that we often first think of the conclusion and then find premises which imply it.

Let us consider this relation of implication a little more closely.

[1] In grammar they are known as the *protasis* and *apodosis* of a subjunctive sentence.

Logical Implication Does Not Depend on the Truth of Our Premises.

The specific logical relation of implication may hold (1) between false propositions or (2) between a false and a true one, and (3) may fail to hold between true propositions.

1. Consider the argument *If Sparta was a democracy and no democracy has any kings, it follows that Sparta had no king.* The falsity of the proposition, *Sparta was a democracy,* does not prevent it from having certain implications nor from determining definite logical consequences.

No argument is more common in daily life than that which draws the logical implications of hypotheses contrary to fact. If there were no death there would be no cemeteries, funeral orations, and so on. All our regrets are based on drawing the consequences of propositions asserting what might have been but did not in fact happen.

> "Had we never loved sae kindly,
> Had we never loved sae blindly,
> Never met or never parted,
> We had ne'er been broken-hearted!"

It is a great error to suppose, as many have unthinkingly done, that in the reasoning we call scientific we proceed only from facts or propositions that are true. This view ignores the necessity for deduction from false hypotheses. In science as well as in practical choices, we are constantly confronted with alternative hypotheses which cannot all be true. Is the phenomenon of burning to be explained by the emission of a substance called phlogiston or by the combination with one called oxygen? Does magnetism act at a distance like gravitation, or does it, like sound, require a medium? We generally decide between such conflicting propositions by deducing the consequences of each and ruling out as false that hypothesis which leads to false conclusions, that is, to results which do not prevail in the field of observable fact. If false hypotheses had no logical consequences we should not thus be able to test their falsity.

That a proposition has definite logical consequences even if it is false follows also from the fact that these logical consequences or implications are part of its meaning. And we must know the meaning of a proposition before we can tell whether it is true. But in all cases (whether a proposition is true or false) the test as to whether there is a logical implication between one proposition and another is the impossibility of the former being true and the latter being false.

2. There is a widespread impression to the effect that false premises must logically lead to propositions that are false. This is a serious error, probably due to a thoughtless confusion with the true principle that if the consequences are false the premises must be false. But that true consequences may be implied by (or logically follow from) false premises can be seen from the following simple examples:

If all Mexicans are citizens of the United States and all Virginians are Mexicans, it logically follows that all Virginians are citizens of the United States. If all porpoises are fishes and all fishes are aquatic vertebrates, it necessarily follows that porpoises are aquatic vertebrates. (The same conclusion follows if all porpoises are mollusks and all mollusks are aquatic vertebrates.) For again the relation between the antecedents and the consequents is such as to rule out the possibility of the former being true and the latter at the same time false.

Of course if a premise is false, the conclusion is *not proved* to be true even though the conclusion is implied by the premise. But it is of the utmost importance to realize that a proposition is not necessarily false, or proved to be so, if an argument in its favor is seen to rest on falsehood. A good cause may have bad reasons offered in its behalf.

3. We have already seen that the proposition *There are five thousand barber shops in New York City,* even if true, is irrelevant to and cannot prove or logically imply the proposition *There are at least two persons in New York City who have the same number of hairs on their heads.* Let us, however, take an instance in which the absence of logical connection or implication is perhaps not so obvious. Does the proposition *Perfect beings can live together without law and men are not perfect* imply *Men cannot live together without law?* Reflection shows that nothing in the premise rules out the *possibility* of there being men who, though not perfect, live together without law. We may be able, on other grounds, to prove that our conclusion is true, but the evidence here offered is not sufficient. There is no necessary connection shown between it and that which is to be proved.

Logical Implication Is Formal.

The fact that the logical implications of a proposition are the same whether it happens to be true or not, and that the validity of such implications is tested by the *impossibility* of the premise being true and its consequences false, is closely connected with what is called the formal nature of logic.

What do we mean by *formal?* The reader has doubtless had occasion to fill out some official blank, say an application for some position, a lease, a draft, or an income-tax return. In all these cases the unfilled document is clearly not itself an application, lease, draft, or tax return; but every one of these when completed is characterized by conforming to the pattern and provisions of its appropriate blank form. For the latter embodies the character or fixed order which all such transactions must have if they are to be valid. A form is, in general, something in which a number of different objects or operations agree (though they differ in other respects), so that the objects may be varied and yet the form remain the same. Thus any social ceremony or act which diverse individuals must perform in the same way if they occupy a given position or office, is said to be formal. Similarly, logical implication is formal in the sense that it holds between all propositions, no matter how diverse, provided they stand to each other in certain relations. Consider any of the foregoing instances of proof, such as *Brown is a minor; all minors are ineligible to vote; therefore Brown is ineligible to vote.* The implication here does not depend on any peculiarity of Brown other than the fact that he is a minor. If any other person is substituted for Brown the argument will still be valid. We can indicate this truth by writing X *is a minor, all minors are ineligible to vote, therefore* X *is ineligible to vote,* where X stands for anyone of an indefinitely large class. Reflection shows that we can also replace the word "minor" with any other term, say, "felon," "foreigner," without invalidating the argument. Thus if X *is a* Y, *and all* Y*'s are ineligible to vote, then* X *is ineligible to vote,* no matter what we substitute for *Y*. We can now take the third step and realize that the logical implication is not only independent of the specific character of the objects denoted by *X* and *Y*, but that the term "ineligible to vote" might be replaced by anything else (*provided it is the same in premise and conclusion*). Thus we get the formula: *If* X *is* Y *and all* Y*'s are* Z*'s, then* X *is a* Z as true in all cases no matter what X, *Y*, and *Z* denote. On the other hand it would be an error to assert that if *All Parisians are Europeans and all Frenchmen are Europeans,* it follows that *All Parisians are Frenchmen.* For if in the generalized form of this argument, *All* X*'s are* Y*'s, and all* Z*'s are* Y*'s, therefore all* X*'s are* Z*'s,* we substitute "Belgians" for "Parisians," we get an argument in which the premises are true but the conclusion false. Similarly we can assert the implication that

If Socrates is older than Democritus and Democritus is older than Protagoras, then Socrates is older than Protagoras. For this will hold no matter what persons are substituted for these three, provided we keep the form, X *is older than* Y *and* Y *is older than* Z *implies* X *is older than* Z. On the other hand, from the proposition: A *is to the right of* B, *and* B *is to the right of* C, it does not necessarily follow that A *is to the right of* C. For if three men are sitting in a circle *A* can be said to be to the left of *C* even though he is to the right of *B* and the latter to the right of *C*. It is the object of logical study to consider more detailed rules for distinguishing valid from invalid forms of argument. What we need to note at present is that the correctness of any assertion of implication between propositions depends upon their form or structure. If any form can be filled by premises which are true and conclusions which are false, that form is invalid, and the assertion of an implication in any such case is incorrect.

Two observations must be added to the foregoing:

1. The more general statement or formula is not a constraining force or imperative existing before any special instance of it. An argument is valid in virtue of the implication between premises and conclusion in any particular case, and not in virtue of the general rule, which is rather the form in which we have abstracted or isolated what is essential for the validity of the argument. For the objects which enter into propositions are related in a certain way, and a form is an arrangement; hence an implication which holds for one arrangement of objects will not hold for another.

2. This formal character of implication (and thus of valid inference) does not mean that formal logic ignores the entire meaning of our propositions. For without the latter we can have only meaningless marks or sounds—not significant assertions or information having logical consequences. However, the fact that logic is concerned with necessary relations in the field of possibility makes it indifferent to any property of an object other than the function of the latter in a given argument. Formal properties must be common to all of a class.

Logical Implication as Determination

We have thus far considered the nature of logical implication from the point of view which regards it as an element in all proof or conclusive evidence. We may, however, also view it as entering into every situation or problem in which certain given conditions are sufficient to determine a definite result or situation. Take, for instance, the familiar problem: How long is the interval between the successive occasions when the hands of a clock are together? When the relative velocities of these hands are given, the value of the resulting interval is uniquely determined by the relation of logical implication—though if we are untrained in algebra, it may take us a long time before we see how to pass from the given conditions as premises to the conclusion or solution which they determine. The process of exploring the logical implications is thus a form of research and discovery. It should be noted, however, that it is not the business of logic to describe what happens in one's mind as one discovers rigorous or determinate solutions to a problem. That is a factual question of psychology. Logic is relevant at every step only in determining whether what *seems* an implication between one proposition and another is indeed such. Logic may, therefore, be also defined as the science of implication, or of valid inference (based on such implication). This may seem a narrower definition of logic than our previous one, that logic is the science of the weight of evidence. For implication as we have discussed it seems restricted to conclusive evidence. Reflection, however, shows that deductive inference, and hence the implication on which it ought to be based, enters into all determination of the weight of evidence.

§ 4. PARTIAL EVIDENCE OR PROBABLE INFERENCE

We have so far discussed the relation between premises and conclusion in case of rigorous proof. But complete or conclusive evidence is not always available, and we generally have to rely on partial or incomplete evidence. Suppose the issue is whether a certain individual, Baron X, was a militarist, and the fact that most aristocrats have been militarists is offered as evidence. As a rigorous proof this is obviously inadequate. It is clearly possible for the proposition *Baron X was a militarist* to be false even though the proposition offered as evidence is true. But it would also be absurd to assert that the fact that most aristocrats are militarists is altogether irrelevant as evidence for Baron X having been one. Obviously one who continues to make inferences of this type (Most *X*'s are *Y*'s, *Z* is an *X*, therefore *Z* is a *Y*) will in the long run be more often right than wrong. An inference of this type, which from true premises gives us conclusions which are true in most cases, is called *probable*. And the etymology of the word (Latin *probare*) indicates that such evidence is popularly felt to be a kind of proof even though not a conclusive one.

The reader will note that where the evidence in favor of a proposition is partial or incomplete, the probability of the inference may be increased by additional evidence. We shall in a later chapter consider with some detail when and how we can measure the degree of probability, and what precautions must be taken in order that our inferences shall have the maximum probability attainable. Here we can only briefly note that deductive inference enters as an element in such determination. To do this, let us consider first the case where a probable argument leads to a *generalization* or *induction,* and secondly, the case where such argument leads to what has been called a *presumption of fact.*

Generalization or Induction

Suppose we wish to know whether a certain substance, say, benzoate of soda, is generally deleterious. Obviously we cannot test this on everybody. We select a number of persons who are willing to take it with their food and whom we regard as typical or representative specimens of human beings generally. We then observe whether the ingestion of this substance produces any noticeable ill effect. If in fact all of them should show some positive ill effect we should regard that as evidence for the general proposition *Benzoate of soda is deleterious.* Such generalizations, however, frequently turn out to be false. For the individuals selected may not be typical or representative. They may have all been students, or unusually sensitive, or used to certain diets, or subject to a certain unnoticed condition which does not prevail in all cases. We try to overcome such doubt by using the inferred rule as a premise and deducing its consequences as to other individuals living under different conditions. Should the observed result in the new cases agree with the deduction from our assumed rule, the probability of the latter would be increased, though we cannot thus eliminate all doubt. On the other hand, should there be considerable disagreement between our general rule and what we find in the new cases, the rule would have to be modified in accordance with the general principle of deductive reasoning. Thus while generalizations from what we suppose to be typical cases sometimes lead to false conclusions, such generalizations enable us to arrive frequently at conclusions which are true in proportion to the care with which our generalization is formulated and tested.

Presumption of Fact

The second form of probable inference (which we have called presumption of fact) is that which leads us to deduce a fact not directly observable. Suppose that on coming home we find the lock on the door forced and a letter incriminating a prominent statesman missing. We believe in the general rule that violators of the law will not hesitate to violate it still further to save themselves from punishment. We then infer that the prominent statesman or his agents purloined the letter. This inference is obviously not a necessary one. Our evidence does not preclude the possibility that the letter was stolen by someone else. But our inference is clearly of the type that often leads to a right conclusion; and we increase this probability, when we show that if someone else than the one interested in the contents of the letter had stolen it, other valuables would have disappeared, and that this is not the fact.

Let us take another case. Suppose, for instance, we notice one morning that our instructor is irritable. We may know that headaches are accompanied by irritability. Consequently we may conclude that our instructor is suffering from a headache. If we examine this argument we find that the evidence for our conclusion consists of a proposition asserting the existence of a particular observable state of affairs (the instructor is irritable) and of another proposition asserting a rule or principle which may be formulated either as *All headaches are accompanied by irritability,* or *Many cases of irritability are due to headaches.* In neither case does our conclusion *The instructor has a headache* follow necessarily. His irritability may in fact be due to some other cause. But our inference is of the type that will lead to a true conclusion in a large number of cases, according to the extent to which irritability is connected with headaches. And we test the truth of the latter generalization (or induction) by deducing its consequences and seeing whether they hold in new situations.

This form of inference is so widespread, not only in practical affairs but even in advanced natural science, that illustration from the latter realm may be helpful. Various substances like oxygen, copper, chlorine, sulphur, hydrogen, when they combine chemically do so according to fixed ratios of their respective weights; and when the same amount of one substance, say chlorine, combines with different amounts of another substance, say oxygen, to form different compounds, it does so in ratios that are all small integral multiples of one. (This is the *observed event.*) We know also that if each of these substances were composed of similar atoms, or mechanically indivisible particles, the substances would combine in such integral ratios of their weights. (This is the *general rule.*) We conclude, then, that these substances are composed of atoms. (This is the *inferred fact.*) From the point of view of necessary implication the inference in this case is invalid. For it is possible that the observable facts may be due to an altogether different general cause than the assumed atomic constitution of matter. But our evidence has a very high degree of probability because we have used the general proposition (Matter has an atomic structure) as a premise from which to deduce all sorts of consequences that have been found true by direct observation and experiment—consequences which have also been shown to be inconsistent with other known assumptions.

This is also the character of the evidence for such everyday generalizations as that bread will satisfy our hunger, that walking or taking some conveyance will get us to our destination. These generalizations are not universally true. Accidents unfortunately happen. Bread may disagree with us, and he who walks or rides home may land in the hospital or in the morgue. All of our life, in fact, we depend on using the most probable generalizations. If our friend should refuse to walk on wooden or concrete floors because it has not been absolutely proved that they might not suddenly disintegrate or explode, we should feel some concern about his sanity. Yet it is unassailably true that so long as we lack omniscience and do not know all of the future, all our generalizations are fallible or only probable. And the history of human error shows that a general consensus, or widespread and unquestioned feeling of certainty, does not preclude the possibility that the future may show us to be in error.

§ 5. IS LOGIC ABOUT WORDS, THOUGHTS, OR OBJECTS?

Logic and Linguistics

While it seems impossible that there should be any confusion between a physical object, our "idea" or image of it, and the word that denotes it, the distinction is not so clear when we come to complexes of these elements, such as the government or literature of a country. As the logical inquiry into the implication of propositions is not directly concerned with physical or historical fact, the view naturally arises that it is concerned exclusively with words. The etymology of the word "logic" (as in "logomachy") supports this view. The great English philosopher Hobbes speaks of logic and reason as "nothing but reckoning, that is, adding and subtracting, of the consequences of general names agreed upon."[2] We must not, however, confuse the fact that words or symbols of some sort are necessary for logic (as for all the advanced sciences) with the assertion that valid reasoning is *nothing but* a consequence of the act of naming. For we can change the names of things, as we do when we translate from one language into another, without affecting the logical connections between the objects of our reasoning or "reckoning." The validity of our reasoning depends on the consistency with which we use whatever language we have, and such consistency means that our words must faithfully follow the order and connection of the items denoted by them. Logic, like physics or any other science, starts from the common social fact usually recorded by lexicographers that certain words have certain meanings, that is, that they denote certain things, relations, or operations. But a knowledge of such usage in English, or in any other language, will not enable us to solve all questions as to the adequacy of evidence, for example, the validity of the proof by Hermite and Lindemann that π (the ratio between the circumference and the diameter of a circle) cannot be accurately expressed by rational numbers in finite form.

While the direct subject matter of logic cannot be restricted to words, or even to the meaning of words as distinguished from the meaning or implication of propositions, logic is closely connected with general grammar, and it is not always easy to draw a sharp line between the grammatical and the logical writings of philosophers like Aristotle, Duns Scotus, and C. S. Peirce. We have already mentioned that logic starts by taking the ordinary meaning of words for granted, and we shall later see how just discrimination in the meanings of words helps us to avoid logical fallacies. It must, however, be added that in the general study of the meaning of words (called semantics) we are dependent on logic. The information conveyed by words depends both logically and psychologically on propositions, or the information conveyed by sentences.

Perhaps the most significant distinction between logic and that part of linguistics called grammar can be put thus: The norm or correctness with which grammar is concerned is conformity to certain actual usages, while the norm or correctness of logic is based on the possibilities in the nature of things which are the objects of our discourse. Grammar is primarily a descriptive social science, describing in some systematic manner the way in which words are

[2] *Leviathan,* Pt. I, Chap. 5.

used among certain peoples. It is only incidentally normative, as the description of fashions in clothes is. It is assumed that certain styles, the King's English [3] or the usage of the "best people," is preferable. Many differences of linguistic form may not correspond to any difference of meaning, for example, the differences between the ablative and dative cases in Latin, or such differences as those between "proved" and "proven," "got" and "gotten" in English. But as language is sometimes used to convey significant information, as well as to express emotions, it is impossible for grammarians to ignore logical distinctions. As ordinary experience does not require great accuracy and subtle insight into the nature of things, ordinary language is not accurate. Logic is necessary to correct its vagueness and ambiguity.

In general, though words are among the important objects of human consideration, it is not true that all propositions are about words. Most propositions are about objects like the sun and the stars, the earth and its contents, our fellow-creatures and their affairs, and the like; and the implication between propositions, which is the subject matter of logic, has to do with the possible relations between all such objects. It is only as words are necessary instruments in our statement or expression of a proposition that logic must pay critical attention to them, in order to appreciate their exact function and to detect errors in inference.

Logic and Psychology

An old tradition defines logic as the science of the laws of thought. This goes back to a time when logic and psychology were not fully developed into separate sciences clearly distinguished from other branches of philosophy. But at present it is clear that any investigation into the laws or ways in which we actually think belongs to the field of psychology. The logical distinction between valid and invalid inference does not refer to the way we think—the process going on in someone's mind. The weight of evidence is not itself a temporal event, but a relation of implication between certain classes or types of propositions. Whether, for instance, it necessarily follows from Euclid's axioms and postulates that the area of no square can be exactly equal to that of a circle is a question of what is necessarily involved in what is asserted by our propositions; and how anyone actually thinks is irrelevant to it. Of course thought (and not mere sense perception) is necessary to apprehend such im-

[3] The English kings from the eleventh to the fourteenth century were of course Frenchmen.

plications. But thought is likewise necessary to apprehend that the propositions of any science are true. That, however, does not make physics a branch of psychology—unless we deny that these sciences have different subject matters, in other words, unless we deny that physical objects and our apprehension of them are distinguishable and not identical. Similarly, our apprehension of the logical implication on which our inferences are based may be studied as a psychological event, but the relation directly apprehended is not itself a psychological event at all. It is a relation between the forms of propositions and indirectly one between the classes of possible objects asserted by them.

The realization that logic cannot be restricted to psychological phenomena will help us to discriminate between our science and rhetoric—conceiving the latter as the art of persuasion or of arguing so as to produce the feeling of certainty. People often confuse the two because the word "certainty" is sometimes used as a characteristic of what is demonstrated and sometimes as the feeling of unquestioning assurance. But such feeling of certainty may exist apart from all logic, and the factual persuasiveness of arguments is more often brought about by properly chosen words, which through association have powerful emotional influences, than through logically unassailable arguments. This is not to belittle the art of rhetoric or to accuse it of always using fallacious arguments. The art of persuasion, or getting others to agree with us, is one that almost all human beings wish to exercise more or less. Harmonious social relations depend on it. But strictly logical argumentation is only one of the ways, and not always the most effective way, of persuading those who differ from us. Our emotional dispositions make it very difficult for us to accept certain propositions, no matter how strong the evidence in their favor. And since all proof depends upon the acceptance of certain propositions as true, no proposition can be proved to be true to one who is sufficiently determined not to believe it. Hence the logical necessity revealed in implication, as in pure mathematics, is not a description of the way all people actually think, but indicates rather an impossibility of certain combinations of the objects asserted. The history of human error shows that the assertion, "I am absolutely certain," or, "I cannot help believing," in regard to any proposition is no adequate evidence as to its truth.

In general, the canons of logical validity do not depend upon any investigation in the empirical science of psychology. The latter, indeed, like all other sciences, can establish its results only by conformity to the rules of logical inference. But a study of psychology is of great help to logic, if for no other reason than that only a sound knowledge of psychology can help us to avoid unavowed but false psychological assumptions in logical theory.

Logic and Physics

If logic cannot be identified with linguistics or psychology, neither is it the same as physics or natural science. The propositions whose relations logic studies are not restricted to any special field, but may be about anything at all—art, business, fairy tales, theology, politics; and while the logical relation of implication is involved in physical science, it is not the primary object of the latter.

The fact that propositions may be about nonexistent objects does not militate against the objectivity of the relation of implication. This relation is objective in the sense that it does not depend upon our conventions of language or on any fiat of ours to think in a certain way. This may perhaps become more obvious if we consider the procedure of pure mathematics. In this field, as we have already indicated, we inquire only as to the implication of our initial propositions, without regard to their truth or to whether their subjects are existent or nonexistent, real or imaginary. And yet research in mathematics of the kind that has been going on for over two thousand years is as bound or determined by the nature of the material (logical implication) as is any geographical exploration of the earth or astronomical study of the movements of the stars.

No linguistic fiat or resolution to think differently can change the truths discovered or deduced in such fields as the theory of prime numbers. And this is true of all rigorous logical deduction.

Logic and the Metaphysics of Knowledge

The essential purpose of logic is attained if we can analyze the various forms of inference and arrive at a systematic way of discriminating the valid from the invalid forms. Writers on logic, however, have not generally been content to restrict themselves to this. Especially since the days of Locke they have engaged in a good deal of speculative discussion as to the general nature of knowledge and the operations by which the human mind attains truth as to the external world. We shall try to avoid all these issues —not because they are not interesting and important, but because they are not necessary for the determination of any strictly logical issues. Indeed, the answers to the questions of metaphysics, rational psychology, or epistemology (as they are variously called) are admittedly too uncertain or too questionable to serve as a basis for the science of all proof or demonstration. We wish, however, to dispose of one of these questions, which may trouble the reader: How can false propositions, or those about nonexistent objects, have implications that are objectively necessary?

This seeming paradox arises from a naïve assumption—that only actually existing things have determinate objective characters. It is rather easy to see that the world of science, the world about which there is true knowledge, cannot be restricted to objects actually existing but must include all their possible functions and arrangements. Consider such elementary propositions as *Carbon burns, Ice melts at 32° F., Metals conduct heat and electricity,* and the like. These all refer to classes or kinds of possibilities of the ideally continuous or recurrent existences we call carbon, ice, or metal. Now whatever actually exists is only one of an indefinite number of possibilities. The actual is a flying moment passing from the future which is not yet to the past which no longer is. Logic may be conceived as ruling out what is absolutely impossible, and thus determining the field of what in the absence of empirical knowledge is abstractly possible. History and the sciences of natural existence rule out certain possible propositions as false, for example: There are frictionless engines, free bodies, perfectly rigid levers, and so on. They are ruled out because they are incompatible with propositions which we believe to be true of the actual world. But the actual world at any one time is only one of a number of possible arrangements of things. A proposition proved false on one set of assumptions may be proved true on another. Thus while logical relations alone are not sufficient to determine which existence is actual, they enter into the determination of every arrangement of things that is at all possible. The essential properties which determine the value of $100 remain the same whether we do or do not possess that amount.

§ 6. THE USE AND APPLICATION OF LOGIC

Like any other science, logic aims at attaining truth in its own special field and is not primarily concerned with the values or uses to which these truths can be put. Bad men may be logically consistent. But correct inference is such a pervasive and essential part of the process of attaining truth (which process in its developed form we call *scientific method*) that a study of the way in which logic enters into the latter is a natural extension of our science, just as pure mathematics is extended and developed by its practical application. This will engage our attention in Book II of this volume. Even at this stage, however, we may note some of the ways in which formal, deductive logic aids us in arriving at true propositions.

1. It is obvious that it is often difficult, if not impossible, to determine the truth of a proposition directly, but relatively easy to establish the truth of another proposition from which the one at issue can be deduced. Thus we have observed how difficult it would be to show by actual count that there are at least two people in New York City who have exactly the same number of scalp hairs. But it is fairly easy to show that the number of inhabitants in New York City exceeds the maximum number of hairs on a human head. For the study of the physiology of hair follicles, as well as random samplings of human heads, enables us to establish that there are not more than five thousand hairs to a square centimeter. Anthropological measurements lead to the conclusion that the maximum area of the human scalp is much under one thousand square centimeters. We may conclude, therefore, that no human being has more than five million scalp hairs. And since the population of New York City is close to seven million, there must be more inhabitants in New York City than any human head has hairs. It follows, in virtue of our previous demonstration, that there must be at least two individuals in that city who are precisely alike in the number of hairs on their heads.

2. Many of our beliefs are formed to meet particular problems, and we are often shocked to find these beliefs inconsistent with one another. But they can be integrated, and their bearings on one another made clear, by exploring deductively their mutual relations. Thus it is deductive reasoning which enables us to discover the incompatibility between the following propositions: *Promise-breakers are untrustworthy; Wine-drinkers are very communicative; A man who keeps his promises is honest; No teetotalers are pawnbrokers; All communicative persons are trustworthy; Some pawnbrokers are dishonest.*

3. Deductive reasoning enables us to discover what it is to which we must, in consistency, commit ourselves if we accept certain propositions. Thus if we admit that two straight lines cannot inclose an area, as well as some other familiar geometric propositions, we must also admit, as we soon discover, that the sum of the angles of any triangle cannot be greater than two right angles.

The full meaning of what it is we believe is discovered by us when we examine deductively the connections between the diverse propositions which we consider. For the propositions which we may be inclined to accept almost without question may have implications altogether surprising to us and requiring us to modify our hasty acceptance of them as premises.

In pointing out these uses of deductive inference, it is not denied that men may and do successfully employ it without any previous theoretical study of logic, just as men learn to walk without studying physiology. But a study of physiology is certainly valuable in preparing plans for training runners. Any competent electrician can adjust our electric lights, but we think it necessary that an engineer who has to deal with new and complicated problems of electricity should be trained in theoretical physics. A theoretical science is the basis of any rational technique. In this way logic, as a theoretical study of the kinds and limitations of different inferences, enables us to formulate and partially mechanize the processes employed in successful inquiry. Actual attainment of truth depends, of course, upon individual skill and habit, but a careful study of logical principles helps us to form and perfect techniques for procuring and weighing evidence.

Logic cannot guarantee useful or even true propositions dealing with matters of fact, any more than the cutler will issue a guarantee with the surgeon's knife he manufactures that operations performed with it will be successful. However, in offering tribute to the great surgeon we must not fail to give proper due to the quality of the knife he wields. So a logical method which refines and perfects intellectual tools can never be a substitute for the great masters who wield them; none the less it is true that perfect tools are a part of the necessary conditions for mastery.[4]

[4] More mature readers will do well to go over Appendix A carefully before undertaking Chapter II.

BOOK I

FORMAL LOGIC

CHAPTER II

THE ANALYSIS OF PROPOSITIONS

§ 1. WHAT IS A PROPOSITION?

In the last chapter logic was defined as dealing with the relation of implication between propositions, that is, with the relation between premises and conclusions in virtue of which the possible truth and falsity of one set limits the possible truth and falsity of the other. Both premises and conclusions are thus propositions; and for the purposes of logic, a *proposition* may be defined as anything which can be said to be true or false. But we shall understand this definition more clearly if we also indicate what a proposition is not.

1. A proposition is not the same thing as the sentence which states it. The three sentences, "I think, therefore, I am," "*Je pense, donc je suis,*" "*Cogito, ergo sum,*" all state the same proposition. A sentence is a group of words, and words, like other symbols, are in themselves physical objects, distinct from that to which they refer or which they symbolize. Sentences when written are thus located on certain surfaces, and when spoken are sound waves passing from one organism to another. But the proposition of which a sentence is the verbal expression is distinct from the visual marks or sound waves of the expression. Sentences, therefore, have a physical existence. They may or may not conform to standards of usage or taste. But they are not true or false. Truth or falsity can be predicated only of the propositions they signify.

2. It should be noted, however, that while the proposition must not be confused with the symbols which state it, no proposition can be *expressed* or *conveyed* without symbols. The structure of the proposition must, therefore, be expressed and communicated by an appropriate structure of the symbols, so that not every combination of symbols can convey a proposition. "John rat blue Jones," "Walk-

ing sat eat very," are not symbols expressing propositions, but simply nonsense, unless indeed we are employing a code of some sort. Only certain arrangements of symbols can express a proposition. That is why the study of symbolism is of inestimable value in the correct analysis of the structure of propositions. And that is why, although grammatical analysis is not logical analysis, the grammar of a language will often clarify distinctions which are logical in nature.

3. A proposition, we have said, is something concerning which questions of truth and falsity are significant. Consequently when Hamlet declares, "Oh, from this time forth, My thoughts be bloody, or be nothing worth!" or when he asks, "Why wouldst thou be a breeder of sinners?" he is not asserting propositions *except implicitly*. For wishes, questions, or commands cannot as such be true or false. It should be noted, however, that the intelligibility of wishes, questions, and commands rests upon assumptions that certain states of affairs prevail. And such assumptions involve propositions. For consider the question, "Why wouldst thou be a breeder of sinners?" It obviously assumes, among other propositions, that the person addressed exists, is capable of breeding children, and that such children are certain to be sinners. Similarly, the exhortation, "My thoughts be bloody, or be nothing worth!" assumes that the speaker is capable of having ideas, that these ideas can be murderous, that they may have some kind of value, and so forth. Moreover, a command or wish may be put in the form of a declaration, which generally expresses a proposition, for example, *I wish you would come; I shall be pleased if you come; You will be sorry if you do not come*. To the extent that the declarations state something that may be true or false they are propositions.

4. Propositions are often confounded with the mental acts required to think them. This confusion is fostered by calling propositions "judgments," for the latter is an ambiguous term, sometimes denoting the mental act of judging, and sometimes referring to that which is judged. But just as we have distinguished the proposition (as the objective meaning) from the sentence which states it, so we must distinguish it from the act of the mind or the judgment which thinks it.

5. Nor must propositions be identified with any concrete object, thing, or event. For propositions are at most only the abstract and selected relations between things. When we affirm or deny the proposition *The moon is nearer to the earth than the sun*, neither the moon alone, nor the earth, nor the sun, nor the spatial distance

between them is the proposition. The proposition is the relation asserted to hold between them. The relations which are the objects of our thought are elements or aspects of actual, concrete situations. These aspects, while perhaps not spatially and temporally *separable* from other characters in the situation, are *distinguishable* in meaning. That is why sense experience never yields knowledge without a reflective analysis of what it is we are experiencing. For knowledge is *of* propositions. And propositions can be known only by discriminating within some situation relations between abstract features found therein.

6. While a proposition is defined as that which is true or false, it does not mean that we must *know* which of these alternatives is the case. *Cancer is preventable* is a proposition, though we do not know whether it is true.

A difficulty is nevertheless suggested: We may not be able to tell whether a given sentence does or does not express a proposition. Consider, for example, the expression, "Three feet make a yard." Are we raising questions of truth and falsity in asserting it? It must be acknowledged that the sentence has the appearance of expressing a proposition. But if we analyze what is usually meant by it, we find it expresses a *resolution* rather than something which is capable of being true. We *resolve* to use a unit of measure so that it will be made up of three feet. But the resolution as such cannot have truth or falsity predicated of it. Such resolutions, which often take the form of definitions, are expressed in ways analogous to the way propositions are expressed; but they must be distinguished from the latter.

The question whether the word "yard" is actually used in the sense defined is of course a factual one, and the answer may be true or false. But such propositions are about linguistic usages and not about the objects denoted by the words.

7. A related difficulty arises from the fact that we popularly speak of propositions as sometimes true and sometimes false, whereas our definition of propositions excludes this possibility and assumes that if a proposition is true it must always be true. Nothing is more common in discussion between candid people than their remark, "What you say is sometimes true, but not always so." This applies to such statements as "Religion preaches love of one's fellow men"; "It is difficult to resist temptation"; "A gentle answer turneth away wrath." We may, however, remove this difficulty by recognizing that if these propositions assert that something is *universally* the case, then the existence of an exception only proves

that they are false. Such an assertion as "Religion sometimes preaches hate of one's neighbors" does not assert the absurdity that a universal proposition *Religion always preaches hate of one's neighbors* is sometimes true.

We can perhaps see this more clearly in the following case. *The present governor of Connecticut is Dr. Cross* seems to be a proposition true for certain years, but surely not always. This, however, is an inadequate analysis. For the phrase "the present governor" clearly presupposes a date; and as we complete our expression by including explicitly the date, we obtain expressions for different propositions, some of which are true and some false. In general, our everyday statements seldom contain all the qualifications necessary to determine whether what we say is true or false. Some of their qualifications we understand, others are not thought of. The incomplete expression is neither true nor false. And when we say that it is sometimes true and sometimes false, we can mean only that our expressions may be completed in some ways which express true propositions and in some ways which express false ones.

§ 2. THE TRADITIONAL ANALYSIS OF PROPOSITIONS

Terms: Their Intension and Extension

According to Aristotle, all propositions either assert or deny something of something else. That about which the assertion is made is called the *subject,* and that which is asserted about the subject is called the *predicate.* The subject and predicate are called the *terms* of the proposition; the proposition is the synthesis or unity of the terms by means of the *copula,* which is always some part of the verb "to be."

This analysis cannot readily be applied to very simple propositions such as *It is raining; There was a parade* and the like. The "It" and the "There" clearly do not denote subjects, of which "raining" and "a parade" are attributes. Nevertheless, there is a certain amount of truth in Aristotle's analysis, if we hold on to the distinction between *terms* and *propositions,* but drop the requirement that there must be just two terms. *It is raining* or *There was a parade* are properly said to be propositions, because they conform to his test, to wit, they are either true or false. "Raining" or "a parade" are not propositions, because they are not either true or false. When we hear the words "raining" or "a parade" we ask, "What about it?" Only when *assertions* are made about these objects can questions of truth or falsity be raised.

These objects, then, as *terms* enter as elements of propositions. A term may be viewed in two ways, either as a class of objects (which may have only one member), or as a set of attributes or characteristics which determine the objects. The first phase or aspect is called the *denotation* or *extension* of the term, while the second is called the *connotation* or *intension*. Thus the extension of the term "philosopher" is "Socrates," "Plato," "Thales," and the like; its intension is "lover of wisdom," "intelligent," and so on.

Although the intension and extension of a term are distinct aspects, they are inseparable. All words or symbols except pure demonstrative ones (those which serve to point out, like a gesture, or "this") signify some attributes in virtue of which they may be correctly applied to a delimited set of objects; and all general terms are capable of being applied to some object, even though at any given time no object may in fact possess the attributes necessary to include it in the extension of the term. *Why* a term is applied to a set of objects is indicated by its intension; the set of objects *to which* it is applicable constitutes its extension.

With many of the difficult problems of extension and intension we shall not concern ourselves. It will be convenient, however, to distinguish between several senses in which the term *intension* is frequently employed. It is necessary to do so if we wish to avoid elementary confusions.

1. The *intension of a term* is sometimes taken to mean the sum total of the attributes which are present to the mind of any person employing the term. Thus to one person the term "robber" signifies: "taking property not lawfully his, socially undesirable, violent," and so forth; to another person it may signify: "taking property with value greater than ten dollars not lawfully his, physically dangerous person, the result of bad disposition," and so on. The intension of a term so understood is called the *subjective intension*. The subjective intension varies from person to person, and is of psychological rather than of logical significance.

2. The *intension of a term* may signify the set of attributes which are essential to it. And by "essential" we mean the necessary and sufficient condition for regarding any object as an element of the term. This condition is generally selected by some convention, so that intension in this sense is called *conventional intension* or *connotation*. The conventional intension of a term, as we shall see later, constitutes its definition.

3. *The intension of a term* may signify *all* the attributes which the objects in the denotation of a term have in common, whether

these attributes are known or not. This is called the *objective intension* or *comprehension*. Thus, if the conventional intension of "Euclidean triangle" is "a plane figure bounded by three straight lines," a part of the objective intension is: "a plane figure with three angles, a plane figure whose angle sum is two right angles," and so on.

It is the conventional intension of a term which is logically important. The denotation of a term clearly depends upon its connotation. Whether we may apply the word "ellipse" to some geometric figure is determined by the attributes included in the connotation of the term. From the point of view of knowledge already achieved, the understanding of the connotation of a term is prior to its denotative use: we must know the connotation of "amoeba" before we can apply it. In the order of the *development* of our knowledge, it is doubtful whether there is such a priority. Philosophers have been unable to resist the temptation of regarding either the intension or the extension of a term prior in every respect, and much ink has been shed over the question. It is reasonable to suppose, however, that the development of our knowledge concerning intension and extension, since they are inseparable aspects of the meaning of a term, go hand in hand. A group of objects, such as pieces of iron, bronze, tin, may be selected for certain special purposes in virtue of their possessing some prominent features in common, like hardness, opacity, fusibility, luster. Such objects may then be denoted by a common term, "metal." These striking features may then be taken as criteria for including other objects in the denotation of this term. But greater familiarity with such objects may lead us to note qualities more reliable as signs of the presence of other qualities. We may then group objects together in spite of superficial differences, or group objects differently in spite of superficial resemblances. The conventional intension of the term "metal" may thus become gradually modified. The assigning of the satisfactory conventional intension (or definition) to a term denoting objects with familiar common traits is a difficult task, and is a relatively late achievement of human thought.

Consider now the terms: "figure," "plane figure," "rectilinear plane figure," "quadrilateral," "parallelogram," "rectangle," "square." They are arranged in order of subordination, the term "figure" denoting a class which includes the denotation of "plane figure," and so on. Each class may be designated as the *genus* of its subclass, and the latter as a *species* of its genus. Thus the *denotation* of these terms *decreases:* the extension of "parallelogram" includes the extension of "rectangle," but not conversely, and so on. On the other hand, the *intension* of the term *increases:* the intension of "rectangle" includes the intension of "parallelogram," but not conversely. Reflection upon such series of terms has led to the rule: *When a series of terms is arranged in order of subordination, the extension and intension vary inversely.*

But this formulation of the relation of extension to intension is not accurate. In the first place, "vary inversely" must not be understood in a strict numerical sense. For in some cases the addition of a single attribute to the intension of a term is accompanied by a greater change in its extension than in other cases. Thus the extension of "man" is reduced much more by the addition of the attribute "centenarian" than by the addition of the attribute "healthy." And in the second place, variation in the intension may be accompanied by no change in the extension. Thus, the extension of the term "university professor" is the same as the extension of "university professor older than five years." It is clear, moreover, that the relation of inverse variation must be taken between the *conventional* intension of a term, and its denotation in a specified universe of discourse. The law of inverse variation must, therefore, be stated as follows: *If a series of terms is arranged in order of increasing intension, the denotation of the terms will either remain the same or diminish.*

The Form of Categorical Propositions

According to the traditional doctrine all propositions can be analyzed into a subject and a predicate joined by a copula, either from the intensional or from the extensional point of view. *All cherries are luscious* may on the first view be interpreted to mean that the attribute of "being luscious" is part of the group of *attributes* which define the nature of cherries. On the second view, this proposition means that the objects called cherries are included in the denotation of the term "luscious." [1]

The traditional view recognizes other forms of propositions, called *conditional,* which it tries to reduce to the categorical form. We shall examine below these conditional forms, as well as other ways of analyzing propositions. Here we shall only note that on the

[1] The reader should note, however, that in the traditional analysis which we shall follow in this section the emphasis will be on the *extensional* interpretation.

traditional view all propositions are analyzable in just this way and only in this way.[2]

Propositions which do not obviously present a subject-predicate form must, then, be changed to exhibit that form. Thus *Germany lost the war* must be expressed as *Germany is the loser of the last war,* where "Germany" is the subject, "the loser of the last war," the predicate, and "is," of course, the copula. The proposition *Ten is greater than five* must be analyzed into "ten" as subject, "a number greater than five" as predicate, and "is" as copula.

It is not difficult to exhibit any proposition as *verbally* conforming to the subject-predicate type, but such verbal identity often obscures fundamental logical distinctions. The chief criticism which modern logic has made of the traditional analysis is that the latter has lumped together (as categorical) propositions which have significant differences in form.

The reader may perhaps wonder what is the significance of this quarrel over the manner in which propositions should be analyzed. The answer is simple. The analysis of propositions is undertaken for the purpose of discovering what inferences may be validly drawn from them. Consequently, if there is a plurality of propositional forms, and such form or structure determines the validity of an inference, an increased refinement of analysis may help us to attain a more accurate view of the realm of possible inference.

Another reason for analyzing the structure of propositions is to devise some *standard* or *canonical* ways of representing what it is we wish to assert. We wish to find certain canonical formulations of propositions of a given type, in order that the process of inference shall be expedited. Thus in elementary algebra it is extremely convenient to write the quadratic equation $5x^2 = 3x - 5$ in the standard form $5x^2 - 3x + 5 = 0$. For if we do so, since we already

[2] The view that all propositions are of the subject-predicate form has been associated historically with certain philosophical interpretations of the nature of things. The subject, on this view, is regarded as a *substance* in which various qualities inhere, and the task of all inquiry is to discover the inhering predicates in some concrete subject. According to Leibniz, for instance, there is an ultimate plurality of substances or monads, each of which is pregnant with an illimitable number of properties. These monads cannot be said to stand to one another in any relation, for if they did one monad would have to be a *predicate* of another, and therefore not self-subsistent. According to others, like Bradley, there is just *one* substance, so that all predication is the affirmation of something of *all reality* conceived as a single, unique individual. Neither of these extreme positions was adopted by Aristotle. He maintained that the ultimate subject of predication is some concrete, individual substance, and that there is an irreducible plurality of such, but that these substances are systematically related.

know the roots of a general quadratic in the standard form $ax^2 + bx + c = 0$, it is very simple to find the numerical answers to our problems. Moreover, if we adopt a standard form for writing equations, it is much easier to compare different equations and note their resemblances. Similar considerations apply in logic. For if we can once establish criteria of validity for inferences upon propositions stated in a standard form, all subsequent testing of inferences becomes almost mechanical.

Much care must be taken, however, in reducing a proposition expressed in one verbal form to the standard form, lest some part of its original meaning be neglected. In reducing a line of Keats's poetry, for example, into a canonical form it is not easy to believe that every shade of meaning of the original has been retained.

Quantity

Categorical propositions have been classified on the basis of their *quantity* and their *quality*. In the proposition, *All steaks are juicy,* something is affirmed of *every* steak, while in the proposition, *Some steaks are tough,* information is supplied about an *indefinite* part of the class of steaks. Propositions which predicate something of *all* of a class are designated as *universal,* while those which predicate something of an *indefinite part* of a class are *particular.* The particles "all" and "some" are said to be *signs of quantity,* because they indicate of how large a part of the subject the predicate is affirmed. The distinction between them is more accurately stated if "all" is called the sign of a *definite* class, and "some" the sign of an *indefinite* part of a class. For in everyday usage, the signs of quantity are ambiguous. Thus in the proposition *Some professors are satirical* it would ordinarily be understood that a *part, but not the whole,* of the class of professors are satirical; here "some" means "some but not all." On the other hand, *Some readers of this book have no difficulty in understanding it* would generally be understood to mean that a portion of the readers, *not excluding the entire class,* had no difficulty with it; here "some" means "some and perhaps all." We shall obviate such ambiguity by agreeing that in logic "some" will be taken in the latter sense; that is, as not necessarily excluding "all."

A different sort of ambiguity occurs in the use of the word "all." Sometimes it denotes all of a finite and enumerated collection, as in the proposition *All the books on this shelf are on philosophy.* Sometimes, however, as in *All men are mortal* the "all" means "all possible," and cannot be taken, without distortion of intended meaning, to indicate merely an enumeration of the men who do in fact exist or have already existed. We shall find this distinction of paramount importance in the discussion of induction and deduction. Many mistaken views concerning the former are the outcome of ignoring it.

Is the proposition *Thaïs was a courtesan in Alexandria* universal or particular? The reader may be tempted to say it is the latter. But that would be an error, for he would then be using "particular" in a sense different from the one employed in classifying propositions. On the basis of the definition of universal propositions as those which affirm something of *all* of the subject, this proposition must be regarded as a universal. This would be even more obvious if, as we suggest, the terms *definite* and *indefinite* were used instead of *universal* and *particular*. However, since in such propositions we are affirming something of a single individual, traditional logic has sometimes designated them as *singular*. But singular propositions must be classified as universal on the traditional analysis. However, a more subtle analysis cannot be content with such a conclusion. Even untutored people feel dimly that there is a difference *in form* between *Dr. Smith is a reassuring person* and *All physicians are reassuring persons* although traditional logic regards both as universal. Modern logic corroborates this feeling by showing clearly that these propositions do in fact illustrate different logical forms. Nevertheless, for many purposes no harm results if, with traditional logic, we regard both as having the same structure.

Quality

A second classification of categorical propositions is concerned with their *quality*. In the proposition *All snakes are poisonous* the predicate is affirmed of the subject. The proposition is therefore said to be *affirmative*. In *No democracies are grateful* something is denied of the subject. The proposition is therefore said to be *negative*. If we think of a categorical proposition as a relation between classes of individuals, an affirmative proposition asserts the *inclusion* of one class or part of a class in another, while a negative proposition asserts the *exclusion* of one class or part of a class from another. It follows that the negative particle, the sign of quality, must be understood to characterize the copula, not the subject or the predicate.

How should we classify *All citizens are not patriots*? It seems to be a negative proposition, and the sign of quantity seems to indicate it as universal. But while it *may* be interpreted as asserting that *No citizens are patriots* it may also be understood as denying that *All citizens are patriots* or as asserting that *Some citizens are not patriots*. Expressions employing "all . . . not" as in the foregoing, or as in *All that glitters is not gold* are essentially ambiguous. In such cases, we must determine what is meant, and then state it in an unambiguous form.

On the basis of quantity and quality we may therefore distinguish four forms of categorical propositions. *All teetotalers are short-lived* is a universal affirmative, and is symbolized by the letter *A*. *No politicians are rancorous* is a universal negative, and is symbolized by *E*. *Some professors are soft-hearted* is a particular affirmative, and is symbolized by *I*. *Some pagans are not foolish* is a particular negative, and is symbolized by *O*. The letters *A* and *I* have been used traditionally for affirmative propositions: they are the first two vowels in *affirmo;* while *E* and *O* symbolize negative propositions: they are the vowels in *nego*.

Exclusive and Exceptive Propositions

In the propositions *The wicked alone are happy, Only the lazy are poor, None but savages are healthy,* something is predicated of something else in an exclusive fashion. They are therefore called *exclusive propositions*. Traditional logic reduces such propositions to the canonical form for categoricals. For example, *The wicked alone are happy* asserts the same thing as *All happy individuals are wicked. None but the brave deserve the fair* asserts the same as *All who deserve the fair are brave. None but Seniors are eligible* asserts the same as *All those eligible are Seniors.*

In the propositions *All students except freshmen may smoke, All but a handful were killed, No child may enter unless accompanied by a parent* the predicate is denied of some part of the denotation of the subject. They are therefore called *exceptive* propositions. They may also be stated in the standard form for categoricals. For exceptive propositions may always be expressed as exclusive ones. Thus *All students except freshmen may smoke* can be reduced to *Freshmen alone among students may not smoke*. Consequently, the above exceptive proposition can be stated as the following *A* proposition *All students who may not smoke are freshmen.*

Distribution of Terms

We shall now introduce a new technical term. We will say a term of a proposition is *distributed* when reference is made to *all* the individuals denoted by it; on the other hand, a term will be said to be *undistributed* when reference is made to an *indefinite part* of the individuals which it denotes.

Let us now determine which terms in each of four types of categoricals are distributed. It is evident that in universal propositions the subject term is always distributed, while in particular propositions the subject is undistributed. How about the predicate terms? In *All judges are fair-minded* is reference made to all the individuals denoted by "fair-minded"? Clearly not, because the proposition supplies no information whether *all* fair-minded individuals are judges or not. Hence the predicate in *A* propositions is undistributed. A similar conclusion is true for *I* propositions. We may therefore conclude that affirmative propositions do not distribute their predicates.

Does the same state of affairs obtain for negative propositions? Consider *No policemen are handsome.* This proposition asserts not only that every individual denoted by "policemen" is excluded from the class denoted by "handsome," but also that all individuals of the latter class are also excluded from the former. Consequently, the predicate in *E* propositions is distributed. A similar conclusion holds for *O* propositions. Thus in *Some of my books are not on this shelf* an indefinite part of the subject class is excluded from the *entire* class denoted by the predicate. The reader will see this clearly if he asks himself what part of the shelf indicated he would have to examine in order to assure himself of the truth of the proposition. Obviously it is not enough to examine only a *part* of the books on the shelf; he must examine *all* the books on the shelf. The predicate is therefore distributed.

We may summarize our inquiry by noting that universal propositions distribute the subject, while particular propositions do not distribute the subject. On the other hand, the predicate is distributed in negative propositions, but undistributed in affirmative ones.

The concept of distribution of terms plays an important part in traditional logic, and is the fundamental idea in the theory of the syllogism. The reader is therefore advised to familiarize himself with it thoroughly. We may note in passing, however, that the subject-predicate analysis of propositions, together with the idea of distribution, sometimes leads to inelegant results. Thus on the traditional analysis *Socrates was snub-nosed* is a universal; its subject must be distributed, since snub-nosedness is predicated of the entire individual Socrates. Nevertheless, while in other universal propositions like *All children are greedy* a corresponding particular proposition may be obtained in which "children" is undistributed, no such corresponding proposition can be found for singular ones. For under no circumstances may the term "Socrates" be undistributed. We shall find other respects in which universal and singular propositions do not receive symmetrical development in traditional logic.

Diagrammatic Representation

The structure of the four types of categorical propositions can be exhibited in a more intuitive fashion if we adopt certain conventional diagrammatic representations. Many methods of doing this have been devised, some having different purposes in mind. The earliest is due to Euler, a Swiss mathematician of the eighteenth century. We shall first explain a slight modification of his method.

Let us agree to the following conventions. A circle drawn in solid line will indicate a distributed term; a circle drawn (in part or in whole) in dotted line will represent an undistributed term. A circle drawn inside another will indicate the inclusion of one class in another; two circles entirely outside each other will indicate the mutual exclusion of two classes; and two overlapping circles will represent either an indefinite partial inclusion or an indefinite partial exclusion.

The four relations between the classes "street-cleaners" and "poor individuals" which characterize the four categorical propositions (in which the former class is subject) may then be diagrammed as follows.

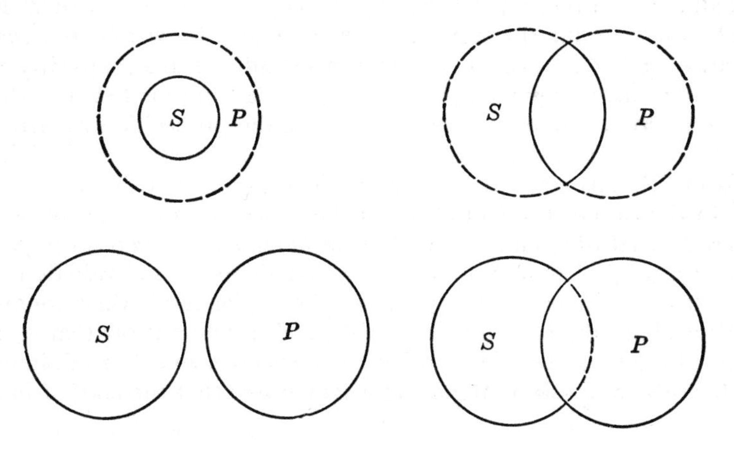

The *S* circle represents the class "street-cleaners" (the subject), and the *P* circle the class "poor individuals" (the predicate).

It is often useful to employ another method of representing the categorical propositions, which is due to the English logician John Venn. We first notice that every proposition tacitly refers to some context within which it is significant. Thus *Hamlet killed Polonius* refers to Shakespeare's play. Let us call the domain of reference the *universe of discourse* and represent it diagrammatically by a rectangle. Now the reader will observe that two classes, together with their negatives, yield four and only four combinations. (By the negative of a class is understood everything in the universe of discourse excluded from that class.) For example, in the universe of discourse restricted to human beings there are the things which are both street-cleaners and poor (symbolized as $S\ P$), or which are street-cleaners and not poor ($S\ \bar{P}$), or not street-cleaners and poor ($\bar{S}\ P$) or neither street-cleaners nor poor ($\bar{S}\ \bar{P}$). The universe of discourse is thus divided into four possible compartments. However, in general not all these possible compartments will contain individuals as members. Which ones do and which ones do not depends upon what is *asserted* by propositions referring to that universe of discourse.

Let us therefore draw two overlapping circles within a rectangle, and so obtain automatically four distinct compartments, one for each of the four logical possibilities indicated. Now since the *A* proposition asserts that all street-cleaners are included in the class of the poor, the class of those who are street-cleaners and not poor cannot contain any members. To show this on the diagram, we shall agree to blot out by shading the corresponding compartment. The diagram for the *A* proposition will thus show that the compartment $S\bar{P}$ is missing. We may also indicate this explicitly by writing under the rectangle $S\bar{P} = 0$, where 0 means that the class in question contains no members. Hence the proposition *All street-cleaners are poor* declares that in its universe of discourse there are no individuals who are both street-cleaners and not poor.

In the case of the *I* proposition the procedure of representing it is somewhat different. If we ask what *Some street-cleaners are poor* asserts, we find that it does *not* say that there *are* individuals who are street-cleaners and not poor (let the reader recall what we said above about the meaning of "some"). Neither does it say that *every* one of the four possible compartments has members. The minimum which the proposition requires for its truth is that the class of indi-

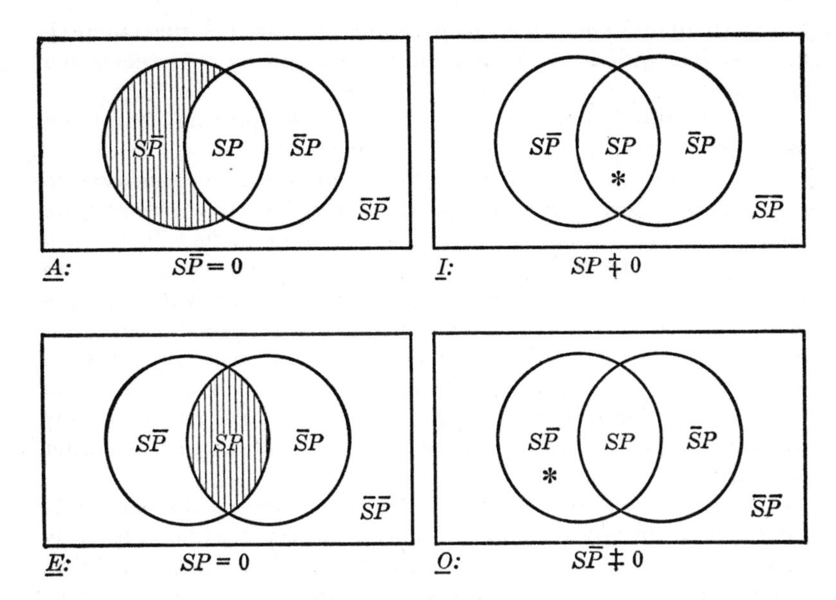

viduals who are street-cleaners and poor be *not* empty. We shall agree to designate this minimum by placing an *asterisk* into the SP compartment, to show that it cannot be empty. We thereby leave indeterminate whether the other compartments have members or not. We may indicate this further by writing $SP \neq 0$ under the rectangle, which signifies that the left-hand member of this inequality is not devoid of members. The reader should study carefully the remaining diagrams. He will find that the analysis of the *E* and *O* propositions is similar to that of the *A* and *I* propositions respectively.

The Existential Import of Categoricals

If we now compare the diagrams, we discover that there is a remarkable difference between universal and particular propositions. Universals do not affirm the existence of any individuals, but simply *deny* the existence of certain kinds of individuals. Particulars do not deny the existence of anything, but simply *affirm* that certain classes do have members. Consequently, the universal *All street-cleaners are poor* means only this: *If any individual is a street-cleaner, why, then he is poor.* It does not say that there actually are individuals who are street-cleaners. On the other hand, the particular *Some street-cleaners are poor* means that there is at least one individual who is at the same time a street-cleaner and poor.

We will anticipate some later discussion by putting the matter as follows. The universal *All street-cleaners are poor* is to be interpreted to assert: *For all instances or values of* X, *if* X *is a street-cleaner, then* X *is poor.* The particular *Some street-cleaners are poor* is to be interpreted to assert: *There is an* X, *such that* X *is a street-cleaner and* X *is poor.* We are thus becoming prepared to understand why modern logic finds fault with classifying a proposition such as *Napoleon was a soldier* with propositions such as *All Frenchmen are soldiers.* The latter, we have seen, means when analyzed: *For all instances or values of* X, *if* X *is a Frenchman then* X *is a soldier.* The former, on the other hand, cannot in any way be interpreted in this manner. We shall return to this matter presently.

The conclusion we have reached, that universals do not imply the existence of any verifying instances, while particulars do imply it, will doubtless seem paradoxical to the reader. (Indeed, a fuller discussion than we can undertake is required to make clear how much of this conclusion is a matter of convention and how much is forced upon us by logical considerations.) The reader will perhaps cite propositions like *All dogs are faithful* and urge that they do imply the existence of dogs. Now it may well be that when the reader asserts *All dogs are faithful* he also intends to assert *There are dogs.* But he should note that then he makes two separate and distinct assertions. But the proposition *All those who are free from sin may cast stones* clearly does not imply that there actually are any individuals free from sin. The universal proposition may be simply a hypothesis concerning a class which we know cannot have any members.

Thus Newton's first law of motion states: All bodies free of impressed forces persevere in their state of rest or of uniform motion in a straight line forever. Will the reader affirm that this proposition asserts the existence of any body which is not under the influence of an impressed force? We need remind him only of the law of gravitation, according to which *all* bodies attract one another. What Newton's first law does assert is the hypothesis that *if* a body were free from impressed forces, it would persevere in its state of rest or in uniform motion in a straight line forever. In the same way, the principle of the lever states what would be the case if the lever were a perfectly rigid body; it does not assert that there is such a body. Indeed, reflection upon the principles of the sciences makes it quite clear that universal propositions in science always function as *hypotheses,* not as statements of fact asserting the existence of individuals which are instances of it. It is true, of course, that if there were no applications of the universal, it would be useless for the science which deals with matters of fact. It is also true that the *meaning* of universal propositions requires at least *possible* matters of fact. But we cannot identify abstract possibilities denoted by universals with actual existences in which these possibilities are annulled by or are combined with other possibilities. Thus, inertia is a phase of all mechanical action, though no instance of inertia by itself can be found in nature. The principle of the lever holds to the extent that bodies are actually rigid, even though no instances of pure rigidity exist in isolation from other properties of bodies.

We may view this matter from another angle. Hitherto we have been discussing propositions as asserting relations between classes of individuals. But we have seen (page 33) that propositions may be interpreted as asserting connections between attributes. When universal propositions are regarded as not implying the existence of any individuals, it is the interpretation in terms of constant relations between attributes which comes to the fore.

It should be noted, finally, that in raising questions of existential import we do not necessarily confine the reference of terms to the physical universe. When we ask, "Did Jupiter have a daughter?" or, "Was Hamlet really insane?" we are not raising questions of *physical* existence, but of the existence of individuals within a universe of discourse controlled by certain assumptions, such as the statements of Homer or Shakespeare. An individual who may thus be said to "exist" in one universe of discourse may not have an existence in another. The proposition *Samson is a pure myth* denies existence to Samson in the universe of authentic history, but obviously not in the field of biblical mythology.

When, therefore, it is said in formal logic that universal propositions do not imply, while particulars do imply, the existence of instances, the reader may find it helpful to interpret this (in part at least) on the basis of the different function each type of proposition plays in scientific inquiry. Just as we cannot validly infer the truth of a proposition concerning some matter of observation from premises which do not include a proposition obtained through observation, so we cannot infer the truth of a particular proposition from universal premises alone; unless indeed we tacitly take for granted the existence of members of the classes denoted by the terms of the universal proposition.

§ 3. COMPOUND, SIMPLE, AND GENERAL PROPOSITIONS

The analysis of propositions thus far has been restricted to those having the categorical form. But logical relations hold between more complicated forms of propositions. Consider the following:

1. The weight of *B* is equal to *G*.
2. The lines *AB* and *CD* are parallel.
3. If the angles *AFG, CGF,* are greater than two right angles, the remaining angles *BFG, DGF,* are less than two right angles.
4. The sum of the interior angles on the same side is equal to, greater than, or less than two right angles.

A comparison of the first two propositions with the last two shows that the second set contains propositions as components or elements, while the first set does not. Thus *The angles* AFG, CGF, *are greater than two right angles* and *The sum of the angles on the same side is equal to two right angles* are themselves propositions which are components in 3 and 4.

Let us apply the characterization *compound* to all propositions which contain other propositions as components. But the reader must be warned that the form of a sentence is not always indicative of the kind of proposition which it expresses. Let him recall the discussion at the end of the last section, where an analysis of categorical propositions was made.

Compound Propositions

Four types of *compound* propositions may be distinguished. Each type relates the component propositions in a characteristic manner.

1. Consider the proposition *If war is declared, then prices go up*. We have agreed to call the proposition introduced by "if" (*War is declared*) the *antecedent,* and the one introduced by "then" (*Prices go up*) the *consequent*. Sometimes the particle "then" is omitted, but it is tacitly understood in such cases. A compound proposition which connects two propositions by means of the relation expressed by "if . . . then" is called *hypothetical* or *implicative*.

When a hypothetical proposition is asserted as true, what do we mean? We clearly do not mean to assert the truth of the antecedent nor the truth of the consequent, although both may in fact be true. What we mean to affirm is that *if* the antecedent is true, the consequent is also true, or in other words, that the two are so connected that the antecedent cannot be true without the consequent's also being true. A hypothetical proposition is sometimes said to express doubt. But this is a misleading way of characterizing such a proposition. We may indeed doubt whether *War is declared,* but when we affirm the hypothetical proposition we do not doubt that *If war is declared, prices go up.*

The antecedent and consequent of a hypothetical proposition may themselves be compound propositions. The analysis of propositions with respect to their logical form will therefore be facilitated if we employ special symbols expressly devised to exhibit logical form. Let us agree that an accent placed over a parenthesis will signify the *denial* or *negative* of the proposition contained in it; hence *It is false that Charles I died in bed* may be written (Charles I died in bed)′. Also, let us replace the verbal symbols "if . . . then" by the ideographic symbol $\supset$. The hypothetical *If Charles I did not die in bed, then he was beheaded* may then be written (Charles I died in bed)′ $\supset$ (he was beheaded).

2. Consider next the proposition *Either all men are selfish or they are ignorant of their own interests.* The relation connecting the compared propositions is expressed by "either . . . or." We shall call the component propositions *alternants,* and the compound proposition an *alternative* proposition.

What do we mean to affirm in asserting an alternative proposition? We do not intend to assert the truth or falsity of any of the alternants: we do not say that *All men are selfish* nor that *All men are ignorant of their own interests.* All that we do assert is that *at least one* of the alternants is true.

Do we mean to say, however, that *both* alternants *cannot* be true? Usage in everyday conversation in this respect varies. An editorial which sums up the economic situation in the alternative proposition *Either this country will adopt national economic planning, or a revolution cannot be avoided* would generally be taken to mean that one, but not both, of the alternatives must be true. On the other hand, when we say *Either he is a fool or he is a knave* we do not mean to exclude the possibility that both alternatives are true. In the interest of unambiguity, we shall adopt the interpretation according to which the minimum is asserted by such a proposition. Henceforth an alternative proposition will be understood to mean one in which *at least one* of the alternatives is true, and *perhaps both.*

It is convenient to introduce a special symbol for the relation expressed by "either . . . or." We shall use the symbol V between the alternants to express this relation. *Either he is a fool or he is a knave* may then be written (He is a fool) V (he is a knave).

3. Consider next the compound proposition *The moon is full, and Venus is a morning star.* The relation connecting the component propositions is the conjunctive "and." We shall call such compound propositions *conjunctives,* and their components *conjuncts.*

What does a conjunctive proposition assert? Obviously, it asserts not only the truth of *The moon is full* and the truth of *Venus is a morning star* taken *singly;* it asserts the truth of the conjuncts *taken together.* Consequently, if either one of the conjuncts is false, the conjunctive proposition must itself be false. The conjunctive proposition must be regarded as a *single* proposition, and not as an *enumeration* of several propositions.

We shall employ a special symbol for the relation expressed by "and." A dot (.) between propositions will hereafter represent a conjunction. Thus *The moon is full and Venus is a morning star* may be written (The moon is full) . (Venus is a morning star).

4. The reader may wonder of what earthly use a conjunctive proposition can be in inference. A conjunctive proposition, he may claim, would never be adduced as evidence for any of its conjuncts. Thus if we are in doubt concerning the truth of *My watch keeps accurate time* would we offer in evidence the conjunctive *My watch keeps accurate time, and all spring-driven mechanisms are subject to climatic influences?* It is obviously more difficult to establish the truth of a conjunctive than to establish the truth of one of its conjuncts.

We may reply, however, that something may be inferred from a conjunctive which cannot be inferred from either of its conjuncts alone. Moreover, the *denial* of a conjunctive proposition yields a proposition that is extremely useful for inference. Indeed, the denial of a conjunctive proposition yields the fourth type of compound proposition. Thus, the denial of the proposition *My watch keeps accurate time, and all spring-driven mechanisms are subject to climatic influences* is *It is not the case that both my watch keeps accurate time and all spring-driven mechanisms are subject to climatic influences.* This means that at least one of the component propositions *My watch keeps accurate time* and *All spring-driven mechanisms are subject to climatic influence* must be false. We shall call the denial or negative of a conjunctive proposition a *disjunctive* proposition and its components *disjuncts.* Since a conjunctive asserts that *both* its conjuncts are *true,* its denial, which is a disjunctive proposition, asserts that *at least one* of the disjuncts is *false.* Both of two disjuncts cannot be true.

We have seen that in everyday conversation the alternants in an alternative proposition may be taken to be mutually exclusive. Thus in *Either he is single or he is married* the truth of one of the alternants excludes the truth of the other. Such alternative propositions *tacitly* assert a *disjunctive* proposition as well. As the usual meaning of *Either he is single or he is married* includes "He cannot be both," it can be expressed as follows: [(He is single) V (he is married)] . [(He is single) . (he is married)]'.

Let us now employ the foregoing distinctions to exhibit the logical form of some compound propositions. Consider the following argument: If every distinct racial group is characterized by a distinct culture, then either all nations differ from one another culturally, or national distinctions do not coincide, in whole or in part, with racial ones. But neither is it the case that the different nations possess distinct cultures, nor is it true that national distinctions do not coincide at any point with racial distinctions. Hence it is not true that each race has a distinct culture.

Let us use the letters p, q, r, to designate the following component propositions in this argument.

$p \equiv$ Every distinct racial group is characterized by a distinct culture.
$q \equiv$ All nations differ from one another culturally.
$r \equiv$ National distinctions do not coincide, in whole or in part, with racial ones.

The premises and conclusion of the argument may therefore be represented as follows:

a. $p \supset (q \vee r)$
b. $q' . r'$
c. p'

Each of the propositions a, b, and c is of a logical form different from those of the other two, a difference which the symbolism helps to bring out. The validity of the argument depends on the structure or form of the proposition a, b, and c, since the conclusion follows from the premises only if the following is true:

d. $(a . b) \supset c$

The reader should note that an important distinction can be drawn between the relation which the antecedent of a hypothetical proposition has to its consequent (as in proposition *a*), and the relation which the premises of a valid argument have to the conclusion (as in *d*). For the former relation material (or factual) evidence must be offered; while for the relation between premises and conclusion such evidence is both irrelevant and impossible, since this latter relation holds only when one of the terms related is logically or analytically contained in the other. The two relations, however, have a common trait, namely, that neither holds when the antecedent or premise is true and the consequent or conclusion is false. It is this common trait which is denoted by "if . . . then" or $\supset$. The reader should be on guard against the fallacy of supposing either that two things in any way alike cannot be unlike in some other way, or that two things in some ways distinct cannot also be alike in other ways.

Simple Propositions

The analysis of compound propositions into propositional elements clearly belongs to logic. But the analysis of a sentence into its verbal elements is an affair of grammar. Logically propositions are prior to words, in the sense that the propositions are not produced by the union of words—but the meaning of the words can be derived only from some propositional context. Ultimately the meaning of words is determined by elementary propositions of the form *This is a truffle, This is the color magenta,* and the like, where the word "this" may be replaced by some gesture of pointing. But while propositions cannot be analyzed into verbal constituents, attention to the latter is often an aid in analyzing or classifying propositions for logical purposes. Consider the following propositions:

1. Archimedes was modest.
2. Archimedes was a mathematician.
3. Archimedes was a greater mathematician than Euclid.

According to the traditional doctrine, each of these is a categorical proposition and its components are a subject, a predicate, and a copula joining them; any proposition such as *Archimedes loved mathematics* or *Archimedes ran naked through the streets crying Eureka* is to be analyzed by transforming it into *Archimedes was one who loved mathematics* or *Archimedes was one who ran naked,* and so on. It may be questioned whether such transformation does not change the meaning somewhat. But in any case it is possible to analyze propositions in other ways than the traditional one. Thus, using proposition 2 given above as a model, we can regard every proposition as asserting that some object is a member of a class. Proposition 1 would then assert that Archimedes was a member of the class of beings called modest, and proposition 3 would assert his membership in the class of mathematicians greater than Euclid. This second mode of analysis is related to the first mode as the point of view of extension is to that of intension.

An entirely different mode of analyzing propositions is to resolve them into the assertion of some relation between at least two objects. Thus our first proposition asserts a relation between Archimedes and modesty (the substance-attribute relation), our second proposition asserts a relation which may be called that of class membership between Archimedes and the class of mathematicians. Such propositions as *Archimedes solved Hiero's problem* may thus be analyzed as *Archimedes stood in the relation of solver to Hiero's problem.*

Now it is quite clear that no one of these modes of analysis can claim to be the only one; nor are they mutually exclusive. Nevertheless, each of these modes of analysis fits some propositions better than others. It seems quite forced to say that in the proposition *The author of* Macbeth *is the author of* Hamlet, "the author of *Hamlet*" is an attribute of "the author of *Macbeth.*" It seems more appropriate to view it as the assertion of a relation of identity in denotation, despite a difference in intension or connotation.

Of more direct logical importance is it to note that if we fail to discriminate between class-membership propositions and those which typify some other relation, we miss something that bears on the nature of implication. Thus, while some relations are transitive, that of class membership is not. *Archimedes was a greater mathematician than Euclid and Euclid was a greater mathematician than Aristotle* implies that *Archimedes was a greater mathematician than Aristotle.* But *Archimedes was a member of the Syracusan city-state and the Syracusan city-state was a member of the Graeco-Carthaginian alliance* does not imply that *Archimedes was a member of the Graeco-Carthaginian alliance.*

In Chapter VI we shall undertake a more systematic study of the relation between classes and of the logical properties of relations in general.

General Propositions

Consider the proposition: *All mathematicians are skilled logicians.* This cannot be fittingly regarded as a proposition of the subject-predicate type, for it does not predicate a character or quality to some individual. Nor does it assert that an individual is a member of a class. Nor would it be accurate to say that it asserts a relation between one individual and one or more others. What it does assert is the specific relation of inclusion between *two classes.* Propositions which deal with the relations between classes, that is, with the total or partial inclusion (or exclusion) of one class in another, are called *general* propositions. We have already indicated what the proper analysis of such propositions should be, in discussing categorical propositions in a previous section. Let us now approach the same conclusion from a slightly different angle.

The series of propositions *Archimedes was a mathematician, Euclid was a mathematician, Ptolemy was a mathematician,* are all of the same form. They differ only in having different terms as subjects. Let us now examine the *expression* "x is a mathematician." This is not a proposition, because it is incapable of being true or false. But propositions may be obtained from it by substituting proper values for x. All the propositions so obtained will have the same form. An expression which contains one or more variables, and which expresses a proposition when values are given to the variables, is called a *propositional function.*

We may vary not only the subject but other terms of the proposition as well. By varying the relation in *Archimedes was killed by a Roman soldier,* we may obtain *Archimedes was praised by a Roman soldier, Archimedes was spoken to by a Roman soldier, Archimedes was a cousin of a Roman soldier,* and so on. If we represent the relation by a variable R, we get the propositional function: Archimedes R a Roman soldier. (It is to be read: Archimedes has the relation R to a Roman soldier.) In such a manner, by letting the terms and relations in a proposition vary, and by representing them by variables, we are able to exhibit logical form or structure in a precise fashion.

When we state that *All mathematicians are trained logicians* what we mean is that if *any* individual is a mathematician he is also a trained logician. We are really affirming an *implication* between being a mathematician and being a trained logician. This may be expressed as an implication between propositions obtained from propositional functions, as follows:

[For all values of x, (x is a mathematician) $\supset$ (x is a trained logician)], where the sign $\supset$ indicates as usual the "if . . . then" relation between the propositions, obtained from the propositional functions by giving values to x.

Propositions of this type, which assert the inclusion (or exclusion) of one class in (or from) another, are analogous in some ways to compound propositions. They should therefore not be confused with class-membership propositions. For the class-membership relation is not transitive, as we have seen, while the relation of class inclusion is transitive. Thus, if *All mathematicians are trained logicians* and *All trained logicians are college professors,* we may validly infer that *All mathematicians are college professors.*

Let us now express each of the four kinds of categorical propositions in this new notation.

1. *All students are independent thinkers* is equivalent to [For all x's, (x is a student) $\supset$ (x is an independent thinker)].
2. *No students are independent thinkers* is equivalent to [For all x's, (x is a student) $\supset$ (x is an independent thinker)′].
3. *Some students are independent thinkers* is equivalent to [There is an x such that (x is a student) . (x is an independent thinker)].
4. *Some students are not independent thinkers* is equivalent to [There is an x, such that (x is a student) . (x is an independent thinker)′].

The two universal propositions (1 and 2) evidently have a logical form quite distinct from the two particulars (3 and 4), while all four propositions are quite distinct in form from the subject-predicate form of propositions.

CHAPTER III

THE RELATIONS BETWEEN PROPOSITIONS

§ 1. THE POSSIBLE LOGICAL RELATIONS BETWEEN PROPOSITIONS

The logician's interest in the structure of propositions arises from his desire to exhibit all possible propositional forms in virtue of which propositions imply one another. Propositions may be related in ways other than by implication. Thus *The exchange value of a commodity is proportional to the amount of labor required for its production* and *The supply of a commodity is proportional to the demand* are related propositions, since they are both about economics. And *Continuous eloquence wearies* and *Thought constitutes the greatness of man* are also related in virtue of the fact that Pascal believed both of them. This kind of relatedness, however, is not the logician's concern. The relations between propositions which are logically relevant are those in virtue of which the possible truth or falsity of one or more propositions limits the possible truth or falsity of others. Let us examine them.

The conclusion of Plato's dialogue *Protagoras* finds Socrates summarizing the discussion on the nature of virtue:

"My only object . . . in continuing the discussion, has been the desire to ascertain the nature and relations of virtue; for if this were clear, I am very sure that the other controversy which has been carried on at great length by both of us—you affirming and I denying that virtue can be taught—would also become clear. The result of our discussion appears to me to be singular. For if the argument had a human voice, that voice would be heard laughing at us and saying: 'Protagoras and Socrates, you are strange beings; there are you, Socrates, who were saying that virtue cannot be taught, contradicting yourself now by your attempt to prove that all things are knowledge, including justice, temperance, and courage,—which tends to show that virtue can certainly be taught; for if virtue were other than knowledge, as Protagoras attempted to prove, then clearly virtue cannot be taught; but if virtue is entirely knowledge, as you are seeking to show, then I cannot but suppose that virtue is capable of being taught. Protagoras, on the other hand, who started by saying that it might be taught, is now eager to prove it to be anything rather than knowledge; and if this is true, it must be quite incapable of being taught.' Now I, Protagoras, perceiving this terrible confusion of our ideas, have a great desire that they should be cleared up." [1]

Let us examine the following propositions in this excerpt:

a. Virtue cannot be taught.
b. If virtue is not knowledge, then virtue cannot be taught.
c. If virtue is knowledge, then virtue can be taught.
d. Virtue can be taught.
e. Virtue is knowledge.
f. Virtue is not knowledge.

Very little reflection is required to show that propositions *a* and *d* are connected *logically,* and not only because both are about virtue. For there is clearly a limitation upon their possible truth or falsity. Both propositions cannot be true, since one affirms what the other denies; and both propositions cannot be false, for the same reason. Precisely the same relation holds between *e* and *f*. Such propositions are *contradictories* of one another.

How about *b* and *c*? The reader may be tempted to regard these as contradictories also. This, however, would be erroneous. Reflection shows that there is no contradiction in saying that virtue can be taught under certain contingencies (if virtue is knowledge) but not under others (if virtue is not knowledge). There is, indeed, no mutual limitation upon the possible truth or falsity of these two propositions. Such propositions, even though they deal with the same subject matter, are logically *independent.* The reader should determine for himself whether there are any other pairs of independent propositions in the set above.

Let us now consider *b* and *f* asserted *jointly,* thus forming a conjunctive proposition, and ask for the relation between this conjunctive and *a*. It is easy to see that if *both b and f* are true, *a* must also be true. But if *a* is true, does anything follow concerning the

[1] Plato, *The Dialogues,* tr. by B. Jowett, 1892. 5 vols., Vol. I, pp. 186-87.

truth of *both b and f?* Evidently not, for *a* may be true on other grounds than those supplied by the conjunctive *b and f;* for instance, human obstinacy, bad habits, or the weakness of the flesh; and the conjunctive may be false even though *a* is true, for it may be true, for example, that virtue is knowledge and yet cannot be taught. Propositions so related that if the first is true the second is also true, but if the second is true the first is *undetermined* or *not thereby limited* in its truth-value, are said to be in the relation of *superaltern to subaltern.* The convenient name of *superimplication* has also been devised for this relation. Can the reader find other combinations of propositions in this set which are in this relation?

We have thus far identified three types of relations between propositions: contradictoriness, independence, relation of superaltern to subaltern. Are these all the possible relations there are? We can obtain an exhaustive enumeration of such relations if we examine all the possible truth-values of a pair of propositions (where the "truth-value" of a proposition is either truth or falsity). Let p symbolize any proposition, and q any other. The following table contains all their possible truth-values. We must allow for the possibility that the truth-value of one or other of the propositions is not limited or determined by the other, by denoting such lack of determination as "Undetermined."

	p	q		p	q
1	True	True	4	False	True
2	True	False	5	False	False
3	True	Undetermined	6	False	Undetermined

Two propositions may be related to each other in any one of these six ways. But to impose only a *single* one of these six conditions upon a pair of propositions is not sufficient to determine their logical relation to one another *uniquely.* Thus the relation called contradictory, such as between *e* and *f,* needs two conditions to determine its properties, namely, the conditions 2 and 4. The relation of superimplication also requires two conditions, namely, 1 and 6. Reflection shows that the other possible logical relations similarly require two of these six conditions to determine them. By pairing each of the first three conditions with each of the last three, we then get nine possible relations between propositions, not all of which, however, are distinct.

1. If p is true, q is true.
   If p is false, q is true.

In this case, the truth-value of q is not limited by the truth-value of p. Two propositions so related are said to be *independent.*

2. If p is true, q is true.
   If p is false, q is false.

Two propositions so connected are said to be *equivalent.*

3. If p is true, q is true.
   If p is false, q is undetermined.

Propositions so related are said to be in the relation of *super-altern* (or *principal*) *to subaltern.* As noted before, we shall also use the designation *superimplication.*

4. If p is true, q is false.
   If p is false, q is true.

The reader will recognize this as the case of *contradictory* relation.

5. If p is true, q is false.
   If p is false, q is false.

Here the falsity of q does not depend on the truth or falsity of p, and the propositions are *independent.*

6. If p is true, q is false.
   If p is false, q is undetermined.

In this case, p and q are said to be *contraries:* both cannot be true, but both may be false.

7. If p is true, q is undetermined.
   If p is false, q is true.

In this case, both propositions cannot be false, but both may be true. Such propositions are called *subcontraries.*

8. If p is true, q is undetermined.
   If p is false, q is false.

In this case, the relation of p to q is the *converse* of the relation in 3, and p is said to stand in the relation of *subimplication* or that of *subaltern to principal* to q.

9. If p is true, q is undetermined.
   If p is false, q is undetermined.

In this case, p and q are also *independent*, since the truth-value of p does not determine the truth-value of q.

There are, therefore, seven distinct types of logical relations between one proposition or set of propositions and another proposition or set. (Note that 1, 5, and 9 are of the same type.) Propositions may be (1) equivalent, (2) related as principal to subaltern, (3) related as subaltern to principal, (4) independent, (5) subcontraries, (6) contraries, or (7) contradictories. These are all the fundamental types of logical relations between propositions, and *every discussion we shall enter into in the present book may be regarded as illustrating one of them.* A full understanding of these seven relations will give the reader an accurate synoptic view of the province of logic.[2]

§ 2. INDEPENDENT PROPOSITIONS

We have agreed to call two propositions independent, if the truth-value of one of them in no way determines or limits the truth-value of the other. Thus, if we were considering whether the proposition *Pericles had two sons* is true or false, we should not regard the truth or falsity of the proposition *Hertz discovered electric waves* as evidence either way. When one proposition is thus no evidence at all for the truth or falsity of the other, we also speak of the former as irrelevant to the latter. One of the duties of a court of law or of any other rational procedure is to rule out irrelevant testimony. This does not deny that propositions which we now do not know to be in any way related may later be discovered to be indirectly connected with each other. No one before the middle of the 18th century could have seen any connection between prop-

[2] There will be seemingly more relations if we introduce the question of symmetry or reversibility of the relations between the propositions p and q. Thus the relation between a hypothesis and its logical consequence will then be described by the tetrad: If p is true, q is true; if p is false, q is undetermined; if q is true, p is undetermined; if q is false, p is false. It is an interesting exercise for the student to try to determine how many tetrads of this kind are logically possible.

ositions about thunder and lightning, about the color of mother of pearl and about the attractive power of a lodestone; and yet they are now all part of electromagnetic theory. But formal logic is not concerned with guaranteeing factual omniscience. The logical test of independence is simply whether it is possible for a given proposition to be a) true, or b) false, or c) have its truth undetermined, irrespective of whether another given proposition is true or false. Thus a) if the proposition, *The angle of reflection of a light ray is always equal to the angle of its incidence,* is true whether the hypothesis *Light consists of corpuscles* is true or false, the former proposition is independent of the latter. Similarly, b) any proposition which can be demonstrated to be false, such as *The sum of two sides of a triangle is not greater than the third side,* will be independent of any proposition which may be considered as true or as false, e.g., *Through a point outside of a straight line only one parallel can be drawn.* c) A third instance of one proposition being independent of another can be seen in the pair, *The greatest contribution to physics in the 18th century was made by England* and *Sir Philip Sidney was the author of the Letters of Junius.*

In all these pairs of propositions, the truth-value of the first member of each pair is not limited or determined by whether the second member is true or false.[3]

§ 3. EQUIVALENT PROPOSITIONS

The realization that there are several ways of saying the same thing is very valuable in the search for truth. Futile controversies flourish not only because everyone prefers his own formulation of the beliefs he holds, but also because this preference makes very few of us willing to analyze apparently opposing expressions of belief in order to discover whether the obvious differences are verbal or whether they are material. In any case, the examination of what propositions are equivalent is an essential part of all rational inquiry.

Several forms of equivalent propositions have been studied by traditional logic. In turning to examine these, the reader will do well to employ either the diagrammatic representation of categorical propositions or the algebraic statements which express their import.

[3] Further consideration of the tests for, and proof of, the independence of propositions will be taken up in Chapter VII.

Conversion

Consider what the proposition *No agricultural country is tolerant on religious questions* asserts. Obviously, it contains the same information as *No countries tolerant on religious questions are agricultural,* for if agricultural countries are excluded from being tolerant, the latter must also be excluded from the former. These two propositions are equivalent—if one is true or false, the other is the same. They have the same terms as subject and predicate, but the subject of the first is the predicate of the second, and the predicate of the first is the subject of the second. The second proposition is said to be the *converse* of the first. The process by which we pass from a proposition to another that has the same truth-value and in which the *order* of subject and predicate is interchanged, is called *conversion.* An *E* proposition can therefore be converted.

Can each of the other categoricals be converted? May we validly infer from *All bald men are sensitive* the proposition *All sensitive men are bald?* We certainly may not. The reader will see this more clearly if he will note that in the first proposition the term "sensitive" is undistributed while in the converse the same term is distributed. This is not permissible, for it would be tantamount to asserting something of all of a class on the evidence of an assertion concerning only an indefinite part of that class. Indeed, we may state as a general principle: *In inferences from categorical propositions no term may be distributed in the conclusion which is undistributed in at least one of the premises.* Consequently, from *All bald men are sensitive* we may infer no more than *Some sensitive men are bald.* Thus, an *A* proposition can be converted only by *limitation* or *per accidens,* that is, its quantity must be changed. But as we saw before, no inference *per accidens* is valid without the assumption that the class denoted by the subject has members, in this case bald men.

The converse of the *I* proposition *Some Republicans are conservatives* is *Some conservatives are Republicans.* Hence the converse of an *A* proposition may be regarded as the converse of its subaltern.

What is the converse of *Some Italians are not dark-haired?* Is the reader tempted to say it is *Some dark-haired individuals are not Italians?* But this is clearly a fallacious inference. From the proposition *Some mortals are not men* it does not follow that

Some men are not mortals. Technically, this is expressed by saying that such an inference violates the principle concerning the distribution of terms. Indeed, the *O* proposition has no converse; but, as we shall see later, we can infer that *Some men not having dark hair are Italians.*

Obversion

We may obtain equivalent propositions in another way. If *All employees are welcome,* what may we infer about the relation of employees to those who are unwelcome? Evidently *No employees are unwelcome* is a valid conclusion. These two propositions are equivalent: the first declares that nobody is both an employee and unwelcome, and the second asserts the same thing. The inference is called *obversion* and each proposition is the *obverse* of the other. The subjects of the propositions are the same, but the predicate term of one is the negative or contradictory of the other; and the quality of the propositions is different. Care must be taken that the predicate of the obverse should be the contradictory of the predicate in the premise. Thus the obverse of *All leaves are green* is not *No leaves are blue,* for "green" and "blue" are not contradictory terms—they are merely contrary. Two terms are *contradictory* in a universe of discourse if they are exhaustive as well as exclusive; two terms are *contrary* if they are simply exclusive. The proper obverse of *All leaves are green* is *No leaves are non-green,* or in more colloquial English *No leaves are other in color than green.*

Each of the four types of categoricals may be obverted simply—that is, without limitation. The reader should verify that the obverse of *No Laplanders are educated* is *All Laplanders are uneducated;* of *Some college presidents are intelligent* is *Some college presidents are not unintelligent;* and of *Some gases are not poisonous* is *Some gases are nonpoisonous.*

Conversion and obversion are two forms of inference for passing from one proposition to another which is equivalent to it. Other types studied in traditional logic can be defined in terms of successive applications of these two.

Contraposition

If *All reasonable petitions are investigated,* what may be inferred concerning the relation of the uninvestigated things either to the reasonable petitions, or to the non-reasonable petitions? The reader may admit that one permissible conclusion is *No uninvestigated things are reasonable petitions* and that another is *All uninvestigated things are non-reasonable petitions.* But if the reader should not see that these conclusions necessarily follow, he will be able to do so if he performs the following series of obversions and conversions. We shall consider all four categorical propositions together.

1.	All reasonable petitions are investigated.	No reasonable petitions are investigated.	Some reasonable petitions are investigated.	Some reasonable petitions are not investigated.
2.	No reasonable petitions are uninvestigated.	All reasonable petitions are uninvestigated.	Some reasonable petitions are not uninvestigated.	Some reasonable petitions are uninvestigated.
3.	No uninvestigated things are reasonable petitions.	Some uninvestigated things are reasonable petitions.		Some uninvestigated things are reasonable petitions.
4.	All uninvestigated things are non-reasonable petitions.	Some uninvestigated things are not non-reasonable petitions.		Some uninvestigated things are not non-reasonable petitions.

The first row contains the four categorical propositions. The second row contains the corresponding *obverses* of the propositions in the first row. The third row contains the *converses* of the propositions in the second row. And the fourth row contains the *obverses* of the propositions in the third row.

The propositions in the third row are called the *partial contrapositives* of the corresponding propositions in the first row. The partial contrapositive of a proposition is one in which the subject is the contradictory of the original predicate, while the predicate is the original subject; it also differs in quality from the original proposition. The *I* proposition has no partial contrapositive, and the *E* has one by limitation. The partial contrapositives of the *A* and *O* are equivalent to the original proposition.

The propositions in the fourth row are the *full contrapositives* of the corresponding propositions in the first row. The full contrapositive is a proposition in which the subject is the contradictory of the original predicate, and the predicate is the contradictory of the original subject. It has the same quality as the original proposition. As before, the *I* proposition has no full contrapositive, and the *E* has one only by limitation.

Obverted Converse

We have obtained equivalent propositions by performing a series of obversions and conversions, *in this order,* upon each of the four types of categoricals. A different set of equivalent propositions are obtained, however, if instead we first convert a given proposition, then obvert the result. The following table summarizes the outcome:

1. All reasonable petitions are investigated.	No reasonable petitions are investigated.	Some reasonable petitions are investigated.	Some reasonable petitions are not investigated.
2. Some investigated things are reasonable petitions.	No investigated things are reasonable petitions.	Some investigated things are reasonable petitions.	
3. Some investigated things are not non-reasonable petitions.	All investigated things are non-reasonable petitions.	Some investigated things are not non-reasonable petitions.	

It will be noted that the *E* and *I* have an obverted converse without limitation, the *A* has a limited obverted converse, while the *O* has none at all.

Inversion

If *All physicists are mathematicians,* what may be inferred about the relation of the non-physicists to the mathematicians, or to the non-mathematicians? Let us discover what may be validly inferred by a successive application of conversions and obversions.

We may begin by first converting the proposition, then obverting, and so on, until we obtain a proposition which satisfies the problem; or we may begin by first obverting, then converting and so on. Let us develop the alternative methods in parallel columns, the first method in the left-hand column, the second in the right-hand one.

All physicists are mathematicians.	All physicists are mathematicians.
Some mathematicians are physicists.	No physicists are non-mathematicians.
Some mathematicians are not non-physicists.	No non-mathematicians are physicists.
	All non-mathematicians are non-physicists.
	Some non-physicists are non-mathematicians.
	Some non-physicists are not mathematicians.

Hence, if we first convert an *A* proposition we are soon brought to a halt, because an *O* proposition cannot be converted. If we first obvert, we get two propositions which are satisfactory. *Some non-physicists are not mathematicians* is called the *partial inverse* of the original proposition. Its subject is the contradictory of the original subject, its predicate is the original predicate. *Some non-physicists are non-mathematicians* is called the *full inverse*. Both its subject and its predicate are the contradictories of the original subject and predicate respectively.

Has each form of categorical proposition an inverse? If the reader will use the method we have indicated, he will discover that from *No professor is unkind* he can infer *Some non-professors are unkind* (the *partial inverse*) and *Some non-professors are not kind* (the *full inverse*). But no inverses can be obtained from either an *I* or an *O* proposition. Hence, only universals have inverses, and in each case inversion is by limitation.

The process of inversion may lead to absurd results as when, from *All honest men are mortal* we seem to get *Some dishonest men are immortal*. Where has the error crept in? The answer is: In our careless use of negatives. The true inverse of our proposition is *Some non-honest-men are non-mortal,* which is not at all an absurd result. For the class of all beings that are not *honest men* is wider than the class *dishonest men* and includes triangles and the like which are certainly non-mortal.

"But look here!" the reader may object. "The partial inverse of *All physicists are mathematicians* is *Some non-physicists are not mathematicians*. In the first proposition the predicate is undistributed, although the same term is distributed in the second. How then can you maintain that the second is a valid consequence of the first? Does it not violate the principle concerning distribution of terms?"

If the reader has understood the discussion of the existential import of propositions, he will have a ready reply. A universal proposition, he will say, asserts *nothing* about the existence or non-existence of instances; particulars, on the other hand, do have existential import. Consequently, a particular can never be validly inferred from a universal or combination of universals, *unless* the premises include a proposition asserting that the classes denoted by the terms of the universal contain at least one member. And specifically, the conversion of *A* is valid only if the predicate denotes such a class.

The source of the trouble in inversion is now apparent. To get the inverse of *All physicists are mathematicians* we are required to convert *All non-mathematicians are non-physicists.* This can be done only if we add the further premise *Some men are non-physicists* or what is the same thing *Some men are not physicists.* If this premise is supplied, the partial inverse does not violate the principle concerning distribution of terms.

If universals always did have an existential import, then not only would the terms of such propositions denote classes with members, but the contradictory terms would do so as well. Thus if *All men are mortal* required that there should be men and mortals, since we may validly infer *All immortals are non-men,* we would be compelled to affirm that there are immortals as well as non-men. That universals do not always have existential import, even in everyday conversation, can be seen from the following. Students of mathematics know that the ancient Greek problem, to construct a square equal in area to a circle with compass and ruler, is demonstrably impossible. We may therefore assert confidently that *No mathematicians are circle-squarers.* Its partial inverse is *Some non-mathematicians are circle-squarers.* But we assuredly did not intend to assert anything whose consequence is that there are some people who can in fact square the circle—for there is a proof that this cannot be done. Hence the original proposition could not have been intended to assert the existence of circle-squarers.

Inference by Converse Relation

If *Chicago is west of New York* we may validly infer that *New York is east of Chicago;* if *Socrates was a teacher of Plato* we may infer that *Plato was a pupil of Socrates;* if *Seven is greater than five* we may infer that *Five is less than seven.* In each of these pairs of propositions the two are equivalent. Such inferences are of the form: If *a* stands to *b* in a certain relation, *b* stands to *a* in the *converse relation.*

Equivalence of Compound Propositions

We must now examine what are the *equivalent* forms of compound propositions.

Consider the hypothetical *If a triangle is isosceles, its base angles are equal.* To assert it means, as we have seen, to assert that the truth of the antecedent involves the truth of the consequent, or that the antecedent could not be true and the consequent false. Hence the hypothetical simply asserts that the conjunctive proposition *A triangle is isosceles, and its base angles are unequal* is false; or, what is the same thing, that the disjunctive proposition *It is not the case that both a triangle is isosceles and its base angles are unequal* is true. It follows that from a hypothetical we may infer a disjunctive proposition.

Moreover, we may infer the hypothetical from the disjunctive as well. For if *It is not the case that both a triangle is isosceles and its base angles are unequal,* then the truth of one of the disjuncts is incompatible with the truth of the other: if one disjunct is true, the other must be false. From this disjunctive proposition we may therefore infer *If a triangle is isosceles, its base angles are equal.* Hence a disjunctive proposition can be found which is *equivalent* to a hypothetical.

If we employ our previous symbols, we may write this as follows:

$$[(\text{A triangle is isosceles}) \supset (\text{its base angles are equal})] \equiv$$
$$[(\text{A triangle is isosceles}) \cdot (\text{its base angles are equal})']'.$$

But this discussion also shows how we can infer an equivalent hypothetical from any hypothetical proposition. For if in the equivalent disjunctive the second disjunct is supposed true, the other disjunct must be false. We may therefore infer *If the base angles of a triangle are unequal, it is not isosceles.* Hence we may write:

$$[(\text{A triangle is isosceles}) \supset (\text{its base angles are equal})] \equiv$$
$$[(\text{The base angles of a triangle are equal})' \supset (\text{the triangle is isosceles})'].$$

These equivalent hypotheticals are said to be the *contrapositives* of each other.

Consider next the alternative proposition *Either a triangle is not isosceles or its base angles are equal.* To assert it means to assert that *at least one* of the alternants is true. If, therefore, one of the alternants were false, the other would have to be true. Hence we may infer from the alternative above the following hypothetical *If a triangle is isosceles, its base angles are equal.* Moreover, the alternative may be inferred from this hypothetical. For the latter is equivalent to *It is not the case that both a triangle is isosceles and its base angles are unequal,* which asserts that *at least one* of the disjuncts must be false. From the disjunctive we may therefore infer *Either a triangle is not isosceles, or its base angles are equal.* We therefore may write the equivalence:

[(A triangle is isosceles)′ ∨ (its base angles are equal)] ≡
[(A triangle is isosceles) ⊃ (its base angles are equal)]

It follows that for every hypothetical there is an equivalent alternative, an equivalent disjunctive, and an equivalent hypothetical. A similar statement holds for every alternative and every disjunctive proposition. On the other hand, a conjunctive proposition is not equivalent to any one of the other three forms of compound propositions.

Let us now state the equivalents of *If he is happily married, he does not beat his wife.* They are *If he beats his wife, he is not happily married, Either he is not happily married or he does not beat his wife,* and *It is not the case that both he is happily married and he beats his wife.* In symbols they will read:

[(He is happily married) ⊃ (He does not beat his wife)] ≡
[(He does not beat his wife)′ ⊃ (He is happily married)′] ≡
[(He is happily married)′ ∨ (He does not beat his wife)] ≡
[(He is happily married) . (He does not beat his wife)′]′

These equivalents may be stated more compactly, and the forms of the equivalent propositions exhibited more clearly, if we adopt further symbolic conventions. Let p represent the antecedent of a hypothetical proposition, and q its consequent. Any hypothetical may then be symbolized by $(p \supset q)$. The equivalences will then be written as follows:

$$(p \supset q) \equiv (q' \supset p') \equiv (p' \vee q) \equiv (p \cdot q')'$$

We shall discuss in Chapter VII equivalences between *systems* of propositions. We may, however, offer at this point an example of two propositions which are equivalent in virtue of their place in a system. Let $p \equiv$ *In Newtonian physics, light is reflected from a surface so that the angle of incidence is equal to the angle of reflection* and let $q \equiv$ *In Newtonian physics light is reflected from a surface so that its path is a minimum:* p and q are equivalent.

§ 4. THE TRADITIONAL SQUARE OF OPPOSITION

Traditionally, the *opposition* between propositions has not been conceived in a manner as general as we have indicated. Since on the traditional view all propositions were analyzable into subject and predicate, only propositions in that form could be opposed. The opposition of compound propositions was not discussed, and the discussion of the opposition of singular propositions was most unsatisfactory.

In this section we shall examine the traditional account of opposition. According to it, two propositions are said to be opposed when they have the same subject and predicate, but differ in quantity or quality or both.

Consider, therefore, the four propositions:

A. All republics are ungrateful.
E. No republics are ungrateful.
I. Some republics are ungrateful.
O. Some republics are not ungrateful.

In discussing the existential import of propositions, we have seen that the universals do not require the existence of republics, while the particulars do. We cannot, therefore, without further assumptions, infer the truth of the *I* proposition from the truth of the *A* proposition. To do so, we require the assumption that there are republics. We shall make that assumption in the present section once for all, and explore the consequences of this hypothesis.

No two of the four propositions above are independent of one another, and no two are equivalent. We may, however, identify the other five relations of the possible nine as follows. The reader will find it helpful to make use of the diagrammatic representation of propositions.

1. *All republics are ungrateful* and *Some republics are not ungrateful* cannot both be true, and they cannot both be false. If one is true, we may validly infer the falsity of the other; and if one is false we may infer the truth of the other. Hence the *A* and *O* propositions are *contradictories;* and the same holds for the *E* and *I* propositions.

2. *All republics are ungrateful* and *No republics are ungrateful* cannot both be true, so that if one were true, we could infer the falsity of the other. But if one were false, the truth-value of the other would be undetermined. Hence the *A* and *E* propositions are *contraries.*

3. Examining *All republics are ungrateful* and *Some republics are ungrateful,* we find that the truth of the second may be inferred from the truth of the first. But if the first is false, we can infer nothing about the truth-value of the second. Hence the *A* proposition is the principal or superaltern to the *I* proposition, which is the subaltern. The same relation holds for the *E* and *O* propositions.

4. On the other hand, from the falsity of the *I* proposition we

may infer the falsity of the *A* proposition. But from the truth of the *I* proposition we cannot infer the truth-value of the *A* proposition. Hence the *I* proposition stands to the *A* proposition as the subaltern to the principal. Similarly for the *O* and *E* propositions.

5. Finally, the truth of *Some republics are ungrateful* is compatible with the truth of *Some republics are not ungrateful,* although when we remember our convention that the word "some" is not to exclude "all" we see that we cannot infer the truth of one from the truth of the other. But if either of them is false, the other must be true. Hence the *I* and *O* propositions are subcontraries. This result also follows from the fact that the *A* and *E* propositions are contraries. For since the *O* and *I* propositions are the contradictories of the *A* and *E* propositions respectively, and since contraries cannot both be true, the *O* and *I* propositions cannot both be false; while since the *A* and *E* propositions may both be false, the *O* and *I* propositions may both be true.

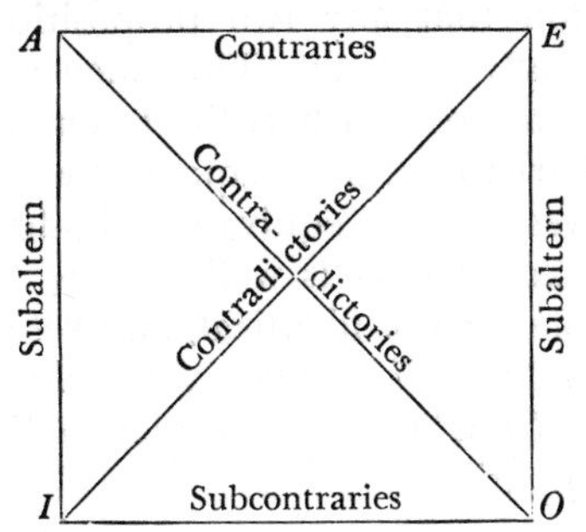

These relations between the categorical propositions have been represented in the ancient square of opposition. We may also construct the following table of valid inferences from each of the four categoricals.

	A	*E*	*I*	*O*
If *A* is true		False	True	False
If *A* is false		Undetermined	Undetermined	True
If *E* is true	False		False	True
If *E* is false	Undetermined		True	Undetermined
If *I* is true	Undetermined	False		Undetermined
If *I* is false	False	True		True
If *O* is true	False	Undetermined	Undetermined	
If *O* is false	True	False	True	

In conclusion, we may point out that without the assumption concerning existence we have made at the beginning of this section, the *I* and *O* propositions cannot be inferred from the *A* and *E* proposition respectively. Moreover, without this assumption, the *A* and *E* propositions would not be contraries, since both might then be true. Thus *All immortal men are in this room* and *No immortal men are in this room* would be both true if there were no immortal men. For if it is false that there are immortal men, then (on the interpretation we have given to particular propositions) both the propositions *Some immortal men are in this room* and *Some immortal men are not in this room* are false. Hence the contradictories of these propositions must be true.

§ 5. THE OPPOSITION OF PROPOSITIONS IN GENERAL

One of the most fruitful sources of intellectual confusion is the too facile assumption that any two propositions which are not equivalent are mutually exclusive. Thus men have debated about the relation of mind to body, of heredity to environment, of selfishness to altruism, of art to nature, frequently without realizing that while the alternatives are not equivalent, it does not follow that they are mutually exclusive. It is well to note that two things related as part to whole are not identical, and yet surely are not contraries. The proposition *All Orthodox Greeks are true believers* is not the same as *All Christians are true believers* but the two are surely not incompatible.

The Contradictory Opposition of Compound Propositions

We have seen this relation illustrated in the traditional square of opposition by such simple cases as *All Mohammedans are true believers* and *Some Mohammedans are not true believers.* But it is a mistake to suppose that it is always easy to tell which propositions do have that relation to one another.

Suppose the reader in studying Rousseau's *Social Contract* found himself in violent disagreement with the opening sentence of the first chapter: "Man is born free; and everywhere he is in chains." How would he contradict this assertion? Or suppose he wished to deny a statement appearing later on: "The strongest is never strong enough to be always the master, unless he transforms strength into right, and obedience into duty." With what proposition would he contradict this? Or, finally, what would he say is the contradictory of this: "Sovereignty, being the exercise of the general will, is the will either of the body of the people or of only a part of it"?

Let us begin with the conjunctive proposition *p: Man is born free; and everywhere he is in chains.* Since to assert this means to assert that both conjuncts are true, to deny it must mean that not both conjuncts are true, or what is the same thing, that at least one of them is false. Hence the alternative *q: Either some men are not born free or man is not in chains everywhere* is the proper contradictory of the original proposition. The reader should convince himself that these two propositions, *p* and *q*, cannot both be true and cannot both be false. Other forms of the contradictory can be obtained, since the alternative is equivalent to *If all men are born free, they are not in chains* as well as to *If man is in chains everywhere, some men are not born free.*

Consider next *The strongest is never strong enough to be always the master, unless he transforms strength into right, and obedience into duty.* This is really a hypothetical proposition, and may be expressed as *If the strongest does not transform strength into right and obedience into duty, he is never strong enough to be always master.* It is therefore equivalent to the disjunctive *It is not the case that both the strongest does not transform strength into right and obedience into duty, and he is sometimes strong enough to be always the master.* Hence its contradictory is *The strongest does not transform strength into right and obedience into duty, and he is sometimes strong enough to be always master.*

Finally, *Sovereignty, being the exercise of the general will, is the will either of the body of the people or of only a part of it* is an alternative proposition, which may be stated explicitly as *Either sovereignty, being the exercise of the general will, is the will of the body of the people, or it is the will of only a part of it.* Since this proposition is equivalent to a disjunctive one, its contradictory is the following conjunctive: *Sovereignty is not the will of the body of the people, and it is not the will of only a part of it.*

It follows that the contradictory of a hypothetical, disjunctive, or alternative proposition can always be stated in the form of a conjunctive proposition. On the other hand, the contradictory of a conjunctive is either a hypothetical, an alternative, or a disjunctive proposition. The precise relations between compound propositions and their contradictories may be compactly stated symbolically. Since

$$(p \supset q) \equiv (q' \supset p') \equiv (p' \vee q) \equiv (p \cdot q')',$$

the contradictory of any one of them will contradict any other. Hence

$$(p \supset q)' \equiv (q' \supset p')' \equiv (p' \vee q)' \equiv (p \cdot q').$$

Or in words, the contradictory of "if p then q" is "p and q'"; the contradictory of "either p or q" is "p' and q'"; the contradictory of "not both p and q" is "p and q."

The reader should note the equivalence $(p' \vee q)' \equiv (p \cdot q')$. This relation is perfectly general, and it does not matter what propositions we substitute for the symbols. Let us therefore replace p' by r; then p will be replaced by r'. We then get

$$(r \vee q)' \equiv (r' \cdot q'),$$

a relation known as De Morgan's theorem. It asserts that the negative (or contradictory) of an alternative proposition is a conjunctive in which the conjuncts are the contradictories of the corresponding alternants. Another form of this theorem is given by

$$(p \cdot r)' \equiv (p' \vee r'),$$

which states that the negative of a conjunctive is an alternative proposition in which the alternants are the contradictories of the corresponding conjuncts.

We have found that specially devised symbols are a great help in exhibiting with great clarity the logical structure of propositions—a structure which is obscured by the unwieldiness of ordinary language. The reader will therefore doubtless agree that symbols are not an obstacle to understanding, but rather an aid. The generalizing power of modern logic, as of modern mathematics, is due in large measure to the adequacy of the symbolism it has adapted for the task.

As a test of his comprehension, we may ask the reader to give the contradictory of *Some men are poor but honest.* It must be understood that the force of the word "but" in this proposition is that the poor are generally dishonest although some also happen to be honest. Hence its explicit meaning is *Some men are poor and honest, and some men are poor and dishonest.* It follows that *Some men are not poor and honest* is not the contradictory of the original statement, and neither is *All men are not both poor and honest.* Application of the previously discussed principles shows that the contradictory is *Either all men are not both poor and honest or all men are not both poor and dishonest.* Similar considerations show that the contradictory of *John came home yesterday on a bicycle* is not *John did not come home yesterday on a bicycle* but rather *Either John did not come home or John did not come home yesterday or John did not come home on a bicycle.*

Contrary Opposition

This relation has also been illustrated in the traditional square of opposition. But clearly other examples of contrary propositions can be found. *I am seven feet tall* and *I am six feet tall. Socrates was the wisest of the Greeks* and *Plato was the wisest of the Greeks. Columbus was the first European to discover America* and *Leif Ericson was the first European to discover America.* These are examples of pairs of contraries which do not fall into the framework of the traditional scheme. It is evident that general propositions may have more than one contrary.

What is the contrary of *Either the book was stolen or I mislaid it?* One contrary is *The book was not stolen and I did not mislay it, and my brother did not borrow it;* these two compound propositions may be both false, for example, when *My brother borrowed it* is true. Symbolically, and in general, the following pair of compound propositions are contraries:

$$(p \vee q) \qquad \text{and} \qquad (p' \cdot q' \cdot r)$$

where p, q, and r are any propositions.

It should be clear from these examples that two propositions may be incompatible with each other even though neither may in fact be true. This elementary point seems to have been often overlooked in some of the great intellectual controversies in the history of human thought. The bitter fights between idealists and realists, between revolutionists and conservatives, between evolutionists and fundamentalists, between theists and deists, have been waged not only on the assumption that the respective points of view are exclusive of one another, but also on the supposition that if one of the parties were proved in the wrong, the other would be proved right. But such is the irony of history that these famous historical oppositions are not today generally acknowledged as dividing the truth between them.

Subcontrary Opposition

In addition to the propositions falling into the traditional schedule, the following pairs illustrate the relation of subcontrary opposition: *There is a page in this book containing misprints* and *There is a page in this book containing no misprints. Hydrogen is not the lightest element* and *Helium is not the lightest element. San Marino is not the smallest country of Europe* and *Andorra is not the smallest country of Europe.* The two propositions in each pair cannot both be false, but both may be true.

Other examples of this important relation are: Under the prevailing organization of governments, the propositions: *Some European countries are monarchies* and *Some European countries are republics* are subcontraries. In a country which is planning to raise taxes, but where the budget must be balanced by increasing either the tariff or the income tax: *The income tax will be raised* and *The tariff will be increased* are again subcontraries. The relation of subcontrariness is thus a very simple one, and it may seem strange that serious errors should ever be made by failing to understand its nature or to recognize it in other examples. Yet the history of human thought shows a powerful tendency to conceive rival hypotheses as contrary or even contradictory when they are in fact subcontraries. Are we to obey the law or be free to change it? Are we to follow our elders or are we to try to improve on their ways? Libraries have been written on these and similar themes on the assumption that the alternatives were mutually exclusive. But it is only to an untutored mind that they appear so. Wisdom consists in seeing ways in which both may be true. Consider, for example, the controversy with respect to free trade and policies of protection. The issues are sometimes stated as if it were incompatible with the best interests of a country to adopt one policy at one time and the other at another time. But under certain conditions tariffs may be necessary for the economic development of a country, while under other conditions they may be detrimental. Hence, while the propositions *Tariffs are detrimental to a country* and *Tariffs are beneficial to a country* are in fact contraries when regarded as formulations of *universal* policies, the actual state of affairs makes it impossible to adopt such universal policies. Hence both propositions may be true if they are regarded as asserting that under *certain qualifying conditions* tariffs may be detrimental, and under others, beneficial. Meat can be both a food and a poison.

Superimplication

Every theory stands to its logical consequences in the relation of principal to subaltern—in the relation of superimplication. For this reason we shall study this and the converse relation in some detail in the sequel. Here we may note a single illustration of this type of opposition that is a little more complicated than the traditional schedule allows. Let p represent the conjunction of postulates and axioms of Euclid, and q represent *The sum of the angles of a triangle is equal to two right angles:* then p is the principal or superimplicant to q.

In this connection, it will be useful to refer to a distinction, canonized by traditional logic, between immediate and mediate inference. An inference is said to be *immediate* when a proposition is inferred from a *single* other proposition; and inference is *mediate* when *at least two* propositions are required in the premises. But this distinction is not significant if any two propositions can be combined into one. We must also remember that some forms of so-called immediate inferences require special assumptions in order to be valid.

Logicians have sometimes been led into drawing a very sharp distinction between equivalent propositions, while on the other hand they have sometimes questioned whether the "immediate inferences" from one to another of an equivalent pair of propositions was a "genuine" case of inference at all. Reflection shows, however, that a part at least of this controversy arises because it is forgotten how arbitrary is the distinction between a proposition and what it implies. Two propositions so related that if the first is true the second is also true, and if the first is false the second is false also, are identical for certain purely logical purposes. For this reason it is not very significant whether we call the contrapositive of a proposition an immediate inference from the latter or whether we regard it as being its equivalent. None the less, although two equivalent propositions are identical with regard to their truth-value, the conventional meaning of the inferred proposition is often an elaboration of the meaning of the premise. It is true, however, that the dividing-line between equivalent propositions which have the same meaning and equivalent propositions which are not precisely alike in meaning is not a sharp one.

Two special cases of immediate inference which fall under the relation of subalternation should also be mentioned. These illustrate the logic of relations, which has been systematically studied only within recent times.

(a) Inference by Added Determinants

We may infer one proposition from another, if we *limit* the subject and predicate of the premise by the *same determinant.* Thus, from *Users of snuff are consumers of tobacco* we may infer *American users of snuff are American consumers of tobacco.* From *All popes are Italians* we may infer *All tall popes are tall Italians.* These propositions, however, are not equivalent. We cannot infer from *All American professors are American scholars* that *All professors are scholars.*

Care must be taken to qualify the subject and predicate with a determinant having the same meaning. Thus, if we argue that because *All husbands are wage-earners* therefore *All unsuccessful husbands are unsuccessful wage-earners* the determinant "unsuccessful" does not have the same meaning when applied to the subject as when it is applied to the predicate, although the same word is used. A husband is unsuccessful relative to his functions as husband; a wage-earner is unsuccessful relative to his wage-earning functions. Hence determinants whose meaning involves a reference to different standards cannot be employed for inference by added determinants.

(b) Inference by Complex Conception

We may infer one proposition from another if we employ the subject and predicate as parts of a more complex conception Thus if *New York is the world's largest city* then *The center of New York is the center of the world's largest city.* If *A horse is an animal* then *The head of a horse is the head of an animal.* The inference consists in inferring that if one term stands to another in a certain relation, then whatever is related to the first term in some specific way stands to whatever is related to the second term in the same specific way in that same certain relation. Such inferences, however, do not yield equivalent propositions. We cannot infer from *The color of his nose is the color of a beet* that *His nose is a beet.*

We must observe the same caution as in the preceding type of inference. Thus it is fallacious to argue that because *All radicals are citizens* therefore *The wealthiest of radicals is the wealthiest of citizens.* For the person standing in the relation of "being the wealthiest of radicals" is measured by different standards of wealth than the person who stands in the relation of "being the wealthiest of citizens."

The Relation of Subimplication or Converse Subaltern

If p stands for *The angle sum of an isosceles triangle is equal to two right angles* and q for *The angle sum of any triangle is equal to two right angles,* p is the subimplicant or subaltern to q. For if p is true, nothing follows as to the truth of q, while if p is false, q

must be false. As we shall see, the relation of a verifying instance of a theory to the theory is the relation of subaltern to principal.

It will become apparent later that no number of verifying propositions relating to a theory can demonstrate the theory, although strictly speaking only one contrary instance of a theory is needed to refute it. But great care must be exercised in making sure that what appear to be contrary instances are really so in fact.

CHAPTER IV

THE CATEGORICAL SYLLOGISM

§ 1. THE DEFINITION OF CATEGORICAL SYLLOGISM

Consider the proposition *Tom Mooney is a danger to society.* What would constitute adequate evidence for this proposition? One may perhaps argue as follows: *All social radicals are a danger to society; and Tom Mooney is a social radical; it follows that Tom Mooney is a danger to society.* The reader would then have to admit that the first two propositions do imply the third, and *if* the first two were true, the third would be true necessarily. The conclusion is therefore a subaltern to the premises, since if the conclusion were true it would not follow that the premises are. It is also clear, however, that although the premises *if true* would be adequate evidence for the truth of the conclusion, the question *whether* the premises are true in fact is not determined by the logical relation in which they stand to the conclusion.

Arguments of this type are frequently employed. Some of them appear to be perfectly sound, although on reflection they are discovered to be faulty. An inference such as *All Parisians are Frenchmen, no Bostonians are Parisians, therefore no Bostonians are Frenchmen* or *All radicals are foreign-born, no patriotic citizen is a radical, therefore no patriotic citizen is foreign-born* is often regarded as sound by those who do not reflect. That neither argument is valid can be shown easily if we employ precisely the same type of inference, but about a different subject matter. Thus *All triangles are plane figures, no squares are triangles, therefore no squares are plane figures* is an argument of the same type as either of the preceding ones, but it will deceive practically no one.

Can we not discover some general rules, easy to apply, to which arguments of this type must conform to be valid? The matter was investigated by Aristotle, who laid the foundation for all subsequent logical inquiry, and his results have constituted the substance of logical doctrines for two thousand years. It is only in comparatively recent times that Aristotle's researches have received extension. The following discussion, however, makes little use of more modern logical techniques, although our procedure will follow the traditional analysis of the syllogism and not Aristotle's own. By departing in some ways from Aristotle's discussion of the categorical syllogism we shall at the same time exhibit the nature of a logical or mathematical system.

A *categorical syllogism* is defined as a form of argument consisting of three categorical propositions which contain between them three and only three terms. Two of the propositions are premises, the third is the conclusion. The premises *All football coaches are well paid* and *All baseball players are popular* cannot yield a syllogistic conclusion, since the premises alone contain four terms. The two propositions have no common term, while the premises of every syllogism have a common term. We may, in fact, interpret a syllogistic inference as a comparison of the relations between each of two terms and a third, in order to discover the relations of the two terms to each other. In the illustration with which we began this chapter, the common term is "social radical"; by examining the relations of the other two terms, "Tom Mooney" and "a danger to society," to this common term, the relation between "Tom Mooney" and "a danger to society" was found. For this reason, the syllogism is classified as a *mediate inference*. From the point of view of a generalized logic, however, the syllogism is a particular instance of an inference by *elimination* of one or more terms contained in the premises.

Is the following argument a categorical syllogism? A *is older than* B; B *is older than* C; *therefore* A *is older than* C. It certainly *looks* like one. But since every categorical proposition is to be analyzed into a subject term, a predicate term, and a copula which is some part of the verb "to be," it follows that this argument, although valid, is not a syllogism, since it contains *four* terms. The following argument, due to C. L. Dodgson (Lewis Carroll), is also not a syllogism as it stands, although it may be transformed into one: "A prudent man shuns hyaenas; no banker is imprudent; therefore no banker fails to shun hyaenas."

The term which is contained in both premises is the *middle term;* the predicate of the conclusion is the *major term;* and the subject of the conclusion is the *minor term*. The premise which contains the major term is the *major premise,* and the premise containing the minor term is the *minor premise.* The order in which the premises are stated does not, therefore, determine which is the major premise. In the syllogism *All mystery tales are a danger to health, for all mystery tales cause mental agitation, and whatever is a cause of mental agitation is a danger to health* the conclusion is stated first, and the major premise last. It is usual, however, to state the major premise first.

§ 2. THE ENTHYMEME

Although syllogistic reasoning occurs frequently in daily discourse, often its presence is not noticed because the reasoning is incompletely stated. A syllogism that is incompletely stated, in which one of the premises or the conclusion is tacitly present but not expressed, is called an *enthymeme.*

The following are familiar illustrations of enthymemes: *This medicine cured my daughter's cough; therefore this medicine will cure mine.* The inference is valid on the tacit admission of the major premise: *Whatever is a cure for my daughter's cough is a cure for mine.* An enthymeme in which the major premise is unexpressed is of the *first order.*

All drunkards are short-lived; therefore John won't live long. Here the missing premise is the minor: *John is a drunkard.* Enthymemes suppressing the minor premise are of the *second order.*

Usury is immoral, and this is usury. The conclusion *This is immoral* is here left unexpressed. Such an enthymeme is of the *third order.* The value of such enthymemes for purposes of innuendo is doubtless well known to the reader.

Although enthymemes do not introduce any new form of inference, their recognition is of very great importance practically. As we shall see, so-called inductive inferences are often believed to constitute a special mode of reasoning, when in fact they are simply enthymemes of the first order.

§ 3. THE RULES OR AXIOMS OF VALIDITY

We have, so far, merely defined what we shall understand by a categorical syllogism. We have not yet stated the conditions under which such an argument is valid. We shall do this by enumerating five propositions which jointly express the determining factors of any valid categorical syllogism. They are referred to as the *rules* or *axioms.* We shall state them without any attempt to prove them, since they will be our "first principles," in terms of which we shall demonstrate other propositions. Nevertheless, although we make no attempt to prove the axioms, we assert them as expressing the conditions of valid syllogistic inference. Reflection on the syllogism as a form of inference in which a connection between two terms may be asserted because of their relations to a common third term, may enable us to "see" that these axioms do in fact express the conditions of validity. But such "seeing" must not be mistaken for *proof.* Since the axioms are principles of logic, we touch upon a fundamental characteristic of logical principles: not all logical principles can be demonstrated logically, since the demonstration must itself employ some principles of logic; and in particular, no proof of the principle of identity (If anything is *A* it is *A*) is possible, without assuming that the "anything that is *A*," which occurs in one part of such an alleged proof is identical with the "anything that is *A*" occurring in another part.

The axioms of the categorical syllogism fall into two sets, those which deal with the quantity or distribution of terms, and those which deal with the quality of the propositions.

Axioms of Quantity

1. The middle term must be distributed at least once.
2. No term may be distributed in the conclusion which is not distributed in the premises.

Axioms of Quality

3. If both premises are negative, there is no conclusion.
4. If one premise is negative, the conclusion must be negative.
5. If neither premise is negative, the conclusion must be affirmative.

These axioms, together with the principles of hypothetical inference, are sufficient to develop the entire theory of the categorical syllogism. The axioms are not independent of each other, since we can derive some of them from the others. However, we shall take all of them as the axiomatic basis for our analysis.

§ 4. THE GENERAL THEOREMS OF THE SYLLOGISM

We shall now demonstrate four theorems.

Theorem I. The number of distributed terms in the conclusion must be *at least one less* than the total number of distributed terms in the premises.

Proof: The number of terms distributed in the conclusion cannot be greater than the total number distributed in the premises (Axiom 2).

The middle term, which must be distributed at least once in the premises (Axiom 1), does not appear in the conclusion (*definition of middle term*).

Therefore the conclusion must contain at least one distributed term less than the premises.

Theorem II. If both premises are particular, there is no conclusion.

Proof: The two particular premises may be either (a) both negative, (b) both affirmative, or (c) one affirmative and one negative.

a. If both premises are negative, there is no conclusion (Axiom 3).
b. A particular affirmative proposition distributes no terms. If both premises are particular affirmative, the premises contain no distributed terms. Hence, there is no conclusion (Axiom 1).
c. An affirmative particular has no distributed terms, and a negative particular has only one. Hence the premises contain one and only one distributed term. Therefore, *if* there is a conclusion, it cannot contain *any* distributed terms (Theorem I). But since one premise is negative, the conclusion must be negative (Axiom 4). Therefore at least *one* term of the conclusion must be distributed. The assumption that there is a conclusion requires us to maintain that it contains at the same time *no* distributed terms and *at least one*. This is absurd. Hence there is no conclusion.

Theorem III. If one premise is particular, the conclusion must be particular.

Proof: The premises cannot both be particular (Theorem II). They must therefore differ in quantity, and may be either (a) both negative, (b) both affirmative, or (c) one affirmative and one negative.

a. If both premises are negative, there is no conclusion (Axiom 3).
b. One premise is an affirmative universal, the other an affirmative particular. The universal distributes only one term, the particular none. Hence the premises contain no more than one distributed term. Therefore the conclusion, if there is any, contains no distributed term (Theorem I). But a universal proposition contains at least one distributed term. Hence the conclusion, if there is any, is particular.
c. We may distinguish two cases: (α) the universal is negative, the particular is affirmative; (β) the universal is affirmative, the particular is negative.
α. The universal distributes two terms, the particular none. Hence the premises distribute two terms.
β. The universal distributes one terms, the particular also one. Hence the premises distribute two terms. In either case, therefore, the premises contain two and only two distributed terms. The conclusion, if there is any, cannot contain more than one distributed term (Theorem I). But the conclusion must be negative (Axiom 4), and its predicate must therefore be distributed. Hence its subject cannot be distributed, and it must be particular.

Theorem IV. If the major premise is an affirmative particular and the minor is a negative universal, there is no conclusion.

Proof: Since, by hypothesis, the minor is negative, the conclusion, if there is any, must be negative (Axiom 4), and its predicate, which is the major term, must be distributed. Hence the major term must be distributed in the major premise (Axiom 2). But the particular affirmative distributes none of its terms. There is therefore no conclusion.

The five axioms and these four theorems which we have demonstrated rigorously by their aid enable us to enumerate all possible valid syllogisms. The reader will do well to notice the nature of the demonstration: it has been shown that the theorems are necessary consequences of the axioms, so that if the axioms are accepted, the theorems must be accepted also, on pain of contradiction.

§ 5. THE FIGURES AND MOODS OF THE SYLLOGISM

But before we enumerate all the valid syllogistic forms, let us consider some syllogisms:

1. No musicians are Italians.
   All barbers are musicians.
   ∴ No barbers are Italians.

2. All gentlemen are polite.
   No gamblers are polite.
   ∴ No gamblers are gentlemen.

3. Some books are not edifying.
   All books are interesting.
   ∴ Some interesting things are not edifying.

4. All business men are self-confident.
   No self-confident men are religious.
   ∴ No religious men are business men.

Although these are all valid syllogisms, they differ from one another in two ways: (1) in the position of the middle term; and (2) in the quality and the quantity of the premises and the conclusion. In the first example, the middle term is the subject of the major and the predicate of the minor: in the second example, the middle term is the predicate of both premises; in the third, the middle term is the subject of both premises; and in the fourth, the middle term is the predicate of the major and the subject of the minor. The position of the middle term determines the *figure* of the syllogism, and on the basis of this distinction there are four possible figures. Letting *S, P, M,* denote the minor term, major term, and middle term respectively, we may symbolize the four figures as follows:

$M-P$ $S-M$ $\therefore S-P$ *First Figure*	$P-M$ $S-M$ $\therefore S-P$ *Second Figure*	$M-P$ $M-S$ $\therefore S-P$ *Third Figure*	$P-M$ $M-S$ $\therefore S-P$ *Fourth Figure*

Aristotle recognized only the first three figures. The introduction of the fourth is generally attributed to Galen, and is therefore called the Galenian figure. Logicians have disputed whether the fourth figure represents a type of reasoning distinct from the first three, and whether Aristotle was or was not mistaken in not recognizing it. If the distinction between figures is made on the basis of the *position* of the middle term, there can be no dispute that there are four distinct figures. But Aristotle did not distinguish the figures in this way. His principle of distinction was the *width* or extent of the middle term as compared with the other two. On this basis there are just three figures: the middle may be wider than one and narrower than the other, wider than either, and narrower than either.

The second way in which syllogisms may differ is with respect to the quantity and quality of the premises and the conclusion.

This determines the *mood* of the syllogism. The first of the four syllogisms above is in the first figure, in mood *EAE*. The syllogism

5. All wholesome foods are cleanly made.
   All doughnuts are wholesome food.
   ∴ All doughnuts are cleanly made.

is a syllogism in the first figure, in mood *AAA*. Syllogisms may therefore differ in both figure and mood (for example, 1 and 3 above) or in figure alone (2 and 4) or in mood alone (1 and 5). However, not all moods are valid in every figure.

Let us count the total number of syllogistic forms, whether valid or not, taking account of differences in mood and figure. Since there are four types of categorical propositions, the major premise may be any one of the four types, and similarly for the minor premise and the conclusion. There are therefore $4 \times 4 \times 4$ or 64 syllogistic moods in each figure, and 64×4 or 256 syllogistic forms in all four figures. Most of these, however, are invalid. But how shall we discover the valid forms? It would be an appalling task to examine each of 256 forms. Such a procedure is not necessary, however, since the invalid forms may be eliminated by applying the axioms and theorems.

Let us write down every possible combination of premises, where the first letter indicates the major premise, and the second letter the minor:

AA	*EA*	*IA*	*OA*
AE	*EE*	*IE*	*OE*
AI	*EI*	*IO*	*OI*
AO	*EO*	*II*	*OO*

But Axiom 3 shows that the combinations *EE, EO, OE,* and *OO* are impossible; Theorem II eliminates *II, IO, OI;* and Theorem IV eliminates *IE*. We are therefore left with the following eight combinations of premises, each of which will yield a valid syllogism in some or all figures: *AA, AE, AI, AO, EA, EI, IA, OA*. The eight combinations which have been eliminated yield no conclusion in *any* figure.

There now remains the task of discovering the valid moods in each figure. This may be done in either of the following ways:

1. Write premises in each of the figures having the quantity and quality indicated by each of the permissible combinations, and by inspection find those which yield a valid conclusion. This method has the disadvantage of being long.

2. Establish special theorems for each figure, and eliminate by their aid the invalid combinations of premises. This method is very elegant, and we shall employ it.

In what follows we shall assume once for all that the classes denoted by the terms of the propositions are not empty, and explore the consequences of this assumption. It will permit us to perform immediate inferences by limitation.

§ 6. THE SPECIAL THEOREMS AND VALID MOODS OF THE FIRST FIGURE

Since the form of the first figure is symbolized by

$$\begin{array}{c} M - P \\ S - M \\ \hline \therefore S - P \end{array}$$

we prove:

Theorem I. The minor premise must be affirmative.

Suppose the minor is negative: then the conclusion must be negative (Axiom 4), and *P* must be distributed. Hence *P* must be distributed in the major premise (Axiom 2), so that the major must be negative. However, both premises cannot be negative (Axiom 3), and the minor must therefore be affirmative.

Theorem II. The major premise must be universal.

Since the minor premise must be affirmative, its predicate *M* cannot be distributed. Hence *M* must be distributed in the major (Axiom 1), making the latter universal.

By means of the special Theorem I, we may therefore eliminate the combinations *AE, AO,* and by means of the second theorem, the combinations *IA* and *OA*. Only the following four yield valid conclusions in the first figure: *AA, AI, EA* and *EI*. The six valid moods are therefore *AAA*, (*AAI*), *AII*, *EAE*, (*EAO*), *EIO*.

The moods we have encircled are called *subaltern* or *weakened moods,* because although the premises warrant a universal conclusion, the actual conclusion is only particular, and therefore "weaker" than it could be. Four of these six valid moods have been given special names, the vowels of which correspond to the quantity and quality of the premises and conclusion. Thus, *AAA* is *Barbara, AII* is *Darii, EAE* is *Celarent,* and *EIO* is *Ferio*. The names have been devised to form mnemonic verses by which the different moods in each of the figures may be recalled, and the moods in figures other than the first reduced to the first. We shall return to the problem of reduction.

§ 7. THE SPECIAL THEOREMS AND VALID MOODS OF THE SECOND FIGURE

The form of the second figure is symbolized by

$$\begin{array}{c} P-M \\ S-M \\ \hline \therefore S-P \end{array}$$

We prove:

Theorem I. The premises must differ in quality.

If both premises are affirmative, the middle term *M* is undistributed in each. Hence one of the premises must be negative (Axiom 1). But both premises cannot be negative (Axiom 3). Hence they differ in quality.

Theorem II. The major premise must be universal.

Since one of the premises is negative, the conclusion is negative (Axiom 4), and *P,* the major term, must be distributed. Hence *P* must be distributed in the major premise (Axiom 2), so that the latter is universal.

Theorem I eliminates the combinations *AA, AI,* and Theorem II eliminates *IA* and *OA.* We are left with four combinations in this figure: *AE, AO, EA,* and *EI,* from which we obtain six valid moods. *AEE* (*Camestres*), (*AEO*), *AOO* (*Baroco*), *EAE* (*Cesare*), (*EAO*), *and EIO* (*Festino*). The moods encircled are weakened syllogisms.

§ 8. THE SPECIAL THEOREMS AND VALID MOODS OF THE THIRD FIGURE

Employing the symbolic form of the third figure

$$\begin{array}{c} M-P \\ M-S \\ \hline \therefore S-P \end{array}$$

we can prove:

Theorem I. The minor must be affirmative.

Suppose the minor negative: the conclusion would then be negative (Axiom 4) and *P*, its predicate, would be distributed. Hence *P* would be distributed in the major premise (Axiom 2), so that the latter would be negative. But this is impossible (Axiom 3), so that the minor cannot be negative.

Theorem II. The conclusion must be particular.

Since the minor must be affirmative, *S* cannot be distributed in the premises. Hence *S* cannot be distributed in the conclusion (Axiom 2), and the latter must be particular.

The first theorem eliminates the combinations *AE* and *AO*, and we are left with the six: *AA, AI, EA, EI, IA, OA*. Keeping in mind the second theorem, we obtain six valid moods: (*AAI*) (*Darapti*), *AII* (*Datisi*), (*EAO*) (*Felapton*), *EIO* (*Ferison*), *IAI* (*Disamis*) and *OAO* (*Bocardo*). In this figure there are no weakened moods. The two moods which we have encircled are called *strengthened syllogisms* because the same conclusion may be obtained even if we substitute for one of the premises its subaltern.

§ 9. THE SPECIAL THEOREMS AND VALID MOODS OF THE FOURTH FIGURE

With the aid of the symbolic representation of the fourth figure

$$\begin{array}{c} P - M \\ M - S \\ \hline \therefore S - P \end{array}$$

we can prove:

Theorem I. If the major premise is affirmative, the minor is universal.

If the major is affirmative, its predicate *M* is not distributed. *M* must therefore be distributed in the minor premise (Axiom 1) and the latter is universal.

Theorem II. If either premise is negative, the major must be universal.

If either premise is negative, the conclusion is negative (Axiom 4) and its predicate *P* will be distributed. Hence *P* must be distributed in the major premise (Axiom 2), which must, therefore, be universal.

Theorem III. If the minor is affirmative, the conclusion is particular.

If the minor is affirmative, its predicate *S* is not distributed. Hence *S* cannot be distributed in the conclusion (Axiom 2) and the latter must be particular.

The first theorem eliminates the combinations *AI, AO,* the second eliminates *OA*. We are left with the five combinations: *AA, AE, EA, IA* and *EI*. Remembering the third theorem, we obtain six valid moods: *AAI* (*Bramantip*), *AEE* (*Camenes*), *AEO*, *IAI* (*Dimaris*), *EAO* (*Fesapo*), and *EIO* (*Fresison*). *AEO* is a weakened syllogism, while *AAI* and *EAO* are strengthened ones.

We thus find that there are just twenty-four valid syllogistic forms in the four figures, each figure containing six moods. The weakened and strengthened forms, however, are legitimate only on the assumption of existential import which we have explicitly made. Where such an assumption is not made, only fifteen valid moods can be obtained.

§ 10. THE REDUCTION OF SYLLOGISMS

We have discovered the valid moods by eliminating all forms incompatible with the axioms of validity and with the theorems derived from them. The only justification of the validity of the remaining forms we have given is that they are in conformity with the axioms. But a different approach to the justification of valid moods was taken by Aristotle, the original writer on the syllogism. According to him, the moods of the first figure were to be tested by applying to them directly a principle known since as the *dictum de omni et nullo*. This principle was, and frequently still is, believed to be "self-evident." It has been stated variously, one form of the *dictum* being: "Whatever is predicated, whether affirmatively or negatively, of a term distributed, may be predicated in like manner of everything contained under it." (Keynes) It is not difficult to show that the *dictum* is equivalent to the axioms and theorems relevant to the first figure. *It cannot, however, be applied directly to syllogisms in other figures.* The first figure was accordingly called the *perfect* figure, the others being *imperfect.*

Let us see how the *dictum* may test the validity of a syllogism in

Barbara: All Russians are Europeans; all communists are Russians; therefore all communists are Europeans. "Europeans" is predicated affirmatively of the distributed term "Russians"; hence, according to the *dictum,* it may be predicated affirmatively of "communists," which is contained under "Russians." But the syllogism *All Parisians are Frenchmen; no Bostonians are Parisians; therefore no Bostonians are Frenchmen* does not conform to the *dictum.* "Frenchmen" is predicated affirmatively of the distributed term "Parisians"; it cannot, however, be predicated in any manner of "Bostonians," since the latter term is not contained under "Parisians."

Now if, with Aristotle, we regard the *dictum* as a "self-evident" principle, and if we believe with him that it is the *sole* "self-evident" principle which can test the validity of syllogistic forms, the only way in which moods in figures other than the first can be justified will be by showing that these imply valid moods in the *first* figure.

This process of exhibiting the connection of moods in other figures with moods in the first is called *reduction.* There are two varieties: (1) *direct reduction,* which is performed by conversion of propositions or transposition of premises; and (2) *indirect reduction,* which requires either obversion and contraposition of propositions, or a form of hypothetical inference known as *reductio ad absurdum.*

Although many logicians have regarded the process of reduction as unnecessary and even as invalid, there is no doubt that, given the basis upon which Aristotle develops the theory of the syllogism, reduction is an essential part of the theory. However, if the doctrine of the syllogism is developed on other bases, which do not assume that the first figure has any intrinsic superiority over the others, reduction cannot have the importance with which it has been regarded traditionally. Our own discussion has shown that the first figure need not be taken as central, and as we shall see later, the syllogism may be developed from a point of view even more general than the one we have taken. Moreover, it is possible to state *dicta* for each of the figures, all of which possess the same degree of "self-evidence" as the *dictum de omni.* Nevertheless, in spite of the diminution in the theoretical importance of reduction, it still serves as a valuable logical exercise. We shall now show how the process may be carried out by reducing to the first figure several moods in figures other than the first.

Direct Reduction

Consider the *AEE* syllogism in the second figure:

All almsgiving is socially dangerous.	↗	No socially dangerous institution is an educational institution.
No educational institution is socially dangerous.	converse ↘	All almsgiving is socially dangerous.
∴ No educational institution indulges in almsgiving.		∴ No almsgiving is undertaken by an educational institution.
		∴ No educational institution indulges in almsgiving.

We have stated its equivalent syllogism on the right, which may be tested directly by means of the *dictum*. The reduction was effected by transposing the premises, converting the minor premise, and finally converting the conclusion. Therefore if we are not in doubt about the validity of the second syllogism, there can be no doubt concerning the validity of the original.

Consider next the *AII* mood in the third figure. It may be reduced to the valid syllogism in the first figure, stated at the right, as indicated:

All brokers are wealthy.	→	All brokers are wealthy.
Some brokers are intelligent.	converse →	Some intelligent men are brokers.
∴ Some intelligent men are wealthy.		∴ Some intelligent men are wealthy.

Finally, the *IAI* in the fourth figure may be reduced as follows:

Some horses are spirited animals.	↗	All spirited animals are difficult to manage.
All spirited animals are difficult to manage.	↘	Some horses are spirited animals.
∴ Some creatures difficult to manage are horses.		∴ Some horses are difficult to manage.
		∴ Some creatures difficult to manage are horses.

Indirect Reduction

The reader will discover that the two syllogisms which contain a particular negative premise, *AOO* in the second figure and *OAO* in the third, cannot be reduced to the first by conversion and transposition of premises alone. If obversion is also permitted, however, the reader will have no difficulty in effecting the reduction.

But obversion was not included by Aristotle in the permissible means of effecting reduction. He discovered, however, a very important logical principle, which is in fact a generalization of the idea of contraposition of hypothetical propositions. Let us illustrate it before we state it.

Suppose that the premises in the following syllogism are true. We wish to demonstrate that the conclusion is necessarily true.

Some steel is not magnetic.
All steels are metals.
∴ Some metals are not magnetic.

Now the conclusion is either true or false. If it is false, its contradictory: All metals are magnetic: is true. Combining this proposition with the minor of the syllogism, we get:

All metals are magnetic.
All steels are metals.
∴ All steels are magnetic.

This, however, is a valid mood in the first figure. But since, by hypothesis, both premises of the original syllogism are true, the conclusion of this second syllogism cannot be true. For it contradicts the original major premise. Consequently, the major of this second syllogism cannot be true, or—what is the same thing—the conclusion of the first syllogism cannot be false. It must, therefore, be true.

The validity of the *OAO* syllogism in the third figure is thus demonstrated by means of a valid syllogism in the first figure and the principle known as the *reductio ad absurdum.* The validity of the *AOO* mood in the second figure may be shown in the same way. This method may be used for other moods as well.

Let us now exhibit the principle of the *reductio ad absurdum* in more abstract form. Let p represent *Some steel is not magnetic,* q represent *All steels are metals,* and r represent *Some metals are not magnetic.* And let p', q', r' symbolize the *contradictories* of each of these propositions respectively. Then the original syllogism asserts that p and q together imply r. Or symbolically, $(p \cdot q) \supset r$. Now what we have shown was that the contradictory of r, together with q, implies the contradictory of p. Symbolically this may be stated: $(q \cdot r') \supset p'$. And the reduction of the first syllogism depends on the *equivalence* between these two implications. This equivalence is an easy extension of the equivalence between a hypothetical proposition and its contrapositive, for we have shown that if a and b are any two propositions $(a \supset b) \equiv (b' \supset a')$. We now have:

$$[(p \cdot q) \supset r] \equiv [(q \cdot r') \supset p'] \equiv [(p \cdot r') \supset q']$$

The principle of indirect reduction may therefore be analyzed as follows: *The syllogism is a form of inference in which two propositions* p *and* q *jointly imply a third* r, *where the three propositions contain three and only three terms.* If, however, we deny the implication $[(p \cdot q) \supset r]$, we must also deny a second implication which is equivalent to it: $[(q \cdot r') \supset p']$. But this second implication in our illustration above was a valid syllogism in *Barbara,* which cannot be denied. Therefore the first implication, which represents an *OAO* syllogism in the third figure (*Bocardo*) cannot be doubted either. For the denial of the validity of *Bocardo* commits us to the denial of the validity of *Barbara,* which is absurd.

If weakened and strengthened forms are not permitted (that is, if we do not assume existential import for universal propositions), reduction enables us to see that all syllogistic arguments can be reduced to two forms: one in which both premises are universal, and the other in which one premise is particular. The former is an argument in which both propositions may be pure hypotheses; the latter involves statements of fact ultimately dependent on observation.

§ 11. THE ANTILOGISM OR INCONSISTENT TRIAD

The principle involved in indirect reduction has been extended by Mrs. Christine Ladd Franklin in such a way as to provide a new and very powerful method for testing the validity of any syllogism. We shall, however, in discussing this method drop the assumption we have made concerning the existence of the classes denoted by the terms of the syllogism. As a consequence, the weakened and strengthened moods must be eliminated as invalid.

Consider the valid syllogism:

All musicians are proud.
All Scotchmen are musicians.
∴ All Scotchmen are proud.

If we let *S, M,* and *P* symbolize the terms "Scotchmen," "musicians," and "proud individuals," and if we make use of the analysis we have given of what asserted is by categorical proposition in

Chapter IV, this syllogism must be interpreted to assert the following:

$$M\bar{P} = 0$$
$$S\bar{M} = 0$$
$$\therefore\ S\bar{P} = 0$$

Now if the premises *All musicians are proud* and *All Scotchmen are musicians* necessarily imply *All Scotchmen are proud,* it follows that these premises are incompatible with the *contradictory* of this conclusion. Hence the three propositions:

1. All musicians are proud.
2. All Scotchmen are musicians.
3. Some Scotchmen are not proud:

are *inconsistent* with one another. They cannot all three be true together. Symbolically stated,

$$M\bar{P} = 0$$
$$S\bar{M} = 0$$
$$S\bar{P} \neq 0$$

are inconsistent. A triad of propositions two of which are the premises of a valid syllogism while the third is the contradictory of its conclusion, is called an *antilogism* or *inconsistent triad.*

An examination of the antilogism above reveals, however, that any two propositions of the triad necessarily imply the *contradictory* of the third. (This can be shown to be true in general, and is a further extension of the equivalence between a hypothetical proposition and its contrapositive.) Thus, if we take the first two of the triad as premises, we get:

All musicians are proud.	$M\bar{P} = 0$
All Scotchmen are musicians.	$S\bar{M} = 0$
∴ All Scotchmen are proud:	$\therefore S\bar{P} = 0$

which is the original syllogism from which the triad was obtained. If we take the first and third of the triad as premises, we get:

All musicians are proud.	$M\bar{P} = 0$
Some Scotchmen are not proud.	$S\bar{P} \neq 0$
∴ Some Scotchmen are not musicians:	$\therefore S\bar{M} \neq 0$

which is a valid mood in the second figure. Finally, if we take the second and third of the triad as premises, we get:

All Scotchmen are musicians.	$S\bar{M} = 0$
Some Scotchmen are not proud.	$S\bar{P} \neq 0$
∴ Some musicians are not proud:	$\therefore M\bar{P} \neq 0$

which is a valid mood in the third figure.

The reader is advised to take a different valid mood of the syllogism, and obtain from it the inconsistent triad and the other two valid syllogisms to which it is equivalent.

The Structure of the Antilogism

Let us now examine the structure of the antilogism. The reader will note, in the first place, that it contains two universal propositions and one particular proposition. This is the same as saying that in the symbolic representation of the members of the triad, there are two *equations,* and one *inequation,* because a universal proposition is interpreted as denying existence, while a particular proposition asserts it. Confining his attention to the symbolic representation, the reader will find, in the second place, that the two universals have a common term, which is once positive and once negative. Finally, the particular proposition contains the other two terms. It can be shown without difficulty that these three conditions are present in every antilogism, and the reader should not hesitate to prove that this is so.

Now since every valid syllogism corresponds to an antilogism, we can employ the conditions we have discovered in every antilogism as a test for the validity of any syllogism. Hence it is possible to develop the theory of the categorical syllogism on the basis of the conditions for the antilogism. The single principle required is: *A syllogism is valid if it corresponds to an antilogism whose structure conforms to the three conditions above.*

The theory of the antilogism represents an attempt to discover a more general basis for the syllogism and other inferences studied in traditional logic. The reader will note the elegance and the power which result from the introduction of specially designed symbols. We shall indicate in the following chapter the close connection between advances in logical theory and improvement in symbolism. We will conclude this discussion, however, by indicating how the antilogism may be used to test syllogisms for their validity.

Is the following valid?

Some Orientals are polite.
All Orientals are shrewd.
∴ Some shrewd people are polite.

Letting S, P, O stand for the minor, major, and middle terms respectively, the symbolic equivalent of this inference is: $OP \neq 0$, $O\bar{S} = 0$, $\therefore SP \neq 0$. The equivalent antilogism is: $OP \neq 0$, $O\bar{S} = 0$, $SP = 0$. This contains two universals and one particular; the universals have a common term which is once positive and once negative; and the particular contains the other two terms. The syllogism is therefore valid.

Is the following valid?

Some professors are not married.
All saints are married.
∴ Some saints are not professors.

Letting S, P, M stand for the minor, major, and middle terms, this may be stated symbolically as: $P\bar{M} \neq 0$, $S\bar{M} = 0$, $S\bar{P} \neq 0$. The equivalent antilogism is: $P\bar{M} \neq 0$, $S\bar{M} = 0$, $S\bar{P} = 0$. This contains two universals and one particular, but the common term in the former is not positive once and negative once. Hence the syllogism is invalid.

§ 12. THE SORITES

It sometimes happens that the evidence for a conclusion consists of more than two propositions. The inference is not a syllogism in such cases, and the examination of all possible ways in which more than two propositions may be combined to yield a conclusion requires a more general approach to logic than the traditional discussions make possible—or an elementary treatise permits. In certain special cases, however, the principles of the syllogism enable us to evaluate such more complex inferences. Thus, from the premises:

All dictatorships are undemocratic.
All undemocratic governments are unstable.
All unstable governments are cruel.
All cruel governments are objects of hate:

we may infer the conclusion:

All dictatorships are objects of hate.

The inference may be tested by means of the syllogistic rules, for the argument is a *chain* of syllogisms in which the conclusion of one becomes a premise of another. In this illustration, however, the conclusions of all the syllogisms except the last remain unexpressed. A chain of syllogisms in which the conclusion of one is a premise in another, in which all the conclusions except the last one are unexpressed, and in which the premises are so arranged that any two successive ones contain a common term, is called a *sorites.*

The above illustration is an *Aristotelian sorites.* In it, the first premise contains the subject of the conclusion, and the common term of two successive propositions appears first as a predicate and next as a subject. A second form of sorites is the *Goclenian sorites.* The following illustrates it:

All sacred things are protected by the state.
All property is sacred.
All trade monopolies are property.
All steel industries are trade monopolies.
∴ All steel industries are protected by the state.

Here the first premise contains the predicate of the conclusion, and the common term of two successive propositions appears first as subject and next as predicate.

Special rules for the sorites may be given. We shall state them and leave their proof as an exercise for the reader.

Special Rules for the Aristotelian Sorites.

1. No more than one premise may be negative; if a premise is negative, it must be the last.

2. No more than one premise may be particular; if a premise is particular, it must be the first.

Special Rules for the Goclenian Sorites.

1. No more than one premise may be negative; if a premise is negative, it must be the first.

2. No more than one premise may be particular; if a premise is particular, it must be the last.

CHAPTER V

HYPOTHETICAL, ALTERNATIVE, AND DISJUNCTIVE SYLLOGISMS

§ 1. THE HYPOTHETICAL SYLLOGISM

In the introductory chapter we saw that the province of logical study is the classification and examination of the evidential value of propositions. Subsequently we discovered that a set of propositions may stand to another set in *six* relations of *dependence,* in virtue of which the first set may be taken as adequate evidence for the other set. In the third chapter we examined with some care some of these relations of dependence. We must now study in greater detail the oppositional relation of superimplication. The various forms of this relation occupy a central rôle in the demonstrative sciences as well as in daily discourse.

In George Moore's historical romance *Héloïse and Abélard* the following discussion is reported between a medieval realist and a nominalistic disciple of Abélard:

"It was plain to all that the Nominalist was not fighting fairly by thrusting theology into Dialectics, but since he had chosen to do so he must take the consequences, and everybody knew that the consequences were that the Realist would do likewise. 'Ah, you are quick, pupil and disciple of Pierre du Pallet . . . you are quick to turn what I offered as an analogy into an argument of heresy against my person. I will meet you on the same ground and with the same weapon. Will you tell us if this concept, this image in the mind of man, of God, of matter, for I know not where to seek it, be a reality?' 'I hold it as, in a manner, real.' 'I want a categorical answer.' 'I must qualify—' 'I will have no qualifications, a substance is or is not.' 'Well, then, my concept is a sign.' 'A sign of what?' 'A sound, a word, a symbol, an echo of my ignorance.' 'Nothing then! So truth and virtue of humanity do not exist at all. You suppose yourself to exist, but you have no means of knowing God; therefore to you God does not exist except as an echo of your ignorance! And what concerns you most, the Church does not exist except as your concept of certain individuals whom you cannot regard as a unity, and who suppose themselves to believe in a Trinity which exists only as a sound or symbol. I will not repeat your words, pupil, disciple, whatever you are pleased to call yourself, of Le Sieur Pierre du Pallet, outside of this house, for the consequences to you would be deadly; but it is only too clear that you are a materialist, and as such your fate must be settled by a Church Council, unless you prefer the stake by judgment of a secular court.' "[1]

Let us distinguish the steps of this argument, supplying the propositions which are taken for granted:

1. If a concept in the mind of man is a sign, it is a sign of a sound, and an echo of ignorance.
   If a concept is a sign of a sound, it is a sign of nothing real.
   ∴ If a concept in the mind of man is a sign, it is a sign of nothing real.

2. If a concept in the mind of man is a sign, it is a sign of nothing real.
   But truth, virtue, God, the Church, the Trinity, are concepts in the mind of man.
   ∴ These things are signs of nothing real.

3. If anyone believes that these things represent nothing real, such a one is a materialist.
   The Nominalist is such a one.
   ∴ The Nominalist is a materialist.

The substance of the discussion is thus analyzable into three distinct steps. Inferences 2 and 3 are of the same logical form, as we shall point out in a moment; inference 1 is of different form. Inferences of the type of 2 and 3 are said to be *mixed hypothetical syllogisms.* They contain three propositions: the first, or *major,* premise is a hypothetical, the second, or *minor,* premise is a categorical, and the conclusion also is a categorical proposition. Inferences of the type of 1 are called *pure hypothetical syllogisms.* They contain two hypotheticals as premises, and a hypothetical as conclusion.

Why is the conclusion in either of the mixed syllogisms above validly inferable from the premises? The reader is prepared by now to give an answer. The correct answer is that if we assert a hypo-

[1] Vol. I, p. 70.

thetical proposition and also assert the truth of its antecedent, we necessarily commit ourselves to assert the consequent.

We may state this differently. The conjunctive proposition *If war is declared prices go up, and war is declared* implies the proposition *Prices go up.* In fact, the conjunctive proposition is the principal or superalternant to *Prices go up.* If then we assert the truth of the conjunctive, we must also assert the truth of its subaltern.

The value of this type of reasoning is very evident. For we may often be able to establish the truth of a hypothetical proposition and also the truth of its antecedent more easily than we can the truth of the consequent. The consequent may then be established *indirectly,* as the conclusion of such an inference. Thus, all attempts to trisect any angle with compass and ruler must be regarded today as useless, because the following two propositions are known to be true: If a geometric construction is expressible as an irreducible algebraic equation of degree greater than two, it cannot be constructed by compass and straight edge alone: and: The trisection of an angle is expressible by an irreducible cubic equation. The trisection of an angle by elementary methods is therefore impossible; this is a result which could have been established in no way other than as the conclusion of a mixed hypothetical syllogism.

The schematic form of the argument is: If *A* is *B,* then *C* is *D; A* is *B;* therefore *C* is *D.* If we employ the symbols previously explained this takes the form: $p \supset q;\ p; \therefore q$. The inference is said to be in the *mood,* or *modus, ponendo ponens.* This expression signifies that by affirming (in the minor premise) we affirm (in the conclusion). It is derived from the Latin *ponere,* which means to take a stand or to affirm.

Suppose we know that *If there is a total eclipse of the sun, the streets are dark* is true. May we then offer as conclusive evidence for *There is a total eclipse of the sun* the proposition *The streets are dark*? If we did, the inference would be fallacious. For the hypothetical simply asserts that if the antecedent is true, the consequent must also be true; it does not assert that the consequent is true *only on the condition* that the antecedent is true. Thus the streets may be dark at night or on cloudy days, as well as during a total eclipse. It is therefore a fallacy to affirm the consequent and infer the truth of the antecedent. We shall have frequent occasions to call the reader's attention to this fallacy. It is sometimes committed by eminent men of science who fail to distinguish between necessary and probable inferences, or who disregard the distinction between demonstrating a proposition and verifying it. For example, if the theory of organic evolution is true, we should find fossil remains of extinct animal forms; but the discovery of such remains is not a proof, is not conclusive evidence, for the theory.

But while it is fallacious to affirm the consequent of a hypothetical, we can obtain a valid conclusion by denying it. Suppose we wished to know whether Tom Mooney is guilty of placing the bomb in the Preparedness Day Parade of 1916 in San Francisco. We may discover, after studying the nature of the bomb, that *If Mooney is guilty, he was on the street corner within ten minutes of the time when the explosion occurred.* But suppose now Mooney has an alibi, and that he can show he was a mile away on an impassable street fifteen minutes before the explosion. We would then have to deny the consequent of the hypothetical, and this denial commits all students of logic, if not all politicians, to the denial of the antecedent. For the hypothetical asserts that it is not the case that the antecedent is true and the consequent false.

The contradictory of the antecedent is, in fact, a subaltern to the conjunctive proposition *If Mooney is guilty, he was on the scene of the explosion within ten minutes of the event* and *He was not on the scene of the explosion within ten minutes of the event.* If we assert the truth of this superaltern, we must assert the truth of the subaltern, because the former implies the latter.

Inferences of this type are the chief methods used to *disprove* suggested theories. Is every point of the earth's surface equidistant from a point internal to it? When we take certain physical principles for granted, it can be demonstrated that *If the earth is a sphere, then a pendulum of specified length will make two swings per second at every point on the earth's surface.* But it can be shown experimentally that there is a variation in the number of swings per second such a pendulum makes as it is moved along a meridian circle. It follows that the earth is not spherical in shape.

This type of argument may be represented schematically as follows: If *A* is *B*, then *C* is *D; C* is not *D;* therefore *A* is not *B*. Also as $p \supset q;\ q';\ \therefore p'$. The inference is in the *modus tollendo tollens,* for by denying (in the minor) we deny (in the conclusion). The expression is derived from the Latin *tollere,* meaning to lift up or to deny.

Can the consequent of a hypothetical proposition be disproved by offering as evidence the falsity of the antecedent? Suppose we inquire whether two rectangular plots of ground have equal areas. We may perhaps know that *If two rectangular plots have their corresponding sides equal, then their areas are equal.* But suppose we discover by actual measurement that the corresponding sides are unequal. Can we validly infer that the areas are unequal also?

Assuredly we cannot. For example, the sides of one of the plots may be four and five units long respectively, and the sides of the other two and ten units long; nevertheless, the areas of the plots are the same. It is a fallacy, therefore, to deny the antecedent and infer the falsity of the consequent. This fallacy is committed not infrequently by men in public affairs and other hard-headed individuals. Thus it may be claimed that *If Congress interferes with the plans of the President, the industrial depression will not be conquered.* And it is frequently believed, as if it were a logical consequence from this, that if only Congress would adjourn or not meet, the depression would be overcome.

§ 2. THE ALTERNATIVE SYLLOGISM

We next consider the valid inferences which can be drawn from an alternative proposition as major premise and a categorical as minor premise. Such arguments are called *mixed alternative syllogisms.* Thomas Paine discussed the nature of the British Constitution in his *Rights of Man* in the form of such an inference: "... Governments arise either *out* of the people or *over* the people. The English Government is one of those which arose out of conquest, and not out of society, and consequently it arose over the people. . . ."

Under what conditions may a conclusion be validly drawn from an alternative proposition as the major premise of a syllogism? Suppose the reader thinks he has been overcharged by the telephone company for services rendered during some month, and he decides to take action. He may decide *Either I will write the company or I will call at its offices.* If this proposition is true, *at least one* of the alternants must be true, but the possibility of both being true is not excluded. Hence if the reader should write his letter of protest, we cannot infer validly that he will not also visit the company's offices; or if he should make his intended visit, we cannot infer validly that he will not also write the letter. We reason fallaciously from an alternative major premise when we conclude that one of the alternants is false because the other is true.

On the other hand, if the reader should suddenly be called out of town, so that he cannot make a personal call on the company, it follows that he will write the letter. Or if he finally decides that letter-writing is of no use, it follows that he will visit the offices. Hence an alternative syllogism is valid when the minor denies one of the alternants, and the conclusion affirms the truth of the other.

We shall restate this result in order to exhibit the oppositional relation of the premises to the conclusion. The conjunctive *Either I will write the company, or I will call at its offices, and I will not write* implies the proposition *I will call at its offices.* The conjunctive is thus the superaltern to the conclusion.

The alternative syllogism, we shall see in detail later, is frequently employed to *eliminate* suggested explanations or solutions of problems. Which factor in the total environment explains the variations in the local weather? We may suppose tentatively that the possible factors are the distance from the sun, the duration of exposure to the sun's rays, the variations in air currents. If we can eliminate one of these possibilities, it will follow, *on the supposition that the alternative proposition is true,* that the cause of local weather is to be found among the remaining suggested explanations. But the reader should note that we do *not* eliminate any of the other factors if we can show, for example, that local weather varies with the behavior of the air currents. For it may be that each of the other factors contributes something to the character of the local weather.

The schematic form of this type of inference is: Either A is B or C is D; A is not B; therefore C is D. Or: $p \vee q;\ p';\ \therefore q$. The inference is said to be in the *modus tollendo ponens,* because by denying (in the minor) we affirm (in the conclusion).

§ 3. THE DISJUNCTIVE SYLLOGISM

"But," the reader may protest, "an inference you have classified as invalid is often admitted as quite correct. For suppose you did not know the exact month of Shakespeare's birth, although you could assert that *Either he was born in April or in May.* And suppose that you subsequently discovered that he was born in April. Would you not infer that Shakespeare was not born in May? And in that case, would you not go counter to the rule you had set up that in alternative syllogisms the minor must *deny* one of the alternants?"

The reader is quite right. But does his illustration really violate the rule we have stated? Examination shows that it does not. For the inference is valid only if we assume an *unexpressed* premise—namely, that one of the alternants *excludes* the other. The major premise in the reader's example when stated in full is as follows: *Either Shakespeare was born in April or in May, and he was not born both in April and in May.* This is a conjunctive proposition, and the part of it which is the real premise for the reader's argument is the conjunct stating a *disjunction*. Such disjunctive propositions are often taken for granted and therefore not expressed explicitly. Nevertheless, they should be made explicit to make clear all the premises involved.

It is advisable, therefore, to consider separately *disjunctive syllogisms*. In such arguments the major is a disjunctive proposition and the minor a categorical. Let us find the conditions under which they are valid.

Suppose we know that *It is not the case that both my watch is an accurate timekeeper* and *the behavior of all mechanisms is influenced by climatic changes.* To assert a disjunctive proposition means to assert that at least one of the disjuncts is false. Consequently, if we assert *The behavior of all mechanisms is influenced by climatic changes* we must also assert as an inference that *My watch is not an accurate timepiece.* But it would be fallacious to conclude that because one of the disjuncts is false the other is true. A disjunctive syllogism is valid if the minor asserts one of the disjuncts and the conclusion denies the other.

The schematic form of the argument is: Not both *A* is *B* and *C* is *D; A* is *B;* therefore *C* is not *D*. Or $(p \cdot q)'; p; \therefore q'$. The inference is said to be in the *modus ponendo tollens,* since by affirming (in the minor) we deny (in the conclusion).

We may now summarize the valid and invalid inferences we have been examining so far in the present chapter.

	Valid	*Invalid*
Ponendo ponens	If p then q p $\therefore q$	If p then q q $\therefore p$
Tollendo tollens	If p then q q' $\therefore p'$	If p then q p' $\therefore q'$
Tollendo ponens	Either p or q p' $\therefore q$	Not both p and q p' $\therefore q$
Ponendo tollens	Not both p and q p $\therefore q'$	Either p or q p $\therefore q'$

§ 4. THE REDUCTION OF MIXED SYLLOGISMS

Some of these principles are doubtless very familiar to the reader, and all of them may have been employed by him all his life, without his having explicitly formulated the principles according to which he had been drawing his inferences. Neither their familiarity nor their simplicity takes away from their importance. These principles are fundamental to all logical theory. They emerge repeatedly in every inquiry into the evidential value of propositions advanced to support other propositions.

These principles, however, are not independent of one another. If the reader bears in mind the equivalences which have been established between compound propositions, he will have no difficulty in reducing an argument in one modus to an argument in any other. We shall therefore simply state the equivalences between the four modi in the following table without further comment:

	Equivalent Syllogisms	*Symbolic Form*
Ponendo ponens	If a man is civilized, he has questioned his first principles. This man is civilized. ∴ This man has questioned his first principles.	$p \supset q$ p $\therefore q$
Tollendo tollens	If a man has not questioned his first principles, he is not civilized. This man is civilized. ∴ This man has questioned his first principles.	$q' \supset p'$ p $\therefore q$
Tollendo ponens	Either a man is not civilized, or he has questioned his first principles. This man is civilized. ∴ This man has questioned his first principles.	$p' \vee q$ p $\therefore q$
Ponendo tollens	It is not the case that both a man is civilized and he has not questioned his first principles. This man is civilized. ∴ This man has questioned his first principles.	$(p \cdot q')'$ p $\therefore q$

§ 5. PURE HYPOTHETICAL AND ALTERNATIVE SYLLOGISMS

We must now turn to the pure hypothetical syllogisms to which we were introduced at the beginning of this chapter.

Consider the following argument:

If the production cost of a commodity is reduced, a greater economic demand sets in for it.
If a commodity is produced in large quantities, the production cost per unit is reduced.
∴ If a commodity is produced in large quantities, a greater economic demand sets in for it.

The first two premises conjoined imply the conclusion, so that the latter is a subaltern to the premises. The premise which contains as component the consequent of the conclusion is called the *major* premise; the premise which contains the antecedent of the conclusion is called the *minor*. The validity of the syllogism clearly depends on the *transitivity* of the "if . . . then" or implication relation. It may be schematized as follows: If *C* is *D*, then *E* is *F*; if *A* is *B*, then *C* is *D*; therefore, if *A* is *B*, then *E* is *F*. Or more abstractly, $q \supset r; p \supset q; \therefore p \supset r$.

Let us now examine:

If a man is healthy, his body is not undernourished.
If a man is poor, his body is undernourished.
∴ If a man is healthy, he is not poor.

In this syllogism the minor is written first. It is valid argument, although it is not quite of the same form as the first illustration. It may be schematized as follows: If *A* is *B*, *C* is not *D*; if *E* is *F*, *C* is *D*; therefore if *A* is *B*, *E* is not *F*. Or as: $r \supset q'; p \supset q; \therefore r \supset p'$. The dependence of the validity of the argument upon the transitivity of the "if . . . then" relation is not as evident now as before. Nevertheless, this condition is still the basis for the valid inference. We can show this by reducing it to a syllogism having the same form as the first one. We need only substitute the equivalent contrapositive of the major premise, as follows:

If a man's body is not undernourished, he is not poor.
If a man is healthy, his body is not undernourished.
∴ If a man is healthy, he is not poor.

Not all pure hypothetical syllogisms are valid. The following is not:

If workmen unite, they enjoy satisfactory conditions of labor.
If workmen are not conscious of their common interests, they do not unite.
∴ If workmen are not conscious of their common interests, they do not enjoy satisfactory conditions of labor.

The reader should discover for himself by actual trial that the form of this invalid syllogism is different from either of the valid forms above, and that it cannot be reduced to either one of the valid forms.

Inferences in which the premises are two alternative propositions are called *pure alternative syllogisms*. They occur rather rarely. Consider the following:

Either men are cowards, or they protest against unjust treatment.
Either men are not cowards, or they do not protect their own economic interests.
∴ Either men do not protect their own economic interests, or they protest against unjust treatment.

This syllogism is valid. But we need not examine, in this place, the conditions of validity, since the reader may *test* the validity of any pure alternative syllogism by reducing it to a pure hypothetical form. The one above may be reduced as follows:

If men are not cowards, they protest against unjust treatment.
If men protect their own economic interests, they are not cowards.
∴ If men protect their economic interests, they protest against unjust treatment.

The conclusion of this pure hypothetical is obviously equivalent to the conclusion of the pure alternative syllogism above.

§ 6. THE DILEMMA

Hypothetical and alternative propositions may be combined in various ways to yield arguments more complex than any considered so far. We cannot study them all, and shall confine ourselves to the argument known as the dilemma. But the reader should note that the dilemma introduces no new logical principle, and that except for the dictates of tradition and the interesting uses to which it may be put, there is no good reason why we should discuss it rather than other complex forms of reasoning.

The *dilemma* is an argument in which one premise, the major, is the conjunctive assertion of two hypothetical propositions, and in which a second premise, the minor, is an alternative proposition. The minor either affirms alternatively the antecedents of the major, or denies alternatively its consequents.

Dilemmas in which the minor alternatively affirms the antecedents of the major are said to be *constructive.* Those in which the minor alternatively denies the consequents of the major are *destructive.* In constructive dilemmas, the antecedents of the major must be different propositions, while its consequents may be either different or identical. In the former case, the dilemma is *complex constructive,* in the latter, *simple constructive.* When the dilemma is destructive, the consequents of the major must be different, while the antecedents may be either different or identical. In the former case, the dilemma is *complex destructive,* in the latter, *simple destructive.* There are, therefore, four distinct kinds of dilemmas. Let us illustrate each.

1. Complex Constructive Dilemma

If women adorn themselves for ostentation, they are vain; and if women adorn themselves in order to attract men, they are immoral.

Either women adorn themselves for ostentation or in order to attract men.

∴ Either women are vain, or they are immoral.

2. Simple Constructive Dilemma

If it is assumed that the angle sum of a triangle is two right angles, Euclid's Postulate 5 can be demonstrated; and if it is assumed that these are two similar triangles with unequal areas Postulate 5 can be demonstrated.

It is assumed that either the angle sum of a triangle is two right angles, or that there are two similar triangles with unequal areas.

∴ Euclid's Postulate 5 can be demonstrated.

3. Complex Destructive Dilemma

If the country goes to war, the unemployment problem can be solved; and if the country does not change its industrial structure, a revolution will take place.

Either the unemployment problem cannot be solved, or a revolution will not take place.

∴ Either the country will abstain from war, or it alters its industrial structure.

4. Simple Destructive Dilemma

If you have a habit of whistling, you are a moron; if you have a habit of whistling, you are not musical.

Either you are not a moron or you are musical.
∴ You have not a habit of whistling.[2]

The Value of Dilemmas

Dilemmatic reasoning is of special value in those cases where we are unable to assert the truth of *any one* of the antecedents or the falsity of *any one* of the consequents in a set of hypothetical propositions, but where we can assert *alternatively* their respective truth or falsity. Thus, although we may not know for what purposes some particular woman adorns herself, on the basis of the premises in the first example above we can conclude that at any rate she is either vain or immoral. And for certain purposes, that may be all the information we need.

The dilemma has become associated in the minds of many with sharp intellectual practices common in debates and polemical literature. This has led to the view that the dilemma is a fallacious form of reasoning. But such an opinion has as little foundation as a similar opinion would have concerning any other valid form of inference, which may none the less be fallaciously used. It is worth while, therefore, because of this persistent misunderstanding, to examine the possible ways of avoiding the admission of the conclusion of a dilemma. But it should be clear that this can be done only by taking exception to the *material truth* of one or other of its premises.

Escaping Between the Horns

The opponent against whom a dilemma is directed may destroy the force of the argument by pointing out that the alternatives enumerated in the minor premise are not exhaustive. Since the alternatives are said to be the "horns of the dilemma" upon which the adversary is to be impaled, he is said "to escape between the horns" by this method.

How may this be done for the first dilemma stated above? St. Thomas Aquinas, who has discussed the question "Whether the Adornment of Women is Devoid of Mortal Sin?" in his *Summa Theologica,* points out that women may adorn themselves not only

[2] It might properly be urged that the common meaning "dilemma" is an argument in which there are two alternative hypotheses, and that what we have called the simple destructive dilemma contains no such alternative, but is only a hypothetical syllogism with a disjunctive consequent in the major premise, for example *If you have a habit of whistling you are a moron or you are not musical.* But this objection is only verbal, since we have defined the dilemma to include the last form.

for the reasons given, but also to hide their physical defects, in which case he believes their action is meritorious. Furthermore, one of the alternatives in the minor may be broken into several alternatives which do not all have the same consequences. Thus St. Thomas distinguishes between the women who adorn themselves to attract men who are not their husbands, and those who adorn themselves to attract their husbands. In the latter case, according to the saint, a woman's action is meritorious.

Taking the Dilemma by the Horns

The opponent may challenge the truth of the major premise. He can do this by flatly contradicting the major, or by showing that some one or other of the antecedents leads to different consequents from those stated. He is then said to "take the dilemma by the horns," that is, by the alternatives offered. Thus, in the first of the above dilemmas we may accept the alternatives offered but deny the major by maintaining that if women adorn themselves to attract men, they do so only in order to save men from worse follies; in that case they should be regarded as kindly.

Rebutting a Dilemma

Finally, the adversary may answer a dilemma with another argument whose conclusion *contradicts* the original conclusion. A rebutting dilemma, however, often only *appears* to contradict the first conclusion. It may then convince an untrained audience, but its logical value is meretricious.

An Athenian mother is said to have cautioned her son from entering public life:

If you tell the truth, men will hate you; if you tell lies, the gods will hate you.
But you must either tell the truth or tell lies.
∴ Either men or the gods will hate you.

The son is reported to have replied with a rebutting dilemma:

If I tell the truth, the gods will love me; if I tell lies, men will love me.
But I must either tell the truth or tell lies.
∴ Either men or the gods will love me.

The reader is now equipped to see that the conclusion of the second dilemma does not contradict the conclusion of the first. It only *appears* to do so, although in fact the two conclusions are perfectly compatible. For the genuine contradictory of the first conclusion is *Men will love me and the gods will love me.*

We may symbolize a dilemma and its specious rebuttal as follows: The first dilemma is:

$$(p \supset q) \cdot (r \supset s)$$
$$p \vee r$$
$$\therefore q \vee s$$

The rebutting dilemma is: $(p \supset s') \cdot (r \supset q')$

$$p \vee r$$
$$\therefore s' \vee q'$$

It is clear that $(q \vee s)$ and $(s' \vee q')$ are not contradictories.

CHAPTER VI

GENERALIZED OR MATHEMATICAL LOGIC

§ 1. LOGIC AS THE SCIENCE OF TYPES OF ORDER

In the previous chapters we have seen that the validity of a demonstration depends not on the truth or falsity of the premises, but upon their form or structure. We have therefore been compelled to recognize that the fundamental task of logic is the study of those objective relations between propositions which condition the validity of the inference by which we pass from premises to conclusions.

Logic has sometimes been defined as the normative science which studies the norms distinguishing sound thinking from unsound thinking. The reader is prepared now to appraise such a definition, and consequently to regard such a characterization of logic as inadequate. For since the study of logic aims to discover the structure of propositions and their objective relations to one another, the capacity of such structures to serve as norms of thought is not their exclusive function, however important it may be. Too great an emphasis on the normative capacity of the structures may incline the student to neglect many of their essential properties in favor of those which may seem to have direct bearing upon normative considerations. Because traditional logic has stressed this side of logical forms, it has failed to consider such forms with sufficient generality and has neglected to undertake a study of all possible formal structures.

Until very recently it was commonly held that Aristotle had explored once for all the entire subject matter of logic. Thus, Kant declared of logic: ". . . Since Aristotle it has not had to retrace a single step, unless we choose to consider as improvements the removal of some unnecessary subtleties, or the clearer definition of its matter, both of which refer to the elegance rather than to the solidity of the science. It is remarkable also, that to the present day, it has not been able to make one step in advance, so that, to all appearance, it may be considered as completed and perfect." [1]

If, however, the reader will turn to some of the illustrations of mathematical reasoning we have cited, he will soon discover that traditional logic has fallen short of its self-appointed task. For types of inference are employed in them which cannot, except with high artificiality or inconsistency of analysis, be reduced to any of the traditional forms. Thus, implications like the following: *If* A *is taller than* B, *and* B *is taller than* C, *then* A *is taller than* C do not fall into any of the types traditionally discussed. From the point of view of logic as an organon which is to guide and test inferences, traditional logic has been remiss in not studying systematically those logical relations which are the basis for the complicated inferences in the mathematical and natural sciences.

In one sense, however, Kant's evaluation of traditional logic is sound. For it has successfully analyzed certain kinds of inferences, and has made clear the formal factors upon which their validity depends. Much of what it has done is of permanent value. Its chief drawbacks are less with what it has done and more with what it has failed to do. Thus it discovered the subject-predicate form of propositions, but failed to note that an identical grammatical construction may cloak propositions of very different types. It stressed the necessity of a copula, but overlooked the logical properties of the copula upon which the validity of an inference rests. It was therefore hindered from developing a more general theory of inference and a more satisfactory calculus of reasoning than the syllogism. The theory of compound propositions was neglected by it, and the important topic of the existential import of propositions was not explicitly considered. Finally, it did not explore logical principles systematically, and so lacked a method for obtaining all the propositions which may be asserted on logical grounds alone.

These limitations may serve to indicate what a program of logical study should include when logic is conceived more adequately. Dissatisfaction with the limited contents of ancient logic is not entirely modern. Thus the Port Royal logicians discussed certain non-syllogistic inferences, such as *The sun is a thing insensible; the Persians worship the sun; therefore the Persians worship a thing insensible.* But they did not make a systematic study of them.

The scope of a more general logic was clearly indicated by

[1] *Critique of Pure Reason*, tr. by Max Müller, p. 688.

Leibniz. In his *New Essays concerning the Human Understanding* (1704) he discussed various kinds of asyllogistic inferences, and pointed the way to a "universal mathematics" which would provide the intellectual instruments for exploring any realm of order whatsoever. Elsewhere he indicated more clearly its main features. On the one hand, there should be constructed a "universal language" or "universal characteristics," in order to express by means of specially devised symbols the fundamental, unanalyzable concepts (the "alphabet of human thought") of all the sciences.[2] The modes of combining these symbols must be clearly determined, and all the sciences may then be restated by their means in order to exhibit clearly the logical structure of their subject matter. On the other hand, there should be invented a "universal calculus," as an instrument for operating on the system of ideas expressed symbolically in the universal characteristics. The relations between propositions would then be systematically discovered, and labor and thought would be economized in investigating rationally any subject matter. Leibniz's ideas have been realized in some measure in recent years, through the work of mathematicians and philosophers. But his own contributions were fragmentary, and had little influence on the history of logical studies until his ideas were discovered independently by others.

A renaissance of logical studies began in the first half of the nineteenth century, due almost entirely to the writings of two English mathematicians, Augustus De Morgan and George Boole. Reflection on the processes of mathematics convinced them that an indefinitely larger number of valid inferences were possible than had been hitherto recognized. De Morgan's most notable contribution was to lay the foundation for the theory of relations. Boole discovered that logical processes might be both generalized and expedited if proper symbolic conventions were made. His book *An Investigation of the Laws of Thought* (1854; the title is a misnomer) marked an epoch in the history of logic. It showed, with undeniable power and success, that the methods of mathematics are applicable not only to the study of quantities, but to *any ordered realm whatsoever,* and in particular to the relations between classes and between propositions. The view that the study of logic was the study of *types of order* gradually forced itself upon men. Since Boole's

[2] The idea of a universal mathematical language had already been envisioned by Raymond Lully in the thirteenth century.

time, mathematicians like Weierstrass, Dedekind, Cantor and Peano, and philosophers like Peirce, Frege, Russell, and Whitehead, have made clear the close connection between logic and mathematics.

§ 2. THE FORMAL PROPERTIES OF RELATIONS

An analysis of the general ideas employed in mathematics shows that one of the most pervasive is that of *relation.* A clear understanding of its nature is, moreover, indispensable in the study of propositional structure.

Relations can be illustrated easily, although difficult to define. "Being greater than," "being colder than," "being as old as," "being the father of," are examples of some relations in which objects of various kinds may stand to one another. An object is said to be in a relation if in our statement about it explicit reference must be made to another object. The term *from* which the relation goes is called the *referent,* the term *to* which it goes is the *relatum.* In *Napoleon was the husband of Josephine* the relation is "being the husband of," which connects "Napoleon" and "Josephine." "Napoleon" is the *referent,* "Josephine" the *relatum.* Such a relation is *dyadic.* In *Borgia gave poison to his guest* the relation is "giving," which connects "Borgia," "poison," and "guest." Such a relation is *triadic.*[3] A *tetradic* relation is illustrated in *The United States bought Alaska from Russia for seven million dollars.* Examples of other polyadic relations may be given, although relations with more than four terms are not common.

Further examples of dyadic relations are: *Mussolini is an Italian:* here the relation is "being a member of a class." In *Italians are Europeans* the relation is "being included in the class." The concept of relation replaces the concept of the copula: the copula of traditional logic is a special type of dyadic relation. We shall discuss some of the properties of dyadic relations upon which valid inference depends. But all polyadic relations may also be classified on the basis of the distinctions we shall make.

[3] We have said that the proposition *Borgia gave poison to his guest* illustrates a *triadic* relation. What it asserts is of course also expressible by *Borgia poisoned his guest,* which seems to illustrate a *dyadic* relation. It would be a mistake, however, to regard the distinction between dyadic and triadic relations as merely a verbal matter. Our example only shows that the same situation may be analyzed from *different* though related aspects. The relation of "giving" is triadic, the relation of "poisoning" is dyadic. The two relations are not identical.

Symmetry

In *Napoleon is the husband of Josephine* the relation is "being the husband of"; in *Josephine is the wife of Napoleon* the relation is "being the wife of." This latter relation is said to be the *converse* of the former. If Napoleon stands to Josephine in the relation of "husband of," Josephine does not stand to Napoleon in that same relation. Consequently, the relation "being the husband of" is said to be *asymmetrical.*

In *John is as old as Tom* the relation "being as old as" is symmetrical, for if John has that relation to Tom, Tom has the same relation to John. A symmetrical relation is one which is the same as its converse; an asymmetrical relation is incompatible with its converse. But if *Gentlemen prefer blondes* it may be that *Blondes prefer gentlemen,* or it may also not be. Relations like "loving," which are sometimes symmetrical and sometimes not, are said to be *nonsymmetrical.*

Transitivity

If A *is the father of* B and B *is the father of* C, *then* A *is* not *the father of* C. A relation like "being the father of" is said to be *intransitive.* But if *John is older than Tom* and *Tom is older than Harry,* then *John is older than Harry.* The relation of "being older than" is *transitive.* Some relations are sometimes transitive, sometimes intransitive. If *Caesar is a friend of Brutus* and *Brutus is a friend of Cassius,* Caesar may be a friend of Cassius or he may not. Such relations are said to be *nontransitive.*

The distinctions based on symmetry and transitivity are independent of one another, and we may therefore obtain any one of the following nine types of relations. a. Transitive symmetrical, for example, "being as old as." b. Transitive asymmetrical, for example, "ancestor of." c. Transitive nonsymmetrical, for example, "not older than." d. Intransitive symmetrical, for example, "spouse of." e. Intransitive asymmetrical, for example, "father of." f. Intransitive nonsymmetrical, for example, "nearest blood-relative of." g. Nontransitive symmetrical, for example, "cousin of." h. Nontransitive asymmetrical, for example, "employer of." i. Nontransitive nonsymmetrical, for example, "lover of."

Correlation

A third principle of classification pays attention to the number of objects to which the referent or relatum may be connected by the given relation.

If *Mr. A is a creditor of Mr. B,* other men besides Mr. A may stand in such a relation to Mr. B, and other men besides Mr. B may stand in that relation to Mr. A. Such a relation is said to be *many-many.*

If *Johann Christian Bach is a son of Johann Sebastian Bach,* other individuals besides Johann Christian may stand in this relation to Johann Sebastian, but there is only one individual to whom Johann Christian may stand in this relation. The relation "son of" is called a *many-one* relation.

The converse of a many-one relation is a *one-many* relation. Thus, in *J. S. Bach is the father of J. C. Bach,* J. S. Bach may stand in this relation to other individuals besides J. C. Bach, but only one individual can stand in this relation to J. C. Bach.

Finally, in *Ten is greater by one than nine,* there is only one number to which "ten" may have this relation, and only one number which has this relation to "nine." Relations like "greater by one than" are called *one-one,* and they play a fundamental rôle in the theory of correlations.

Connexity

A fourth principle of classification depends on whether a relation holds between *every pair* of a collection or not. Consider the integers and the relation "greater than." Any two integers either stand to each other in the relation "greater than" or in the converse relation "less than." A relation with this property is said to have *connexity,* otherwise not. The relation "greater than by two" does not possess connexity.

§ 3. THE LOGICAL PROPERTIES OF RELATIONS IN SOME FAMILIAR INFERENCES

Many of the inferences we have studied in previous chapters can be interpreted to depend on the nature of the relations of class inclusion or exclusion. We must indicate briefly in what way the logical properties of these relations, as well as of others, is significant for the study of valid inferences.

1. The conversion of categorical propositions depends on the *symmetry* or *nonsymmetry* of the class inclusion (or exclusion) relation. *All firemen are muscular* may be interpreted to assert that the class "firemen" is included in the class "muscular men." This proposition cannot be converted simply, because total inclusion of one class in another is a nonsymmetrical relation. But *Some firemen are muscular* may be converted, because partial inclusion of

classes is a symmetrical relation. Also *No firemen are muscular* may be converted simply, because total exclusion of one class from another is symmetrical.

2. The validity of categorical syllogisms depends on the *transitivity* of the relation of class inclusion. Consider the syllogism *All men are cowards; all professors are men; all professors are cowards.* It may be interpreted to assert that if the class "men" is included in the class "cowards," and the class "professors" included in the class "men," then the class "professors" is included in the class "cowards." The relation is obviously transitive. Valid syllogisms in other moods and figures may be shown to depend on the same logical property of the copula.

But in those syllogisms where one of the premises is singular, a different analysis is required. Consider the syllogism: *All men are cowards; Mussolini is a man; Mussolini is a coward.* It asserts that if the class "men" is included in the class "cowards," and if "Mussolini" is *a member of* the class "men," then he is *a member of* the class "cowards." In the minor a different type of relation is asserted from that asserted in the major; for the relation "is a member of" is nontransitive (see p. 49), while the relation "is included in" is transitive. The validity of the inference illustrates a modified form of the *dictum de omni.*

3. All the so-called relational (or *a fortiori*) syllogisms depend on the transitivity of the relations. Thus in *John is older than Tom, Tom is older than Henry, John is older than Henry,* the relation "is older than" is transitive.

4. Consider next the sorites:

> All professors are handsome.
> All handsome men are married.
> All married men are well fed.
> ∴ All professors are well fed.

It is clear that the transitivity of the relation of class inclusion is the basis for the inference.

5. Let us finally examine the pure hypothetical syllogism:

> If it rains, I will take my umbrella.
> If I take my umbrella, I am sure to lose it.
> ∴ If it rains, I am sure to lose my umbrella.

Each of the three propositions asserts an implication. The conclusion is a valid consequence because the implication relation is transitive.

§ 4. SYMBOLS: THEIR FUNCTION AND VALUE

If the value of distinguishing various kinds of relations were confined to laying bare the basis of inferences already familiar, the reader might regard such investigations as without much profit. In fact, however, the study of the logical properties of relations is the gateway to the systematic study of more complex inferences. But the more complex forms of inference cannot be studied with any care until a specially devised symbolism is introduced. Because of the prominence of such special symbols in the generalized study of logic, the latter has also been called *symbolic* or *mathematical* logic.

The importance of symbols in the development of modern logic cannot be overestimated. According to Peirce: "The woof and warp of all thought and research is symbols, and the life of thought and science is the life inherent in symbols; so that it is wrong to say that a good language is *important* to good thought, merely, for it is the essence of it." [4] We must therefore inquire into the function and value of symbols.

1. *The Generic Traits of Language.* It will be best to begin with the generic traits of all languages. Languages differ from one another in two ways: different phonetic or ideographic elements are employed in them; and different groups of ideas are expressed by fixed phonetic and graphic elements. But the experiences which a language intends to express and communicate vary in an unlimited manner, while the language employs only a finite number of fundamental linguistic elements, which may be called word stems. It follows that every language, as we know it, must be based upon a far-reaching classification or categorization of experience. Sense impressions and emotional states are grouped on the basis of broad similarities, although no two of them are identical in every respect. Such groupings indicate, therefore, the presence of a large number of characteristics common to each group. Furthermore, various kinds of relations between these several groupings also come to be noted, and require to be expressed in language.

What groupings of experiences are made depends upon the interests of those using the language, as well as upon the subject matter which comes to be expressed. Consequently, the categorizations of experience which are satisfactory for one purpose may be inept for another; and the language which functions as a satisfactory symbolic scheme on one occasion may be too clumsy or too subtle for

[4] Collected Papers, Vol. II, p. 129.

another. In any case, the reader must note that *all* language, and not merely symbolic logic, is symbolic. All communication and inquiry must take place by means of words, sounds, graphic marks, gestures, and so on. Words *refer* to something, whether it be to the emotions, to ideas, or to that which ideas are about. When we say, "There is a fire next door," the sounds we utter *are* not what they *signify*. The sounds, for example, do not burn, they are not a definite number of feet distant, and so on. They do signify something to someone.

Since the groupings of experiences upon which languages are based are generally not sharply demarcated from one another, all languages suffer in some degree from vagueness. Thus, although the number of distinguishable shades of color is very large, only a few colors receive special names. In different languages different colors receive special names, and the limits of what constitutes one color are also frequently different. Even in scientific languages there is a fringe of uncertainty as to the limits of the denotation or application of names.

In virtue of the fact that on the one hand the number of word stems in any language is small, while on the other hand the number of varieties of experience is unlimited, it is necessary to represent many of the latter by combining the word stems in some manner. These modes of combination are the *formal elements* in the language, and are the theoretical basis for grammar. But it is clear that the formal aspects of a language are not altogether arbitrary. For the combination of word stems must represent experiences so made up that they are analyzable into relations between what is denoted by the word stems taken in isolation. Although grammar cannot be a substitute for logical analysis, there is a kinship between logic and grammar. For grammatical structure also represents certain abstract relations which a subject matter possesses when expressed in language.

Ordinary languages have been devised to meet practical needs, for which fine distinctions may be useless or even a hindrance. Most language is emotive or ceremonial in character, aiming to convey or provoke some emotion. These facts, coupled with the foregoing considerations, make it clear that such languages cannot be used to represent adequately, and without serious modification, those distinctions which are the outcome of subtle analyses, or which arise from a classification of experience made for different purposes. Therefore, if we wish to avoid the distortion which the emotional and intellectual overtones of ordinary words introduce when we are making careful analyses; if we wish to restrict as much as possible the vagueness of common symbols; if we desire to prevent the often subtle transformations which the meaning of verbal symbols undergoes—then it is essential to devise a specially constructed symbolism.

Linguistic Changes

How words change their meanings is a fascinating study. We shall have occasion to point out in the next chapter some of the confusions which have taken place in the philosophy of mathematics because this fact of change has been overlooked. We shall restrict ourselves here to noting two ways in which the original meaning of words may become totally lost and other meanings substituted for it.

One such way has been called *generalization*. The same symbol may come to denote a *more extended class* of objects, so that it no longer denotes with accuracy the more specific things it may once have symbolized. Thus "paper" once denoted papyrus; it was next employed to denote writing material made of old rags; today it symbolizes not only the product of rags, but also the product of chemically treated wood pulp. The history of the word "number" also illustrates such a process of progressive generalization. It once denoted only integers, then gradually included fractions, irrationals, transcendentals like π, and even determinants. Other such words are "force," "energy," "geometry," and "equality."

A second way in which meanings become altered is by *specialization*. The same symbol may become restricted in its application to a smaller range of objects, so that it comes to denote more specific characters than it once did. Thus the word "surgeon" once meant anyone who worked with his hands; today it is restricted to those with special medical training. Other words whose history illustrates the process of specialization are "minister," "physician," and "artist."

A fertile and interesting source of change in the meaning of words arises when their application is broadened because of a *metaphorical extension* of their meaning. Thus "governor" originally meant a steersman on a boat, "spirit" meant breath; a bend in a pipe is called an "elbow," the corresponding parts of a pipe-fitting are called "male" and "female," and so on.

The Value of Special Symbols

Let us now state summarily the advantages which we can expect from a specially devised symbolism. In the first place, such symbols enable us to distinguish and keep distinct different meanings. We need only agree to employ a different symbol for each distinct notion and to have no symbol represent more than one notion. The ambiguity with which ordinary language is infected is then at a minimum.

In the second place, a convenient symbol enables us to concentrate upon what is essential in a given context. When in mathematics we substitute a single letter like R for a complex expression like $(a + b + c + d)$; or when we use letters like S, P, M for the terms "Socrates," "mortal," "man," in a syllogism, we make clear that the results of our reasoning depend not on the *special* meaning of these expressions, but on the abstract relations which connect them with others.

A third important function of symbols is to exhibit clearly and concisely the *form* of propositions. This has long been recognized in mathematics. Thus, to take an elementary illustration, the difference in form between $4x^2 = 5x - 1$ and $4x^3 = 5x^2 - 1$, and the identity in form of $x + y = 1$ and $4x = 3y$, can be perceived at a glance. In the first pair of equations one is quadratic, the other cubic; both equations of the second pair are linear. It would be well-nigh humanly impossible to carry out a long series of inferences if such equations were stated in words. Thus the verbal statement of what is represented by Maxwell's equations would easily fill several pages and the essential relations between the various factors involved would thus be hidden. An adequate symbolism makes clear just what it is that is constant or invariable in a proposition, and what is variable. The invariable features are its form or structure.

A fourth and no mean advantage of such symbols is their labor-saving and thought-saving function. When once a symbolism has been perfected, much that hitherto required concentrated attention can be performed mechanically. Very often a good symbolism may suggest conclusions which otherwise would have escaped the thinker completely. The discovery of negative and imaginary numbers, Maxwell's introduction of the dielectric displacement and the subsequent discovery of ether waves, are directly due to the suggestiveness of symbols. For this reason, it has been said that "in calculation the pen sometimes seems to be more intelligent than the user." It is the ability of a properly constructed symbolism to function as a *calculus* which makes evident their importance in a striking manner.

§ 5. THE CALCULUS OF CLASSES

The development of an adequate symbolism, together with the discovery of the formal properties of relations, made it possible both to generalize traditional logic and to establish a powerful calculus.

For example, the operations of addition, multiplication, and so on of the mathematical sciences may be viewed in terms of the theory of relations. Thus the operation of addition is based upon a triadic relation. The relation $a + b = c$ connects the two summands a and b with c. It is a many-one relation, for to every pair of summands there corresponds one and only one sum, while to one sum there corresponds an indefinite number of pairs of summands. But if the sum and one of the summands is fixed, the other summand is uniquely determined. Such triadic relations, which are exemplified in various kinds of operations, can be studied in great detail.

It is not necessary, however, that the operations should be those of ordinary algebra. Operations, of an altogether *nonquantitative* type have been found for combining classes, when these are taken in extension.

We shall sketch briefly the general theory of classes and of propositions. We may, however, preface what follows with this advice taken from C. L. Dodgson: "When you come to any passage you don't understand, *read it again:* if you *still* don't understand it, *read it again:* if you fail, even after *three* readings, very likely your brain is getting a little tired. In that case, just put the book away, and take to other occupations, and next day, when you come to it fresh, you will very likely find that it is *quite* easy."

In the history of symbolic logic, the theory of classes was developed first, because it was first noted that the Aristotelian logic may be interpreted as dealing with the mutual relations of classes. In a *systematic* exposition of logical principles, however, the logic of classes is not prior to other logical principles. For to assert that two classes are related in any way is to assert a *proposition.* Every investigation of the theory of classes employs principles that belong to the theory of propositions. The theory of propositions is, therefore, required for every other logical inquiry, and should be developed first. But in an elementary discussion, such as ours, where the intent is to suggest the direction in which traditional logic may be extended rather than to provide a systematic analysis of such a generalized logic, we may neglect this point. No great harm will be done if we reverse the logical order, and follow the temporal order of development.

Operations and Relations

We shall understand by a "class" a group of individuals each having certain properties, in virtue of which they are said to be members of the class. Thus the class "man" is the set of individual men, the class "even numbers" is the set of even integers, and so on. In other words we shall treat classes in extension. The domain of possible classes is called the *universe of discourse,* or simply the *universe.* It will be symbolized by 1. It may happen that a class has no members. Thus the class of twenty-foot men has no members, although it does have a defining characteristic, namely, "twenty-foot men." Such a class will be called the *null* or *zero* class, and will be symbolized by *0*. The notion of a null class, although puzzling to beginners, has many technical advantages.

Classes may be operated upon in three ways, each operation or combination receiving a special notation. Consider the class "males" in the universe of human beings. Exclude from the universe this class, and we get the class "females." The individuals who are members of the universe but are not members of the class "males," will be said to belong to the *negative* of the class "males." Hence "females" is the negative of "males" in this universe of discourse. A class and its negative are mutually exclusive, and they exhaust the universe. If a represents a class, $\bar{a}$ (to be read *"not-a"*) represents its negative.

Consider next the two classes "English books" and "French books." The class which contains *either French or English books* will be said to be the *logical sum* of these classes. The operation of combining them in this way will be called *logical addition.* If a and b are classes, their logical sum is represented by $a + b$. This may be read as "a plus b," or as "either a or b." The alternation is, as usual in logic, *not* disjunctive. The symbol $+$ is employed because logical addition has certain formal analogies to the addition of ordinary arithmetic.

Consider next the classes "professors" and "bad-tempered individuals." Suppose we wish to pick out all the individuals who are members of *both* classes, and so obtain the class "bad-tempered professors." Such an operation is called *logical multiplication,* and the result is called the *logical product* of the classes. If a and b are classes, their product is symbolized by $a \times b$, or (more conveniently) by ab.

How the idea of a null class arises may now be evident. We assume that the results of multiplying classes are classes. The logical product of "women" and "locomotive engineers" is "female locomotive engineers." Hence this is a class even though it may have no members.

We have been considering operations upon classes. But we cannot establish a calculus unless *relations* between classes are also symbolized. The difference between operations upon classes and relations between classes is this: operations upon classes yield *classes;* the assertions of relations between classes are *propositions,* not classes. The relation we shall regard as fundamental is that of class inclusion. One class will be said to be included in another if every member of the first is also a member of the second. If a and b are classes, we symbolize the proposition a *is included in* b by $a < b$.

The relation $<$ is transitive and nonsymmetrical, for if $a < b$, and $b < c$, then $a < c$; but if $a < b$, it does not follow that $b < a$. We may define the *equality* of two classes in terms of mutual inclusion. Class a is equal to b if a is included in b and b is included in a—in other words, if they have the same members. In symbols, $(a = b) \equiv (a < b) . (b < a)$, where the sign $=$ indicates *equality between classes,* the sign $\equiv$ indicates *equivalence* between propositions, and the dot (.) indicates the *joint assertion* of two propositions.

Principles of the Calculus of Classes

In order to get the calculus started, we must state a number of fundamental principles which define very explicitly the nature of the operations and relations we have just been discussing. The following set of principles are those which are usually assumed.

1. *Principle of identity:* for every class, $a < a$.

This principle asserts that every class is included in itself. From the definition of equality and this principle, it follows that $a = a$.

2. *Principle of contradiction:* $a\bar{a} = 0$

Nothing is a member of both a and *not-a.*

3. *Principle of excluded middle:* $a + \bar{a} = 1$

Every individual in the universe is either a member of a or of *not-a.*

4. *Principle of commutation:* $ab = ba$

$$a + b = b + a$$

We may illustrate this as follows: The class of individuals who are both German and musical is the same as the class of those who are both musical and German; the class of individuals who are either German or musical is the same as the class of those who are either musical or German.

5. *Principle of association:* $(ab)\,c = a\,(bc)$
$$(a + b) + c = a + (b + c)$$

6. *Principle of distribution:* $(a + b)\,c = ac + bc$
$$ab + c = (a + c)\,(b + c)$$

The first expresses what is analogous to the well-known properties of ordinary numbers. The second, however, introduces a significant difference between the present algebra and the ordinary (quantitative) kind.

7. *Principle of 'autology:* $aa = a$
$$a + a = a$$

Both of these principles mark a radical difference between ordinary (quantitative) algebra and the present one.

8. *Principle of absorption:* $a + ab = a$
$$a\,(a + b) = a$$

9. *Principle of simplification:* $ab < a$
$$a < a + b$$

In virtue of these two principles it follows that the zero class is included in every class ($0 < a$) and that every class is included in the universe ($a < 1$). To see this, all that needs to be done is to let $b = 0$ in the first, and $b = 1$ in the second expression.

10. *Principle of composition:* $[(a < b)\,.\,(c < d)] \supset (ac < bd)$
$$[(a < b)\,.\,(c < d)] \supset [(a + c) < (b + d)]$$

Here we are employing, as usual, the symbol $\supset$ for the relation of implication, and the dot (.) for the joint assertion of two propositions. The first expression reads: If *a* is included in *b*, and *c* is included in *d*, then the logical product of *a* and *c* is included in the logical product of *b* and *d*.

11. *Principle of syllogism:* $[(a < b)\,.\,(b < c)] \supset (a < c)$

If *a* is included in *b*, and *b* in *c*, then *a* is included in *c*. The relation "included in" is herewith declared to be transitive.

The Representation of the Traditional Categorical Propositions

Let us now represent symbolically each of the four categorical propositions.

All a's *are* b's may be represented as $(a < b)$. It can be shown, moreover, that this is equivalent to $(a\bar{b} = 0)$, so that we have $(a < b) \equiv (a\bar{b} = 0)$.

No a's *are* b's is equivalent to *All* a's *are* non-b's. Hence, it may be symbolized as $(a < \bar{b})$. But this expression is equivalent to $(ab = 0)$, so that $(a < \bar{b}) \equiv (ab = 0)$.

Since the particular propositions are the contradictories of the universals, they deny what the latter affirm. Hence *Some* a*'s are* b*'s* must deny *No* a*'s are* b*'s* (symbolically, $a < \bar{b}$). It may be represented as $(a < \bar{b})'$ or as $(ab \neq 0)$.

Some a*'s are not* b*'s* must contradict $(a < b)$. Hence, it may be represented as $(a < b)'$ or as $(a\bar{b} \neq 0)$.

Each of these four symbolic forms should be familiar to the reader from our previous analysis of the categorical propositions.

Proof of De Morgan's Theorem

In the space at our disposal we cannot develop the calculus of classes to show its great power. But we wish to illustrate the nature of proofs in this calculus by offering a demonstration of De Morgan's theorem as applied to classes.

We wish to find the negative of the class $(a + b)$.

In virtue of the principle of excluded middle, $a + \bar{a} = 1$, and $b + \bar{b} = 1$. Also, by the principle of simplification, $1 \times 1 = 1$, and hence $(a + \bar{a})(b + \bar{b}) = 1$. Applying the distributive and associative principles this may be written: $(ab + a\bar{b} + \bar{a}b) + \bar{a}\bar{b} = 1$.

Consider now the two classes $(ab + a\bar{b} + \bar{a}b)$ and $\bar{a}\bar{b}$. They exhaust the universe, since their sum is 1; and they are mutually exclusive, since their product is 0. Hence either one is the negative of the other.

But $ab + a\bar{b} + \bar{a}b = ab + a\bar{b} + \bar{a}b + ab$, by applying the principle of tautology. The right-hand member by the distributive principle is equal to $a\,(b + \bar{b}) + b\,(a + \bar{a}) = a + b$. Therefore, since $\bar{a}\bar{b}$ is the negative of $(ab + a\bar{b} + \bar{a}b)$, which is equal to $(a + b)$, it is also the negative of this latter.

Hence we get $(\overline{a + b}) = \bar{a}\,\bar{b}$, one form of the De Morgan's theorem.

Let us now obtain the negative of ab.

By an argument identical with the above, (ab) and $(a\bar{b} + \bar{a}b + \bar{a}\bar{b})$ are negatives of one another. But we also have:

$$
\begin{aligned}
a\bar{b} + \bar{a}b + \bar{a}\bar{b} &= a\bar{b} + \bar{a}b + \bar{a}\bar{b} + \bar{a}\bar{b} \\
&= (a\bar{b} + \bar{a}\bar{b}) + (\bar{a}b + \bar{a}\bar{b}) \\
&= \bar{b}\,(a + \bar{a}) + \bar{a}\,(b + \bar{b}) \\
&= \bar{b} + \bar{a} = \bar{a} + \bar{b}
\end{aligned}
$$

Hence $(\overline{a \times b}) = \bar{a} + \bar{b}$, a second form of De Morgan's theorem. These results may be generalized for any finite number of classes thus: $\overline{(a+b+c+d+\ldots)} = \bar{a}\,\bar{b}\,\bar{c}\,\bar{d}\ldots$ and $\overline{(abcd\ldots)} = \bar{a}+\bar{b}+\bar{c}+\bar{d}+\ldots$

§ 6. THE CALCULUS OF PROPOSITIONS

The calculus of propositions was first developed as another interpretation of the symbolism elaborated in the theory of classes. The two calculi have, up to a certain point, an identical formal structure, and every proposition in the theory of classes has a corresponding proposition, obtainable by a suitable interpretation, in the theory of propositions. The following table can serve as a dictionary for translating the theorems of the calculus of classes into theorems in the calculus of propositions:

$a, b, c, \ldots$ any classes	$p, q, r, \ldots$ any propositions
$\bar{a}$, the negative of class a	p', the contradictory of proposition p
$a+b$, the logical sum of two classes	$p \vee q$, the logical sum of two propositions, or their alternative assertion
$a\ b$, the logical product of two classes	$p \, . \, q$, the logical product of two propositions, or their joint assertion
$a < b$, a is included in b	$p \supset q$, p implies q.
0, the class having no members	The proposition which is false
1, the class containing all classes	The proposition which is true

By means of this dictionary all the principles which are true for classes may be restated, in different symbolism, for propositions.

Nevertheless, such an approach to the theory of propositions has several drawbacks, although it does help to bring out the formal analogies between the two calculi. In the first place, as already mentioned, there are several theorems which are true when the terms are propositions, but false when they are classes. Thus *If* p *implies* q *or* r, *then* p *implies* q *or* p *implies* r *is a true theorem for propositions*. (Symbolically, it is $[p \supset (q \vee r)] \supset [(p \supset q) \vee (p \supset r)]$. It is false when interpreted for classes. For example, it is false that *If all English people are either men or women, then all English people are men or all English people are women.*

A more serious objection arises from the fact that in developing the calculus of propositions, we wish to enumerate *all* the principles of inference we are employing. If we develop the theory of propositions systematically and deductively, thus beginning with a number of undemonstrated principles for propositions, we can demonstrate every other principle. With enough care we can guard ourselves against the danger of employing a principle of inference not yet assumed or demonstrated. Consequently, with such a process we can arrive at a satisfactory systematization of logical principles. If, however, we use the class calculus as the basis for developing the theory of propositions, we can no longer use such a method of obtaining all the principles of inference.

Just as in the calculus of classes where all classes are interpreted in extension, so in the calculus of propositions all propositions are considered only with respect to their *truth-values* and not for the *specific meaning* of what they assert. The reader must be clear about this if he does not want to commit bad blunders.

Let us illustrate this point by analyzing the definition of *implication* which is often given in discussions of symbolic logic. $(p \supset q)$ is defined as equivalent to $(p' \vee q)$ or to $(p \cdot q')'$. In words: p implies q is true if "*not-p* or q" is true.

But "*not-p* or q" is true in any one of the following cases: (1) p is true and q is true; (2) p is false and q is true; (3) p is false and q is false. The only case which would make it false is when p is true and q is false. It follows that "p implies q" is true in any one of the first three cases. But if we examine these cases, we must say that so long as p is false, no matter what q is, "p implies q" is true; and so long as q is true, no matter what p is, "q is implied by p" is true. This may be stated in the paradoxical fashion that a false proposition implies any proposition, and that any proposition implies a true one. Each of the following propositions must therefore be true: *2 + 2 = 5 implies that Sacco and Vanzetti were executed for murder* and *Alfred Smith was defeated for President in 1928 implies that the base angles of an isosceles triangle are equal.*

The paradox disappears, however, if the reader dismisses from his mind the prejudice in favor of the usual meaning of the word "implication" and recognizes that, by definition, we have made it denote something else in the propositional calculus. The distinction is recognized by calling the first *formal* and the latter *material* implication.[5] (Sometimes the former is called *entailment, tautologous implication,* or *strict implication.*) The assertion of formal implication, as we saw in our first chapter, involves no assumption as to the factual truth or falsehood of either of two propositions, but only that they are so connected by virtue of their structure (which they share with all other propositions of the same form) that it is *impossible* for the implicating proposition to be true and the implied

[5] This use of the word "formal" must not be confounded with the use in the *Principia Mathematica* of Whitehead and Russell.

one to be false. *Material implication* is the name we give to the fact that one of a pair of propositions happens to be false or else the other happens to be true.

The two kinds of implications are, however, not unrelated. The formal implication of the syllogism means that in every specific instance of the syllogism there is a material implication between the premises and the conclusion. But this account leaves out the fact that in every syllogism there is a necessity, based on an element of identity not directly present in all other cases of material implication. When we say that *Whales are mammals and all mammals have lungs* implies *Whales have lungs* there is a connection not present when we say that *Dante was born in 1250* implies that *Lithium is a metal* (because the first happens to be false and the second true). This, however, also involves a question of metaphysics, to wit, whether all truths are necessarily connected in the ultimate nature of things.

CHAPTER VII

THE NATURE OF A LOGICAL OR MATHEMATICAL SYSTEM

§ 1. THE FUNCTION OF AXIOMS

Although the Babylonians and Egyptians had much information about the eclipses of the sun and the moon, the measurement of land and the construction of buildings, the disposition of geometric figures in order to form symmetrical designs, and computation with integers and fractions, it is generally recognized that they had no *science* of these matters. The idea of a science was a contribution of the Greeks.

Information, no matter how reliable or extensive, which consists of a set of isolated propositions is not science. A telephone book, a dictionary, a cookbook, or a well-ordered catalogue of goods sold in a general store may contain accurate knowledge, organized in some convenient order, but we do not regard these as works of science. Science requires that our propositions form a logical *system,* that is, that they stand to each other in some one or other of the relations of equivalence and opposition already discussed. Therefore in the present chapter we continue our study of the nature of proof, in order to make clearer some of the generic characteristics of deductive systems. Such a study, we shall find, is identical with the study of the nature of mathematics.

Let us remember that no proposition can be *demonstrated* by any experimental method. The reader is doubtless familiar with the Pythagorean theorem that in a right triangle the square on the hypotenuse is equal to the sum of the squares on the arms. He has, no doubt, "proved" or "demonstrated" it in his school days. Nevertheless every gathering of college-trained men is likely to contain at least one member who, when asked how the theorem may be proved, will suggest protractors, carefully drawn triangles, and finely graduated rulers. In this respect, such an individual has made no essential advance upon the methods of the ancient Egyptian surveyors.

Suppose, for instance, we were to attempt to prove the Pythagorean theorem by actually drawing the squares on the three sides of a right triangle on some uniformly dense tinfoil, then cutting them out and by weighing them seeing whether the square on the hypotenuse does actually balance the other two squares. Would this constitute a proof? Obviously not, for we can never know that the tinfoil is in fact absolutely uniform in density, or that the pieces cut out are perfect squares. Hence, if in a number of experiments we should fail to find a perfect balance in the weights, we should not consider that as evidence against the view that there *would* be a perfect equilibrium *if* our lines were perfectly straight, the angles of the square were perfect right angles, and the mass of the tinfoil were absolutely uniform. A logical proof or demonstration consists, as we have seen, in exhibiting a proposition as the necessary consequence of other propositions. The demonstration asserts nothing about the factual truth of either the premises or their logical consequences.

"But look here!" the reader may protest. "Don't we prove that the theorems in geometry are *really* true? Isn't mathematics supposed to be the most certain of the sciences, in which some property is shown to hold for all objects of a definite type, once and for all? If you examine any statement of a theorem, for example the Pythagorean, you find something asserted about 'all' triangles. Now if you admit that something is in fact proved true of all triangles, why do you refuse to admit that we are establishing the 'material' truth of such a theorem? Doesn't 'all' really mean 'all'?"

This protest, however, simply ignores that fact already noted that a logical proof is a "pointing-out" or "showing" of the implications between a set of propositions called axioms and a set called theorems, and that the axioms themselves are not demonstrated.

The reader may reply: "The axioms are not proved, because they need no proof. *Their truth is self-evident.* Everybody can recognize that propositions like *The whole is greater than any one of its parts* or *Through two points only one straight line may be drawn* are obviously true. They are therefore a satisfactory basis for geometry, because by their means we can establish the truth of propositions not so obvious or self-evident."

Such a reply represents a traditional view. Up to the end of the nineteenth century it was generally believed that the axioms are materially true of the physical world, and that the cogency of the demonstrations depends upon their being thus materially true. Nevertheless, this view of the axioms confuses three different issues:

1. How is the material truth of the axioms established?
2. Are the axioms materially true?
3. Are the theorems the logical consequences of the explicitly stated axioms?

We must consider these separately.

1. The answer generally given to the first question is that the axioms are self-evident truths. But this view is a rather complacent way of ignoring real difficulties. In the first place, if by "self-evidence" is meant psychological obviousness, or an irresistible impulse to assert, or the psychological unconceivability of any contrary propositions, the history of human thought has shown how unreliable it is as a criterion of truth. Many propositions formerly regarded as self-evident, for example: *Nature abhors a vacuum; At the antipodes men walk with their heads beneath their feet; Every surface has two sides,* are now known to be false. Indeed, contradictory propositions about every variety of subject matter, thus including most debatable propositions, have each, at different times, been declared to be fundamental intuitions and therefore self-evidently true. But whether a proposition is obvious or not depends on cultural conditions and individual training, so that a proposition which is "self-evidently true" to one person or group is not so to another.

This view assumes a capacity on the part of human beings to establish universal or general propositions dealing with matter of fact simply by examining the *meaning* of a proposition. But, once more, the history of human thought, as well as the analysis of the nature of meaning, has shown that there is an enormous difference between *understanding the meaning* of a proposition and *knowing its truth.* The truth of general propositions about an indefinite number of empirical facts can never be absolutely established. The fundamental reason, therefore, for denying that the axioms of geometry or of any other branch of mathematics, are self-evidently true is that each of the axioms has at least one significant contrary.

"But doesn't the mathematician discover his axioms by observation on the behavior of matter in space and time?" the reader may ask. "And aren't they in fact more certain than the theorems?"

In order to reply, we must resort to the ancient Aristotelian distinction between the *temporal order* in which the logical dependence of propositions is discovered and the *logical order* of implications between propositions. There is no doubt that many of the axioms of mathematics are an expression of what we believe to be the truth concerning selected parts of nature, and that many advances in mathematics have been made because of the suggestions of the natural sciences. But there is also no doubt that mathematics as an *inquiry* did not historically begin with a number of axioms from which subsequently the theorems were derived. We know that many of the propositions of Euclid were known hundreds of years before he lived; they were doubtless believed to be materially true. Euclid's chief contribution did not consist in discovering additional theorems, but in exhibiting them as part of a system of connected truths. The kind of question Euclid must have asked himself was: Given the theorems about the angle sum of a triangle, about similar triangles, the Pythagorean theorem, and the rest, what are the minimum number of assumptions or axioms from which these can be inferred? As a result of his work, instead of having what were believed to be independent propositions, geometry became the first known example of a deductive system. The axioms were thus in fact *discovered later* than the theorems, although the former are *logically prior* to the latter.

It is a common prejudice to assume that the logically prior propositions are "better known" or "more certain" than the theorems, and that in general the logical priority of some propositions to others is connected in some way with their being true. Axioms are simply assumptions or hypotheses, used for the purpose of systematizing and sometimes discovering the theorems they imply. It follows that axioms *need not* be known to be true before the theorems are known, and in general the axioms of a science are much less evident psychologically than the theorems. In most sciences, as we shall see, the material truth of the theorems is not established by means of first showing the material truth of the axioms. On the contrary, the material truth of axioms is made *probable* by establishing empirically the truth or the probability of the theorems.

2. We must acknowledge, therefore, that an answer to the question, "Are the axioms materially true?" cannot be given on grounds of logic alone, and that it must be determined by the special natural science which empirically investigates the subject matter of such axioms. But it must also be admitted that the material truth

or falsity of the axioms is of no concern to the logician or mathematician, who is interested only in the fact that theorems are or are not implied by the axioms. It is essential, therefore, to distinguish between *pure mathematics,* which is interested only in the facts of implication, and *applied mathematics,* or natural science, which is interested also in questions of material truth.

3. Whether the theorems are logical consequences of the axioms must, therefore, be determined by logical methods alone. This is not, however, always as easy as it appears. For many centuries Euclid's proofs were accepted as valid, although they made use of other assumptions than those he explicitly stated. There has been a steady growth in the logical rigor demanded of mathematical demonstrations, and today considerable logical maturity, as well as special technical competence, is a prerequisite for deciding questions of validity. Indeed, in certain branches of mathematics the cogency of some demonstrations has not yet been established.

We may now state summarily our first results concerning the nature of a logical system. Propositions can be demonstrated by exhibiting the relations of implication between them and other propositions. But not all propositions in the system can be demonstrated, for otherwise we would be arguing in a circle. It should, however, be noted that propositions which are axiomatic in one system may be demonstrated in another system. Also, terms that are undefined in one system may be definable in another. What we have called pure mathematics is, therefore, a *hypothetico-deductive system.* Its axioms serve as hypotheses or assumptions, which are entertained or considered for the propositions they imply. In general, the logical relation of axioms and theorems is that of a principal to its subaltern. If the whole of geometry is condensed into one proposition, the axioms are the antecedents in the hypothetical proposition so obtained. But they also characterize, as we shall see presently, the formal structure of the system in which the theorems are the elements.

§ 2. PURE MATHEMATICS—AN ILLUSTRATION

The reader is probably familiar with some examples of logical systems from his study of mathematics, and we have already considered one example of such a system in our discussion of the syllogism. It will be valuable, however, to start anew. Consider the following propositions, which are the axioms for a special kind of geometry.

Axiom 1. If A and B are distinct points on a plane, there is at least one line containing both A and B.

Axiom 2. If A and B are distinct points on a plane, there is not more than one line containing both A and B.

Axiom 3. Any two lines on a plane have at least one point of the plane in common.

Axiom 4. There is at least one line on a plane.

Axiom 5. Every line contains at least three points of the plane.

Axiom 6. All the points of a plane do not belong to the same line.

Axiom 7. No line contains more than three points of the plane.

These axioms seem clearly to be about points and lines on a plane. In fact, if we omit the seventh one, they are the assumptions made by Veblen and Young for "projective geometry" on a plane in their standard treatise on that subject. It is unnecessary for the reader to know anything about projective geometry in order to understand the discussion that follows. But what are points, lines, and planes? The reader may think he "knows" what they are. He may "draw" points and lines with pencil and ruler, and perhaps convince himself that the axioms state truly the properties and relations of these geometric things. This is extremely doubtful, for the properties of marks on paper may diverge noticeably from those postulated. But in any case the question whether these actual marks do or do not conform is one of *applied* and not of *pure* mathematics. The axioms themselves, it should be noted, do not indicate what points, lines, and so on "really" are. For the purpose of discovering the implications of these axioms, it is unessential to know what we shall understand by points, lines, and planes. These axioms imply several theorems, not in virtue of the visual representation which the reader may give them, but in virtue of their logical form. Points, lines, and planes may be any entities whatsoever, undetermined in every way except by the relations stated in the axioms.

Let us, therefore, suppress every explicit reference to points, lines, and planes, and thereby eliminate all appeal to spatial intuition in deriving several theorems from the axioms. Suppose, then, that instead of the word "plane," we employ the letter S; and instead of the word "point," we use the phrase "element of S." Obviously, if the plane (S) is viewed as a collection of points (elements of S), a line may be viewed as a class of points (elements) which is a subclass of the points of the plane (S). We shall therefore substitute for the word "line" the expression "l-class." Our original set of axioms then reads as follows:

Axiom 1′. If A and B are distinct elements of S, there is at least one *l-class* containing both A and B.

Axiom 2′. If A and B are distinct elements of S, there is not more than one *l-class* containing both A and B.

Axiom 3′. Any two *l-classes* have at least one element of S in common.

Axiom 4′. There exists at least one *l-class* in S.

Axiom 5′. Every *l-class* contains at least three elements of S.

Axiom 6′. All the elements of S do not belong to the same *l-class*.

Axiom 7′. No *l-class* contains more than three elements of S.

In this set of assumptions no explicit reference is made to any specific subject matter. The only notions we require to state them are of a completely general character. The ideas of a "class," "subclass," "elements of a class," the relation of "belonging to a class" and the converse relation of a "class containing elements," the notion of "number," are part of the fundamental equipment of logic. If, therefore, we succeed in discovering the implications of these axioms, it cannot be because of the properties of space as such. As a matter of fact, none of these axioms can be regarded as propositions; none of them is in itself either true or false. For the symbols, S, *l-class*, A, B, and so on are *variables*. Each of the variables denotes any one of a class of possible entities, the only restriction placed upon it being that it must "satisfy," or conform to, the formal relations stated in the axioms. But until the symbols are assigned specific values the axioms are *propositional functions,* and not propositions.[1]

Our "assumptions," therefore, consist in relations considered to hold between undefined terms. But the reader will note that although no terms are *explicitly* defined, an *implicit* definition of them is made. They may denote anything whatsoever, provided that what they denote conforms to the stated relations between themselves. This procedure characterizes modern mathematical

[1] The statement in the text is concerned with forms such as "X is a man," which does not assert anything until some definite value is assigned to the variable X. In this case, the truth of the proposition asserted by the sentence (obtained by substituting a determinate value of X) depends upon the value assigned to X. Propositions, however, of the form *X is a man implies X is mortal, for all values of X* do assert something which is true no matter what value is assigned to X. In this case, X is said to be an apparent variable, since the truth of the proposition does not depend upon the value given to X.

technique. In Euclid, for example, *explicit* definitions are given of points, lines, angles, and so on. In a modern treatment of geometry, these elements are defined *implicitly* through the axioms. As we shall see, this latter procedure makes it possible to give a variety of interpretations to the undefined elements, and so to exhibit an identity of structure in different concrete settings.

We shall now demonstrate six theorems, some of which may be regarded as trite consequences of our assumptions.

Theorem I. If A and B are distinct elements of S, there is one and only one *l-class* containing both A and B. It will be called the *l-class* AB.

This follows at once from Axioms 1′ and 2′.

Theorem II. Any two distinct *l-classes* have one and only one element of S in common.

This follows from Axioms 2′ and 3′.

Theorem III. There exist three elements of S which are not all in the same *l-class.*

This is an immediate consequence of Axioms 4′, 5′, and 6′.

Theorem IV. Every *l-class* in S contains just three elements of S.

This follows from Axioms 5′ and 7′.

Theorem V. Any class S which is subject to Axioms 1′ to 6′ inclusively contains at least seven elements.

Proof. For let A, B, C be three elements of S not in the same *l-class.* This is possible by Theorem III. Then there must be three distinct *l-classes,* containing AB, BC, and CA, by Theorem I. Furthermore, each of these *l-classes* must have an additional element, by Axiom 5′. And these additional elements must be distinct from each other, and from A, B, C, by Axiom 2′.

Let these additional elements be designated by D, E, and G, so that ABD, BCE, and CAG form the three distinct *l-classes* mentioned. Now AE and BG also determine *l-classes,* which must be distinct from any *l-classes* yet mentioned, by Axiom 1′. And they must have an element of S in common, by Axiom 4′, which is distinct from any element so far enumerated, by Axiom 2′. Let us call it F, so that AEF and BFG are *l-classes.*

Consequently, there are at least seven elements in S.

Theorem VI. The class S, subject to all seven assumptions, contains no more than seven elements.

Proof. Suppose there were an eighth element T. Then the *l-class* determined by AT and BFG would have to have an element in common, by Axiom 3′. But this element cannot be B, for the elements AB determine the *l-class* whose elements are ABD, so that $ABTD$ would need to belong to this very same *l-class;* which is impossible by Axiom 7′. Nor can this element be F, for then $AFTE$ would have to belong to the *l-class* AEF; nor G, for then $AGTC$ would need to belong to the *l-class* AGC; these results are impossible for the same reason (Axiom 7′).

Consequently, since the existence of an eighth element would contradict Axiom 7′, such an element cannot exist.

We have now exhibited a miniature mathematical system as a hypothetico-deductive science. The deduction makes no appeal whatsoever to experiment or observation, to any sensory elements. The reader has had a taste of pure mathematics. Whether anything in the world of existence conforms to this system requires empirical knowledge. If this be the case, that portion of the actual world must have the systematic character indicated formally in our symbolic representation. That the world does exemplify such a structure can be verified only within the limits of the errors of our experimental procedure.

§ 3. STRUCTURAL IDENTITY OR ISOMORPHISM

We want to show now that an abstract set such as the one discussed in the previous section may have more than one concrete representation, and that these different representations, though extremely unlike in material content, will be identical in logical structure.

Let us suppose there is a banking firm with seven partners. In order to assure themselves of expert information concerning various securities, they decide to form seven committees, each of which will study a special field. They agree, moreover, that each partner will act as chairman of one committee, and that every partner will serve on three and only three committees. The following is the schedule of committees and their members, the first member being chairman:

Domestic railroads	Adams,	Brown,	Smith
Municipal bonds	Brown,	Murphy,	Ellis
Federal bonds	Murphy,	Smith,	Jones
South American securities	Smith,	Ellis,	Gordon

Domestic steel industry	Ellis,	Jones,	Adams
Continental securities	Jones,	Gordon,	Brown
Public utilities	Gordon,	Adams,	Murphy

An examination of this schedule shows that it "satisfies" the seven axioms if the class *S* is interpreted as the banking firm, its elements as the partners, and the *l-classes* as the various committees.

We exhibit one further interpretation, which at first sight may seem to have nothing in common with those already given. In the following figure there are seven points lying by threes on seven lines, one of which is "bent." Let each point represent an element of *S*, and each set of three points lying on a line an *l-class*. Then all the

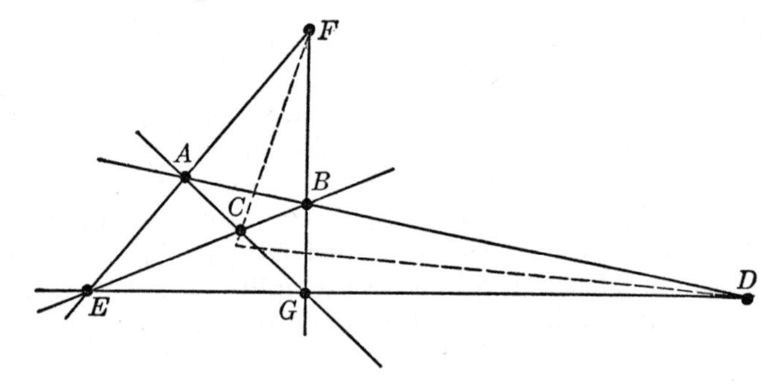

seven assumptions are satisfied. This geometric pattern exemplifies the *same formal relations* as does the array of numbers and the schedule of the banker's committees we have already given. A third representation will be found on page 146.

Let us examine these three representations. We find, in the first place, that we can make correspond in a one-one manner every element of one interpretation with elements of the other two. And in the second place, every relation between the elements in one representation corresponds to a relation with the same logical properties between the *corresponding* elements in the other two. Thus, as an illustration, the element 0 in the numerical interpretation given below, can be placed in one-one correspondence with the point *A* in the geometrical interpretation, and also with Mr. Adams in the banking firm; the element 1 corresponds to point *B*, and also to Mr. Brown, and so on. And the three termed relation between the numbers 0, 1, 3 (page 146) in virtue of which they belong to the same group, corresponds to the relation between the points *ABD* in virtue of which they lie on the same line, and also corresponds to the relation between Messrs. Adams, Brown, and Smith in virtue of which they are on the same committee; and so on.

Two or more systems which are related in this manner are said to be *isomorphic* or to have an *identical structure* or *form*. We may now give a general definition of *isomorphism*. Given two classes S, with elements $a, b, c, \ldots$ and S', with elements $a', b', c', \ldots$; suppose the elements of S can be placed in one-one correspondence with those of S', so that, say, a corresponds to a', b to b', and so on. Then, if for every relation R between elements of S (so that, for example, $a\ R\ b$) there is a relation R' between the corresponding elements of S' ($a'\ R'\ b'$), the two classes are *isomorphic*.

We are now prepared to understand the great importance of the mathematical method as a tool in the natural sciences. In the first place, a hypothesis or set of assumptions may be studied for its implications without raising questions of material truth or falsity. This is essential if we are to understand *to what* a given hypothesis commits us. In the second place, a hypothesis when abstractly stated is capable of more than one concrete representation. Consequently, when we are studying pure mathematics we are studying the possible structures of many concrete situations. In this way we discover the constant or invariable factor in situations sensibly different and undergoing change. Science has been characterized as a search for system (order, constancy) amidst diversity and change. The idea of isomorphism is the clearest expression of what such a system means.

Some examples of isomorphism are well known. An ordinary map is a useful device because the relations between the points on it have a structure identical with the relations between the places in the countryside to which the map points correspond. In physics, we can see how the formula of inverse squares applies to electrical attraction and repulsion as well as to the force of gravitation. This is possible because these different subject matters have an identical formal structure with respect to the properties studied. Physics also discovers that the same set of principles is applicable to the motion of planets, the dropping of a tear, and the swinging of a pendulum. It is the isomorphism found in diverse subject matter which makes possible theoretical science as we know it today. An elementary exposition of a "dictionary" which translates theorems of Euclidean into non-Euclidean geometry the reader will find in Henri Poincaré's *The Foundations of Science,* page 59. From an abstract point of view these different geometries have an identical structure.

It should be noted that two systems may not be identical in structure *throughout* and yet share many common properties. Euclidean and non-Euclidean geometries have many theorems in common, while at the same time some theorems in one system are formally incompatible with some theorems in the other. This suggests the possibility that two systems may be incompatible with each other in their totality and yet possess a common subsystem. We may illustrate this as follows. Consider the system determined by Axioms 1′ to 7′. Consider also the system obtained by replacing 7′ with the assumption 7″: *No* l-class *contains more than four elements of* S. These two systems are not isomorphic, as comparison of the representation of the first system (page 146) with that of the second system (page 147) will show. Nevertheless, all the theorems in both systems which follow from the first six axioms will be the same. The system determined by Axioms 1′ to 6′ is therefore a common subsystem of the incompatible systems determined by 1′ to 7′ on the one hand and by 1″ to 7″ on the other.

This is a very important observation. Research in the natural sciences often tempts us to believe that a theory is true because some consequence of the theory has been verified. Nevertheless, an identical consequence may be drawn from an alternative and incompatible theory. We cannot, therefore, validly affirm either theory. With care, however, we may discover those common assumptions of both theories upon which the identical consequence depends. It may then be possible to ascertain *which* of the assumptions in virtue of which the theories are *different* theories are in disagreement with experimental findings.

One further remark needs to be made about deductive systems. Every system is of necessity *abstract:* it is the structure of certain *selected* relations, and must consequently omit the structure of other relations. Thus the systems studied in physics do not include the systems explored in biology. Furthermore, as we have seen, a system is deductive not in virtue of the special meanings of its term, but in virtue of the universal relations between them. The specific quality of the things which the terms donate do not, as such, play any part in the system. Thus the theory of heat takes no account of the unique sensory qualities which heat phenomena display. A deductive system is therefore doubly abstract: it abstracts from the specific qualities of a subject matter, and it selects some relations and neglects others. It follows that there is a *plurality* of systems, each of which may be explored in isolation from the others. Such a plurality of systems may, indeed, constitute a set of subsystems of a single, comprehensive system, but we have no evidence for such a state of affairs. In any case, it is not necessary to know this comprehensive system in order to explore adequately any one of the many less inclusive systems. It appears that human knowledge of the natural world is possible only because it is capable of being studied as a set of relatively autonomous systems.

§ 4. THE EQUIVALENCE OF AXIOM SETS

It has been pointed out that in every deductive system some propositions are indemonstrable *in that system* and some terms indefinable. We have also suggested, however, that a proposition which is an axiom in one system may be a theorem in another. We wish now to illustrate this.

Consider the following assumptions concerning a class *S;* its elements *A, B, C,* and so on; and its *l-* (or sub-) *classes a, b, c,* and so on:

Axiom 1′. If *a* and *b* are distinct *l-classes* of *S* there is at least one element of *S* in both *a* and *b*.

Axiom 2′. If *a* and *b* are distinct *l-classes* of *S* there is not more than one element of *S* in both *a* and *b*.

Axiom 3′. Any two elements of *S* are elements of some one *l-class* of *S*.

Axiom 4′. There exists at least one element in *S*.

Axiom 5′. Every element of *S* belongs to at least three *l-classes* of *S*.

Axiom 6′. There is no element of *S* belonging to all the *l-classes* of *S*.

Axiom 7′. There is no element of *S* which belongs to more than three *l-classes* of *S*.

None of these axioms are identical with any axioms of the previous set, although some of the new axioms are identical (except for verbal form) with several of the theorems we had previously demonstrated. Thus, Axiom 3′ is identical with Theorem I, Axiom 4′ with part of Theorem III. None the less, the previous set of axioms and the present set characterize the same system of relations. These two sets are *equivalent*. Two postulate sets are equivalent if, and only if, every postulate of the first is either a postulate or a theorem in the second, and every postulate of the second is either a postulate or a theorem in the first. The equivalence for the two sets above may be shown by deducing from the first set those postulates in the second set which have not already been demonstrated, and then by deducing from the second set all the postulates of the first. Thus Axiom 1′ follows at once from Axiom 3′, Axiom 5′ follows from Axioms 1′ to 5′, and so on.

It is quite important that the reader become convinced that there are no *intrinsically undemonstrable propositions.* Failure to realize this fact has been one source for the belief in "self-evidently true" propositions. It is quite easy to fall into the mistaken prejudice that because a proposition cannot be demonstrated on one set of premises, it is altogether undemonstrable. Moreover, the fact that two systems may be equivalent without being identical, axiom for axiom, throws fresh light on the question of logical priority. In a given system, one proposition is logically prior to another if the first is required as a premise, or part of the premise, for the second. In another system, however, the relation of logical priority between two propositions may be reversed.

What has been said about undemonstrated propositions is equally true for undefined terms. In the geometry the reader has studied in his youth, points were taken as fundamental and undefined, and lines, circles, defined in terms of them. That there must be undefined terms is clear from any attempt to define a term. We may try to define "equal distances" as follows: The distance between points *A* and *B* on a straight line is equal to the distance between *C* and *D,* if the segment *AB* can be moved *by a rigid motion of this segment* so that it coincides with the segment *CD.* But obviously, the phrase "a rigid motion" cannot itself be defined in terms of "equal distances," on pain of a circular definition. Nevertheless, it is a mistake to suppose that there are *intrinsically* undefinable terms. *Undemonstrability and undefinability are both relative to a system.* It is not necessary to regard points as undefinable provided we select other undefinables, such as lines, in terms of which points may themselves be defined. Thus different axiomatic foundations have been given for Euclidean geometry. Hilbert has found a set of twenty-one assumptions, requiring five primitive or undefined ideas, from which all the theorems of geometry can be deduced. Veblen, on the other hand, discovered twelve assumptions, requiring only two undefined terms, which perform the same task. We cannot pursue this topic any further, except to point out that the number of undefined terms is closely connected with the number and character of the undemonstrated propositions.

§ 5. THE INDEPENDENCE AND CONSISTENCY OF AXIOMS

We must now consider some fundamental questions connected with a set of axioms. What are some of the essential and desirable properties which a set of axioms must possess?

1. Axioms are studied for the propositions they imply. Consequently, *fertility* is one property which axioms should possess; this means that they should imply many theorems. But there is no criterion as to whether a set of assumptions may give rise to a comprehensive set of theorems. It is very likely that fertility is not an intrinsic character of an axiom set, but reflects the ability of the human reasoner to discover their implications. Moreover, a set of assumptions is regarded as important in proportion to our ability to find *interpretations* for it in terms of investigations in the natural sciences or in other branches of mathematics. We shall return to this point.

2. A very desirable, and historically significant, property of axioms is their *independence.* A set of assumptions is *independent* if it is impossible to deduce any one of the axioms from the others. If a set of axioms is independent, it is possible to make a sharp distinction *in that system* between assumptions and theorems. And unless we know that two propositions are independent, we are unable to say whether we are entertaining different and alternate possibilities or simply the same possibility in a different form.

The question whether the axioms and postulates of Euclid are independent is historically of great interest. To the many attempts to answer it we owe some of the greatest advances that have been made in mathematics, physics, and philosophy. As we have indicated in a previous chapter, mathematicians have tried for more than two thousand years to deduce the parallel postulate from the other assumptions of Euclid. The basis for their doing so was their conviction that all his assumptions except the one about parallel lines were "self-evidently true." Consequently, they believed it was a serious blemish that any non-self-evident proposition should be taken as an axiom. They did not succeed in deducing Postulate 5 from the others without assuming some other proposition not included in the original assumptions of Euclid. But what did their lack of success prove—did it prove that Postulate 5 could not be deduced from the others? It certainly did not. But it turned some inquirers' minds to search for the reason of the lack of success. It led some mathematicians to look for a proof that the parallel postulate was independent of the others.

The proof was finally discovered. We have seen that demonstrative proof consists in pointing out that certain axioms imply certain theorems. Such an alleged implication is denied if we can show that it is possible for the theorems to be false and the axioms to be true. By developing a possible system of geometry in which Euclid's parallel postulate is denied while the other axioms are retained, Lobatchevsky was able to show that the parallel postulate cannot be a logical consequence of the other axioms. It will be seen that this proof illustrates the form of the logical principle we have discussed as the inconsistent triad. If a set of (consistent) propositions *P* imply another proposition *Q*, then the propositions consisting of *P* together with the contradictory (or contrary) of *Q* must be inconsistent with each other. If the inconsistency, shown by finding two contradictory propositions, appears in the set of axioms, the task is completed: *Q* is not independent of *P*. If the inconsistency does not appear in the set of axioms, then it must be possible to deduce by valid reasoning one or more theorems which contradict either some of the axioms or some other theorem validly derived. If, on the other hand, *P* does not imply *Q*, the set of propositions *P* together with the contradictory of *Q* is a consistent set, and no contradictions can ever be discovered.[2]

Let us indicate summarily the essence of one type of non-Euclidean geometry. Euclid's Postulate 5 is equivalent to the assumption that through a point outside of a given line only one line may be drawn parallel to the given line. In the Lobachevskian geometry this is replaced by the assumption that through a point outside of a line more than one parallel may be drawn to it. From this assumption and the other assumptions of Euclid a host of theorems may be obtained, some of which are identical with the theorems of Euclid, while others are contradictories of these. Thus the propositions. *The base angles of an isosceles triangle are equal* and *Two*

[2] The history of non-Euclidean geometry began when a clear perception of this simple logical principle was attained. Saccheri, an Italian mathematician of the eighteenth century, already possessed it, and by making an assumption contrary to that of Euclid's parallel postulate obtained many theorems of what is now known as non-Euclidean geometry. But for some unexplained reason he came to the conclusion that such a geometry was self-contradictory—perhaps because many of the theorems he obtained were formal contradictories of the theorems of Euclid. If that was his reason for rejecting non-Euclidean geometry, he overlooked the possibility that both Euclidean and non-Euclidean geometries may be self-consistent, although the two systems are incompatible with each other. The discovery of non-Euclidean geometry must be attributed, therefore, to Lobachevsky and Bolyai, who wrote in the first half of the nineteenth century. Still another kind of non-Euclidean geometry was discovered by Riemann

lines which are parallel to a third are parallel to each other are common to Euclid and Lobachevsky. On the other hand, the propositions *The sum of the angles of a triangle is equal to two right angles* and *The area of a circle is* πr^2 are correct only in Euclid.

3. Here the reader may protest: "I don't yet see that the parallel postulate has been proved to be independent of the others. You have shown that by assuming a contradictory postulate a host of theorems differing from that of Euclid may be obtained. But you have not yet shown that such a new set of postulates is *consistent*. And unless you do that, you have no good reason for supposing that a non-Euclidean geometry is really possible."

This is quite right. The fact that after any finite number of theorems have been derived from a non-Euclidean set of assumptions no contradiction has turned up proves nothing about the consistency of that set. For a contradiction *may* appear after a larger number of theorems have been obtained. And the same objection can be raised no matter how large is the number of theorems deduced. The reader's protest expresses clearly how closely connected are the problems of the independence and the consistency of a set of propositions.

We may, however, be permitted to ask the reader a question in return. "You think non-Euclidean geometries have not been proved to be consistent, and that since they lack such proof their very possibility is endangered. But what basis have you for believing that Euclidean geometry is self-consistent? It is true that after thousands of years of studying it, mathematicians have not discovered any contradictions in it. But you surely will not accept *that* as a proof. The Euclidean and non-Euclidean geometries seem, in this respect, to be in the same boat."

Let us try to resolve the reader's perplexity by turning once more to our miniature mathematical system, and face similar problems with regard to it. Are the seven axioms independent? Are they consistent with one another?

Mathematicians have found only one way of answering the second question. The method consists in discovering a *set of entities which will embody the relations of our set of abstract axioms*. On the assumption that these entities themselves are free from contradiction and that they in fact fully embody the axioms, the latter are shown to involve no inconsistencies.

We will illustrate the use of this method. Let the integers 0 to 6 inclusive be arranged in distinct groups of three each, as follows:

0	1	2	3	4	5	6
1	2	3	4	5	6	0
3	4	5	6	0	1	2

We now regard these seven integers as the elements of the class *S*. Every column of integers will then represent an *l-class*. On this interpretation, as a little reflection shows, every one of the seven axioms in our set is verified. The axioms are therefore consistent.

It must be emphasized, however, that this method merely shifts the difficulty. For the question still remains whether the set of entities and our method of interpretation are consistent. To this question no completely satisfactory answer seems at present available. We have, however, a certain amount of confidence that since the Euclidean axioms have enabled us to deal so adequately with the properties and relations of physical bodies, Euclidean geometry as a logical system is also consistent, because we assume that nothing occupying spatial and temporal position can be self-contradictory. Since non-Euclidean geometries have been shown to correspond, element for element, to Euclidean geometry in accordance with definite transformation formulae, it follows that if a contradiction could appear in non-Euclidean geometry, a corresponding contradiction would of necessity have to occur in the geometry of Euclid.[3]

We return once more to the problem of independence of axioms, and shall illustrate the problem by means of our miniature system. Is Axiom 7′ independent of the others? The answer is yes if the first six axioms, together with any assumption incompatible with

[3] The assumptions required for other branches of mathematics are shown to be consistent in a similar way. However, complications enter. Mathematics as a system of propositions has advanced much beyond the achievements of Euclid. Mathematicians have shown that all the higher branches of mathematics, such as higher algebra, analysis, geometry, and so on, may be interpreted as studying the relations between whole numbers (integers); and that they require no fundamental notions other than those employed in arithmetic. This achievement has been called the arithmetization of mathematics, and is due largely to such men as Weierstrass, Dedekind, and Hilbert. An even further step in the analysis of mathematical ideas was made when arithmetic itself was shown to require no fundamental notions except those of logic, such as "class," "member of a class," "implies," and so on. This work has been accomplished largely through the efforts of Cantor, Frege, Peano, Whitehead, and Russell, and has received its most adequate expression in the *Principia Mathematica* of the last two men. As a consequence of a century of labor many, though not all, mathematicians are convinced that mathematics can be developed in terms of the ideas of pure logic. If this thesis is sound, the consistency of every branch of mathematics is dependent upon the consistency of the principles of formal logic. The question as to the consistency of any branch of mathematics is then reduced to the question whether logical principles themselves form a consistent system.

the seventh, form a consistent set. This condition is equivalent to finding an interpretation which will satisfy the first six axioms and fail to satisfy the seventh. Such an interpretation may be given in several ways, of which the following is one:

0	1	2	3	4	5	6	7	8	9	10	11	12
1	2	3	4	5	6	7	8	9	10	11	12	0
3	4	5	6	7	8	9	10	11	12	0	1	2
9	10	11	12	0	1	2	3	4	5	6	7	8

These thirteen numbers from 0 to 12 inclusive are the members of *S*. Each column of four numbers represents an *l-class* of *S*. Examination will show that all the axioms except the seventh are satisfied. This axiom is therefore independent of the first six. In a like manner we can show that each of the other assumptions is independent of the rest.

§ 6. MATHEMATICAL INDUCTION

"Aren't you, however, forgetting that induction takes place in mathematics?" the reader may protest. "You have been describing mathematics as a typical deductive science, in which all the theorems are necessary consequences of the axioms. But surely you are not going to overlook the method of proof known as mathematical induction?"

The reader has doubtless been ensnared by a word. There is indeed a method of *mathematical induction,* but the name is unfortunate, since it suggests some kinship with the methods of experimentation and verification of hypotheses employed in the natural sciences. But there is no such kinship, and mathematical induction is a purely demonstrative method.

Is it necessary, however, once more to caution the reader against the common error of confounding the temporal order of our discovering the propositions of a science and the order of their logical dependence? Everybody who has ever worked a problem in geometry knows that there is a preparatory "groping stage," in which we guess, speculate, draw auxiliary lines, and so on until, as the saying goes, we "hit upon" the proof. But no one will confuse that preparatory stage, however essential, with the proof finally achieved. Such an initial "groping" stage has indeed close kinship with human investigations in any field whatever. A process of tested guessing characterizes research in mathematics as well as research in the natural sciences.

The principle of mathematical induction may be stated as follows: If a property belongs to the number 1, and if when it belongs to n it can be proved to belong to $n+1$, then it belongs to all the integers. Let us demonstrate, by its means, the theorem: For all integral values of n, $1+3+5+7+\ldots(2n-1)=n^2$.

This clearly is true for $n=1$. Let us now show that if it holds for the integer n it holds for $(n+1)$.

a. $1+3+5+\ldots(2n-1)=n^2$.

Adding $(2n-1)+2$ or $(2n+1)$ to both sides, we get:

b. $1+3+5+\ldots(2n-1)+(2n+1)=n^2+(2n+1)=(n+1)^2$.

But *b* has the same form as *a*. Hence we have shown that if the theorem is true for the integer n it is true for $(n+1)$. Now it is true for $n=1$. Therefore it is true for $n=1+1$ or 2; therefore it is true for $n=2+1$ or 3, and so on for every integer which can be reached by successive additions of 1. The proof, therefore, is perfectly rigorous, deductive, and altogether formal. It makes no appeal to experiment. And the principle of mathematical induction, as modern researches show, is part of the very meaning of finite or "inductive" numbers.

§ 7. WHAT GENERALIZATION MEANS IN MATHEMATICS

In the preceding chapter we called attention to the changes in the meaning of words by the process of *generalization*. In mathematics, too, such processes take place, and reference is often made to the "modern generalization of number." It is easy to fall into error as to the sense in which "number" has in fact been generalized. Let us examine the matter.

The word "number" was originally restricted to the *integers* 1, 2, 3, and so on. Numbers, so understood, can be added and multiplied, and in some cases subtracted and divided. The abstract nature of integers may be expressed by means of a set of propositions which indicate what operations can be made upon integers, and what the relations are in which the operations stand to one another. For example, the following are some of the abstract properties of integers:

$$a+b=b+a$$
$$(a+b)+c=a+(b+c)$$
$$a\times b=b\times a$$
$$a\times(b+c)=a\times b+a\times c$$

Now on some of the integers, the operations *inverse* to multiplication and addition can be performed. Thus, $4 \times 3 = 12$; hence there is an integer x such that $x \times 3 = 12$: such a number x is the quotient of 12 *divided by* 3. But, unless we enlarge our conception of number, the inverse operation of division cannot *always* be formed. Thus, there is no integer x such that $x \times 3 = 5$. Consequently, in order that there should be no exceptions to the possibility of division, the fractions were introduced. They were also called numbers, and so the domain of number was increased in the interest of continuity and generality.

This was the first generalization of "number." Why were the fractions designated as numbers? The answer is simple, although it has been discovered only recently. It is because operations of addition, multiplication, and even division could be performed on them; and because the *formal relations* of integers to one another with respect to these operations are the same as the formal relations of fractions. In other words, integers and fractions form isomorphic systems.

But it must be pointed out that while addition or multiplication for integers is formally the same for fractions, nevertheless the differences are not thereby denied. Thus, the sign $+$ in $7 + 5 = 12$ and in $\frac{1}{2} + \frac{1}{3} = \frac{5}{6}$, while denoting formal properties common to the two cases, none the less denotes two distinct and different operations. The second is much more complex than the first. It is easy to confuse them because the same symbol is used to denote them both, but neither must we forget that the same symbol is applicable to the two cases because they have common elements of procedure.

Later on other "numbers" were discovered, when it was noticed that some of the previously defined numbers had square roots, cube roots, and so on, but others did not. Thus, the Pythagoreans proved that the diagonal of the square is incommensurable with its sides. In modern notation, this means that $\sqrt{2}$ cannot be expressed as the ratio of two integers. But why should the operation of extracting a root be legitimate only for some numbers (for example 4)? Why not permit the operation to be performed on every one of the previously defined numbers? Hence, in the interest of continuity of treatment and of generality, the *irrationals* were discovered, and they too were regarded as a "species of number."

Why? The answer is again simply: Because the operations upon them possess the formal properties of the operations upon integers and fractions.

Similar remarks, with only few qualifications, apply to the other "species of number" with which modern mathematics is familiar. Negative numbers, imaginary numbers, quaternions, transcendental numbers, matrices, have been introduced into the domain of number because continuity and universality of treatment demanded them. But they have been designated as "numbers" because they share certain abstract properties with the more familiar instances of mathematical entities.

Generality of treatment is thus an obvious goal of mathematics. But it is clearly a mistaken idea to suppose that the definition of "number" as applicable specifically to the cardinals 1, 2, 3, and so on, has in some sense been "extended" or "generalized" to apply to fractions, irrationals, and the rest. There is no generic definition of "number" of which the cardinals, ordinals, fractions, and so on are special instances *except in terms of the formal properties of certain "operations."* It is in virtue of the permanence or *invariance* of these formal properties that these entities are all "numbers."

This conclusion, so obvious when it is once pointed out, has been won only at the expense of tremendous labor by modern philosophers of mathematics. The source of many of the confusions in this subject is the frequent use of the same symbol to denote two essentially different ideas. Thus, the cardinal number 2 and the ratio $2/1$ are usually denoted by the same symbol 2; they denote, however, radically distinct ideas. But this danger from the symbolism of mathematics is undoubtedly outweighed by the great advantages it offers. It enables us to exhibit concisely the *structure* of mathematical propositions, and so makes possible our noting the precise analogies or isomorphisms in contexts that are in other respects very different from one another.

CHAPTER VIII

PROBABLE INFERENCE

§ 1. THE NATURE OF PROBABLE INFERENCE

In daily conversation "probability" is one of the most loosely used words, and in logical theory the correct analysis of the nature of probable inferences is one of the most disputed themes. Nevertheless, we make plans for births, marriages, deaths, holidays, commercial enterprises, friendships, and education on the basis of rational evidence whose weight can be recognized as probable and not conclusive. "Probability," Bishop Butler remarked, "is the very guide of life." The subject, therefore, needs to be examined in detail. The reader will keep his bearings more easily, however, if he recalls the definition of probable inference stated in the introductory chapter. That definition may serve as the Ariadne thread through the mazes of a long discussion. An inference is probable, we said, if it is one of a *class* of arguments such that the conclusions are true with a certain relative frequency when the premises are true.

The following incident occurs in Voltaire's story *Zadig:*

"One day, when he [Zadig] was walking near a little wood, he saw one of the queen's eunuchs running to meet him, followed by several officers, who appeared to be in the greatest uneasiness, and were running hither and thither like men bewildered and searching for some most precious object which they had lost.

" 'Young man,' said the chief eunuch to Zadig, 'have you seen the queen's dog?'

"Zadig modestly replied: 'It is a bitch, not a dog.'

" 'You are right,' said the eunuch.

" 'It is a very small spaniel,' added Zadig; 'it is not long since she has had a litter of puppies; she is lame in the left forefoot, and her ears are very long.'

" 'You have seen her, then?' said the chief eunuch, quite out of breath.

" 'No,' answered Zadig, 'I have never seen her, and never knew that the queen had a bitch.'

"Just at this very time, by one of those curious coincidences which are not uncommon, the finest horse in the king's stables had broken away from the hands of a groom in the plains of Babylon. The grand huntsman and all the other officers ran after him with as much anxiety as the chief of the eunuchs had displayed in his search after the queen's bitch. The grand huntsman accosted Zadig, and asked him if he had seen the king's horse pass that way.

" 'It is the horse,' said Zadig, 'which gallops best; he is five feet high, and has small hoofs; his tail is three and a half feet long; the bosses on his bit are of gold twenty-three carats fine; his shoes are silver of eleven pennyweights.'

" 'Which road did he take? Where is he?' asked the grand huntsman.

" 'I have not seen him,' answered Zadig, 'and I have never even heard anyone speak of him.'

"The grand huntsman and the chief eunuch had no doubt that Zadig had stolen the king's horse and the queen's bitch, so they caused him to be brought before the Assembly of the Grand Desterham, which condemned him to the knout, and to pass the rest of his life in Siberia. Scarcely had the sentence been pronounced, when the horse and the bitch were found. The judges were now under the disagreeable necessity of amending their judgment; but they condemned Zadig to pay four hundred ounces of gold for having said that he had not seen what he had seen. He was forced to pay his fine first, and afterwards he was allowed to plead his cause before the Council of the Grand Desterham, when he expressed himself in the following terms:

" 'Stars of justice, fathomless gulfs of wisdom, mirrors of truth, ye who have the gravity of lead, the strength of iron, the brilliance of the diamond, and a close affinity with gold, inasmuch as it is permitted me to speak before this august assembly, I swear to you by Ormuzd that I have never seen the queen's respected bitch, nor the sacred horse of the king of kings. Hear all that happened: I was walking towards the little wood where later on I met the venerable eunuch and the most illustrious grand huntsman. I saw on the sand the footprints of an animal, and easily decided that they were those of a little dog. Long and faintly marked furrows, imprinted where the sand was slightly raised between the footprints, told me that it was a bitch whose dugs were drooping, and that consequently she must have given birth to young ones only a few days before. Other marks of a different character, showing that the surface of the sand had been constantly grazed on either side of the front paws, informed me that she had very long ears; and, as I observed that the sand was always less deeply indented by one paw than by the other three, I gathered that the bitch belonging to our august queen was a little lame, if I may venture to say so.

" 'With respect to the horse of the king of kings, you must know that as I was walking along the roads in that same wood, I perceived the marks of a horse's shoes, all at equal distances. "There," I said to myself, "went a horse with a faultless gallop." The dust upon the trees, where the width of the road was not more than seven feet, was here and there rubbed off on both sides, three feet and a half away from the middle of the road. "This horse," said I, "has a tail three feet and a half long, which, by its movements to right and left, has whisked away the dust." I saw, where the trees formed a canopy five feet above the ground, leaves lately fallen from the boughs; and I concluded that the horse had touched them, and was therefore five feet high. As to his bit, it must be of gold twenty-three carats fine, for he had rubbed its bosses against a touchstone, the properties of which I had ascertained. Lastly, I inferred from the marks that his shoes left upon stones of another kind, that he was shod with silver of eleven pennyweights in quality.' " [1]

This miniature detective story is a fair representative of the sort of inferences which are employed in many practical problems and in many fields of scientific research. Why do we regard Zadig's conclusions as *reasonably well founded,* even though the evidence for them is not *absolutely complete*?

The argument that the queen's dog had passed by may be formulated as follows:

1. These marks on the sand have a determinate shape. (This is a true proposition, asserting an *observed fact.*)
2. But, if a small dog has passed by, its footprints on the sand would be marks of this shape. (This is a proposition, expressing a *general rule* which is believed to be true.)
3. Therefore, a small dog has passed by. (This is the *inferred proposition.*)

[1] Translation by R. B. Boswell, 1910, pp. 58-60.

Now this argument is formally invalid. It would be valid if we knew not only that proposition 2 is true, but that the proposition 2′ *If a mark of this shape has been made on sand, then a small dog has produced them as its footprints* is also true. (The reader should restate the argument using 2′ instead of 2, and convince himself of the validity of the inference.) Proposition 2′ is the *converse* of 2. But, in general, although we may know the truth of a proposition such as 2, we do not know the truth of its converse. Indeed, we may know that while small dogs produce footprints of this shape, marks very similar to those observed may be produced in other ways. Zadig's conclusion clearly did not necessarily follow from his premises.

Nevertheless, his conclusion is highly *probable* on the premises. For the inference he made *from* the observed fact *to* the inferred proposition *by means* of the general rule, is one of *a class of inferences* in which the number of times true propositions are inferred from true premises is a very large fraction of the number of times such inferences are employed at all. In other words, Zadig *in some instance* may be wrong in inferring the proposition *A small dog has passed by* from the proposition asserting the markings in the sand. But if he, and other observers, were to make a very large number of such inferences, they would be right *much more frequently* than wrong.

We may now state the matter in a slightly different way. Zadig is led, by some reason or other, to study the markings on the road. He is able to establish as true the proposition *These marks on the sand have a determinate shape*. But he also has some ground for asserting the following proposition: *Marks on the sand having such a shape are produced, in a ratio* r *of the time, by small dogs*. Now while nothing follows *necessarily* from these propositions, he may conclude, with *probability*, that *A small dog has passed by*. The conclusion is probable on the evidence, because a fraction r of the time when such an inference is employed, the conclusion is discovered to be true when the premises are true. In any one instance the conclusion (the inferred fact) of such an argument may be false even though the premise is true. But a series of such inferences would yield true conclusions with a perhaps calculable relative frequency r. The numerical value of r cannot always be obtained, and we may have to be satisfied with more or less trustworthy impressions as to its magnitude. But though the evidence for our estimate may be unreliable, the meaning of our judgment of probability is clear. When r is zero, the argument is worthless; when it is 1 the inference is altogether conclusive; when it is greater than $1/2$, the argument leads us aright more often than not.

We must now consider the probability of hypotheses. Observation of magnetized bars of iron, needles, nails, and the like shows that magnets have two unlike poles such that unlike poles attract and like poles repel one another. We also observe that metals are not all equally magnetizable, and that the best metal magnets loose their magnetic properties with lapse of time. How may we explain this? One way of doing so is to assume that all metal substances are composed of small particles, each a *permanent* magnet with two poles, and each capable of rotating around a fixed center. It can be shown that in one arrangement of these particles the opposite poles will neutralize each other completely, so that the entire bar will display no magnetic properties; but in another arrangement, only *some* of these small particles will neutralize one another, so that the entire bar will show magnetic properties. The changes in the relative position of these small permanent magnets, and the ease of their rotation around their fixed centers, will then explain why a metal bar acquires or loses magnetic properties.

Now we have already seen that this hypothesis as to the nature of magnetism is not *demonstrated* when it is shown that its consequences agree with observation. For other theories may also explain the phenomena of magnetism. Moreover, we cannot be sure that some of the logical consequences of our assumption may not turn out to be in disagreement with observation. Nevertheless, our assumption or hypothesis as to the nature of magnetism can be said to be probable in the same sense in which Zadig's conclusions were probable.

We can throw our argument into the familiar form of the hypothetical syllogism *This, that, and the other metal bars exhibit phenomena of magnetism under specified conditions* (true proposition, asserting *observed facts*); *but if each metal is composed of imperceptibly small permanent magnets free to rotate around a fixed axis, then these metals will exhibit magnetic properties under the stated conditions* (*general rule*); *therefore, each metal is composed of permanent magnets,* and so on (*inferred proposition*). We notice again that as it is stated the argument is invalid. Nevertheless, the conclusion is *probable* on the evidence supplied by the premises, because in a series of such inferences the conclusion is true with a considerable relative frequency when the premises are true.

"But hold on!" the reader may protest. "What do you mean by

saying that the conclusion is true a certain fraction of the time that the premises are true? How do you *ever* know that the *conclusion* is true? Aren't the permanent magnets which compose the metal bars imperceptibly small, and doesn't that make it simply impossible ever to establish the truth of the conclusion? And in that case, can we ever determine the relative frequency of its truth?"

The answer to this objection is based on the distinction between the meaning of a proposition and the evidence in favor of its truth. The meaning of probability in general may be definite, even though in some cases we do not have enough evidence to determine its specific numerical value. The difficulty here pointed out has its analogue elsewhere. Thus the question What is a sphere? may and must be answered independently of and before the question whether in a given case the object examined is in fact a sphere.

The analysis of the probability of propositions which cannot be *directly* verified is much more complicated than we have indicated so far. In such cases the argument depends upon the fact that the theory has as its logical consequences propositions which *may* be directly verified, and which lead to the observation of phenomena *other* than those for which the theory was originally proposed. As we shall see in a later chapter, the evidence upon which a theory is probable consists of *samplings* made from all the necessary consequences of the theory. For the present example, the theory that all metals are composed of small permanent magnets has as one consequence, among others, that hammering or heating a magnet should make it lose its magnetic properties, and this inferred fact is directly observable. A more complete form of the argument may therefore be stated as follows *This, that, and other metals exhibit magnetic properties under specified conditions* (the observed fact); *but, if these metals are composed of small permanent magnets,* and so on, *then they should exhibit magnetic properties under specified conditions* (general rule); *therefore, each metal is composed of permanent magnets,* and so on (the inferred fact, or theory, not capable of direct verification). However, if metal bars are composed of permanent magnets, then hammering or heating a magnetized needle will make it lose its magnetic properties (*inferred fact,* directly verifiable). The purpose of the deductive elaboration of the theory is to supply us with as many verifiable consequences of it as possible. The argument *for* the theory thus proceeds *from* a proposition known to be true, *to* other propositions directly verifiable, *via* the theory which is not directly verifiable. The argument is now more complicated, but this does not alter the nature of the inference employed. The theory is probable on the evidence, because the argument for it belongs to a *class* of arguments in which the relative frequency r of the truth of the conclusion when the premise is true is not necessarily 1.

We must, however, analyze still more complicated cases of inferences that are recognized as probable. But before we do so, let us state explicitly the essential characteristics of probable inferences. Some of these characteristics we have already noted; others are stated in anticipation of later discussions.

1. Probable inference, like all inference, is based upon certain relations between propositions. No proposition is probable *in itself*. A proposition is probable in relation to other propositions which serve as evidence for it.
2. Whether a proposition has or has not a degree of probability on definite evidence, does not depend on the state of mind of the person who entertains the proposition. Questions of probability, like questions of validity, are to be decided entirely on objective considerations, not on the basis of whether we *feel an impulse* to accept a conclusion or not.
3. An inference is probable only in so far as it belongs to a definite class of inferences in which the frequency of the conclusions being true is a determinate ratio of the frequency of the premises being true. This is the same as saying that the very *meaning* of probability involves *relative frequencies*.
4. Since the probability of a proposition is not an intrinsic character of it, the same proposition may have *different* degrees of probability, in accordance with the evidence which is marshaled in its support.
5. The evidence which may be marshaled in support of a proposition may have different degrees of relevance. In general, that evidence is chosen for a proposition which will make the degree of probability of that proposition as great as possible. But the relevance of the evidence cannot be determined on formal grounds alone.
6. While the meaning of the *measure* of the probability of an inference is the relative frequency with which that type of inference will lead to true conclusions from true premises, it is true that in most cases, as we shall see, the definite numerical value of the probability is not known.

In comparison to the number of cases in which we judge a proposition to be probable on the basis of certain evidence, it is relatively infrequently that we are in a position to determine the exact numerical degree of such probability. But this does not annul the analysis of probability we have given, since we may very well know what probability is *in general* without having adequate evidence in a given case on the basis of which to determine what the numerical value is.

§ 2. THE MATHEMATICS OR CALCULUS OF PROBABILITY

The modern study of probability began when the Chevalier de Méré, a famous gamester of the seventeenth century, consulted his friend, the saintly Pascal, on how best to lay his bets in games with dice. And most discussions of probability center around questions to which numerical answers can be given: What is the probability of getting three heads in four throws with a coin? What is the probability of getting seven with a pair of dice? Problems of this kind, as well as those of a more complicated nature, have been studied by mathematicians. At present, practically every branch of physics and some branches of chemistry and biology find it necessary to employ the calculus of probability. We must consider the simpler problems of this nature in some detail, and note the limitations of the mathematical approach.

Let us begin with the limitations. Mathematics is a discipline which studies the necessary consequences of any set of assumptions. And so conceived, mathematics is not concerned with the truth or falsity of the premises whose consequences it explores. In this respect, logic and mathematics are indistinguishable.

It follows that no purely mathematical theory can determine the degree of probability of any proposition which deals with actual matters of fact. It can determine the probability of a proposition when certain assumptions are explicitly made concerning it. It can tell us what are the necessary consequences of these assumptions; but it cannot determine, and is not concerned with, the truth or falsity of those assumptions. Consequently, *the theory of probability can be purely mathematical only if it is restricted to questions of necessary inference.* This is the case if the theory of probability is viewed as a part of pure mathematics. We shall make a brief study of its elementary theorems, in part because tradition has become fixed in this respect, and in part because the nature of scientific method is illuminated through the use of the rigorous theorems of the calculus of probability in the study of probable inference.

Let us begin with a very simple problem. What is meant by the "mathematical probability" of obtaining a head when a coin is tossed? Let us employ the usual terminology. Instead of discussing the probability of the *proposition: This coin will fall head uppermost,* we shall talk of the probability of the *event* "getting a head," or simply "heads." "Heads" and "tails" are referred to as the *possible events* or *possible alternatives.* If we are interested in getting a head, "heads" is said to be the *favorable event,* all the others the *unfavorable events.* The mathematical probability is then defined as the fraction whose numerator is the number of possible favorable events and whose denominator is the number of possible events (the sum of the number of favorable and unfavorable events), *provided that all the possible events are equiprobable* (equally probable). If, therefore, a coin has 2 faces which can fall in no other way except to yield heads or tails, and if the 2 faces are also equiprobable, the probability of getting a head is ½. In general, if f is the number of favorable events, u the number of unfavorable ones, and if the events are equiprobable, the probability of the favorable event is defined as $\frac{f}{f+u}$. It is clear that this is always a proper fraction, having values between 0 and 1 inclusive; a probability of 0 indicates that the event is impossible, a probability of 1 that it will necessarily take place.

The condition that the possible events must be equiprobable is of fundamental importance, but very difficult to define. It has been the source of serious errors, some of which we shall examine later. What is meant, roughly, is that one possibility should occur as frequently as any other; and it has often been maintained that two possibilities are equally probable when we know no reason why one rather than the other should occur. Nevertheless, whatever may be the difficulties in determining the equiprobability of a set of alternatives, it is not the mathematician's business to find the criteria of equiprobability, since his concern is with the necessary consequences of such an assumption irrespective of its truth or falsity. The importance of this condition will be clear if we ask for the probability of getting a six with a die. We can argue as follows: there are 2 possibilities, getting a six, and getting other-than-a-six; one of them is favorable; consequently, the probability is ½. This answer, however, may be false unless we make the material assumption that these 2 alternatives are equiprobable. This material assumption is generally not made, because the possibility of getting other than a six is assumed to be made up of 5 subsidiary alternatives—getting a one, getting a two, and so on—each of which is equiprobable with getting a six. Hence if the 6 sides are assumed as equiprobable, the probability of getting a six with a die is $1/6$.

The main burden of the calculus of probability is to determine the probability of a complex event from a knowledge of the probability of its component events. We therefore require further definitions. Two events are said to be *independent* if the occurrence of one is not affected by the occurrence or nonoccurrence of the other. The assertion that two events are in fact independent is a *material assumption* which must be explicitly stated. Many serious errors arise in applying the calculus when the independence of events is assumed without adequate ground, or when this condition is completely neglected.

The Probability of a Joint Occurrence

What is the probability of getting 2 heads in tossing a coin twice (or 2 coins once)? This is a complex event, whose components are: 1 head on the first throw, and 1 head on the second. If the events are independent, and if the probability of getting a head on each throw is $1/2$, the calculus of probability demonstrates that the probability of the *joint occurrence* of heads on each throw is the *product* of the probability of heads on each throw: that is, the probability of getting two heads is $1/2 \times 1/2$, or $1/4$. We can see why this result is a necessary consequence of the assumptions if we enumerate all the possible events in throwing the coin twice. These are: *HH, HT, TH, TT,* in which the order of letters in any one group indicates one possible sequence of heads and tails. There are, therefore, on the assumptions made, 4 equiprobable possibilities; only 1, *HH,* is favorable. Consequently, the probability of getting 2 heads is $1/4$, in agreement with the previous result. In general, if a and b are two independent events, $P(a)$ the probability of the first, and $P(b)$ the probability of the second, the probability of their joint occurrence is $P(ab) = P(a) \times P(b)$.

Care must be taken, when calculating the probability of complex events, to enumerate all the possible alternatives. If we require the probability of getting *at least* 1 head in 2 throws with a coin, the enumeration of the alternatives shows 3 favorable events. Hence the probability of at least 1 head is $3/4$. Eminent scientists have erred by failing to note all the alternatives. According to D'Alembert, for example, the probability of at least 1 head is $2/3$. He enumerated the possibilities as *H, TH, TT,* for he argued that if a head comes uppermost on the first throw, it is not necessary to continue throwing in order to get at least 1 head. But this analysis is faulty, since the possibilities he enumerated are not equiprobable: the first alternative may be regarded as including two distinct cases that are equiprobable with the others.

The probability of the joint occurrence of two events may sometimes be calculated even if the events are not completely independent. Suppose an urn containing 3 white and 2 black balls, and assume all the balls equiprobable in drawing them from the urn one at a time. What is the probability of getting 2 white balls in succession in the first 2 drawings if the balls are not replaced after they have been drawn? Now the probability of drawing a white ball is $3/5$. If a white ball is drawn and not replaced, there remain 2 white and 2 black balls in the urn. Hence the probability of getting a second white ball *if the first ball drawn has been white* is $2/4$. It follows therefore that the probability of getting 2 white balls under these conditions is $3/5 \times 2/4$, or $3/10$.[2] In general, if $P(a)$ is the probability of event a, and $P_a(b)$ the probability of event b when a has occurred, the probability of the joint occurrence of a and b is $P(ab) = P(a) \times P_a(b)$.

The Probability of Disjunctive Events

We sometimes require not the probability of the joint occurrence of two events, but the probability that *either one* of the events will occur. For this purpose we define *strictly alternative* or *disjunctive events.* Two events are disjunctive if both cannot simultaneously occur (if one does occur, the other cannot). In tossing a coin, the possibilities (heads and tails) are assumed to be disjunctive. It is demonstrable that the probability that either one of two disjunctive events occurs is the *sum* of the probabilities of each. What is the probability of getting either 2 heads or 2 tails in tossing a coin twice, on the assumption that the probability of heads is $1/2$ and that the tosses are independent? Now the probability of getting 2 heads is the product of the probabilities of 1 head on the first throw and 1 head on the second, or $1/4$; similarly, the probability of 2 tails is $1/4$. Hence the probability of either 2 heads or 2 tails is $1/4 + 1/4$, or $1/2$. The same result is obtained by applying directly the defini-

[2] This result conforms to the definition of mathematical probability. For the total number of ways of drawing 2 balls from a collection of 5 (that is, the number of combinations of 5 balls, 2 at a time) is $\frac{5 \times 4}{1 \times 2}$, or 10. And the number of ways of drawing 2 white balls from a collection of 3 whites is $\frac{3 \times 2}{1 \times 2}$, or 3; this is the number of favorable events. The probability of getting 2 white balls under these conditions is therefore 3/10, as before.

tion of probability to the four possible events *HH, HT, TH, TT.* Two of these are favorable to either 2 heads or 2 tails; consequently, the required probability is $2/4$, or $1/2$. In general, if $P(a)$ and $P(b)$ are the respective probabilities of two exclusive events a and b, the probability of obtaining either is $P(a+b) = P(a) + P(b)$.

These two theorems, the product theorem for independent events and the addition theorem for exclusive events, are fundamental in the calculus of probability. In terms of them and their extensions more complicated problems can be solved easily. Let us suppose we make one drawing from each of two urns. The first contains 8 white and 2 black balls, the second contains 6 white and 4 black balls. The balls are assumed to be equiprobable. What is the probability that when we draw a ball from each urn, *at least one* of the balls is white? The probability of a white ball from the first urn is $8/10$, and of a white ball from the second urn is $6/10$. It is tempting to add these fractions in order to obtain the probability of a white ball from either urn. But this would be an error. The answer would be greater than 1, which is absurd. And indeed we cannot simply add in this case, because the events are not exclusive. But we may calculate the result as follows: The probability of not getting a white ball (that is, of getting a black) from the first urn is $2/10$; and of not getting a white ball from the second is $4/10$. Therefore, assuming the drawings to be independent, the probability of getting a white neither from the first nor from the second urn is $2/10 \times 4/10$, or $8/100$. Consequently, since we must either get no white ball from either urn or at least 1 white ball from either, the probability of getting at least 1 white ball is $1 - 8/100$ or $92/100$.

What is the probability of getting just 3 heads in tossing 5 coins, assuming that heads and tails are equiprobable on each coin, and that the ways the coins fall are independent of one another? This problem will introduce us to an important formula in the calculus of probabilities. We may perhaps reason as follows: Since 5 coins are thrown, and the probabilities of getting a head on each is $1/2$, the probability of getting heads is $1/2 \times 1/2 \times 1/2$, or $1/8$; but we want just 3 heads, and therefore the other 2 coins must fall tails, the probability of which is $1/2 \times 1/2$, or $1/4$; hence, we may conclude, the probability of getting *just* 3 heads (that is, 3 heads and 2 tails) is $1/8 \times 1/4$, or $1/32$. This result, however, is not correct. This may be seen quite readily if we write out all the possible ways in which the five coins can fall, and then apply directly the definition of probability to these equiprobable alternatives.

The possible alternatives are:

HHHHH 1 possible way of getting 5 heads, 0 tails

HHHHT
HHHTH
HHTHH
HTHHH
THHHH
} 5 possible ways of getting 4 heads, 1 tail

HHHTT
HHTHT
HTHHT
THHHT
HHTTH
HTTHH
TTHHH
HTHTH
THTHH
THHTH
} 10 possible ways of getting 3 heads, 2 tails

HHTTT
HTHTT
HTTHT
HTTTH
THTTH
TTHTH
TTTHH
TTHHT
THHTT
THTHT
} 10 possible ways of getting 2 heads, 3 tails

HTTTT
THTTT
TTHTT
TTTHT
TTTTH
} 5 possible ways of getting 1 head, 4 tails

TTTTT 1 possible way of getting 0 heads, 5 tails.

There are therefore 32 equiprobable possibilities, 10 of which are favorable to the event. The probability of getting 3 heads and 2 tails is $^{10}/_{32}$, a result ten times as large as the one obtained by the incorrect method.

We can understand now, however, why that method is incorrect. For that method failed to take into account the different orders or arrangements in which 3 heads and 2 tails can occur. We must therefore have some way of evaluating the number of different arrangements which can be made of 5 letters, 3 of which are of one kind and 2 of which are of another kind. Readers who are familiar with the theory of combinations will have no difficulty in evaluating this and similar numbers. However, readers who are unacquainted with this branch of arithmetic need not despair, for there is a very simple formula which will yield the required probabilities easily. For the number of possibilities favorable to each category of complex event (that is, 1 for 5 heads, 0 tails; 5 for 4 heads, 1 tail, and so on) is nothing other than the appropriate coefficient in the expansion of the binomial $(a + b)^5 = a^5 + 5a^4b + 10a^3b^2 + 10a^2b^3 + 5ab^4 + b^5$.

In general, then, it is rigorously demonstrable that if p is the probability of an event, q the probability of its *sole exclusive alternative,* then the probability of a complex event with n components is obtained by selecting the appropriate term in the expansion of the binomial $(p + q)^n$. This binomial expansion can be performed quite simply. It is: $(p + q)^n = p^n + \frac{n}{1} p^{n-1}q + \frac{n(n-1)}{1 \times 2} p^{n-2}q^2 + \frac{n(n-1)(n-2)}{1 \times 2 \times 3} p^{n-3}q^3 + \ldots + q^n$.

We shall consider one further illustration of the binomial formula. An urn contains 2 white and 1 red balls, and we are to make 4 drawings from the urn, replacing the ball after each drawing. We are allowed to assume that the balls are equally probable, and that the contents of the urn are thoroughly mixed after each drawing, so that the drawings are independent. What is the probability of getting 3 white balls and 1 red? The probability of drawing a white ball is $p = 2/3$, and of drawing a red ball is $q = 1/3$. To obtain the required answer, we need only expand $(p + q)^4 = p^4 + 4p^3q + 6p^2q^2 + 4pq^3 + q^4$, and substitute the indicated numerical values in the term which represents the probability of getting 3 white and 1 red balls. This term is $4p^3q$, and the required probability is $4 \times (2/3)^3 \times (1/3)$, or $32/81$.

§ 3. INTERPRETATIONS OF PROBABILITY

This brief discussion of the calculus of probability does not exhaust the interesting theorems it contains. We must, however, resume the discussion of the *logic* of probable inference. But we must repeat a former warning. The mathematical theory of probability studies the necessary consequences of our assumptions about a set of alternative possibilities; it cannot inform us as to the probability of any actual event. The question very naturally arises, therefore, how the probability of such events is to be determined. Under what circumstances may the theorems of the calculus of probability be applied?

Probability as a Measure of Belief

The analysis of probable inference which we discussed at the beginning of this chapter is not the usual interpretation. The probability of an event has been commonly identified with the *strength of belief* in the event taking place. Probability means, according to De Morgan, "the state of mind with respect to an assertion, a coming event, or any other matter on which absolute knowledge does not exist." The expression, "It is more probable than improbable," means, according to him, "I believe that it will happen more than I believe that it will not happen." [3] An omniscient being would never employ probable inferences, since every proposition would be known to be certainly true or certainly false. Beings lacking omniscience must rely on probabilities, since their knowledge is incomplete, and probability measures their ignorance. When we *feel altogether sure* that an event will take place, its probability is 1; when our *belief in its impossibility is overwhelming,* its probability is 0; when our *belief is intermediate* between the certainty of its occurrence and the certainty of its nonoccurrence, the probability is some fraction intermediate between 1 and 0.

On this interpretation of what probability is, the calculus of probability may be employed only when our ignorance is equally distributed between several alternatives. As we have seen, the mathematical theory can answer the question, "What is the probability of 3 heads with a coin tossed 3 times?" only when information is supplied concerning (1) the number of alternative ways in which the coin can fall, (2) the equiprobability of these alternatives, and (3) the independence of the different throws. On the psychological theory of probability as a measure of belief or expectation, this information is obtained quite simply. For this theory employs a famous criterion called the *principle of insufficient reason* or the *principle of indifference.* According to this principle: *If there is no known reason for predicating of our subject one rather than another of several alternatives, then relatively to such knowledge the assertions of each of these alternatives have an equal probability.* And if there is no known reason for believing that two events are independent rather than dependent, it is just as probable that they are independent as it is that they are not. Two alternatives are equally probable if there is "perfect indecision, belief inclining neither way." When we are completely ignorant about 2 alternatives, the probability of 1 of them occurring must, on this view, be $1/2$. If, then, we are shown a strange coin, since we can have no reason for believing that one face is more likely to come up than the other,

[3] *Formal Logic,* Chap. IX.

we must say the probability of heads is the same as the probability of tails.

It is difficult to believe, however, that this interpretation does justice to the foundations of probability. In the first place, our ability to predict successfully and control portions of the flux of things by means of probable inferences (for example, in thermodynamics and statistical mechanics) is altogether inexplicable if such inferences rest on nothing but our ignorance or the strength of our beliefs. It would be suicidal for an insurance company to conduct its business on the basis of estimating the feeling of expectation which its officers may have.

In the second place, if probability is a measure of belief, whose belief are we to measure? That beliefs about the same event vary considerably in intensity with different people is notorious. All of us have sufficiently mercurial temperaments so that our beliefs at some time and with respect to some things run the entire gamut from despair to certitude for very trivial reasons indeed. Which state of expectation shall be taken as a measure of probability?

Thirdly, since probabilities can be added and multiplied, beliefs would have to be combinable in a corresponding manner. But in fact no operations of addition of beliefs can be found, and beliefs cannot be measured, as our discussion of the principles of measurement will show.

Finally, the psychological theory of probability can be shown to lead to absurd results—unless indeed the range of its application is considerably restricted. Thus suppose we know that the volume of a unit mass of some substance lies between 2 and 4. On this interpretation of probability, it is just as probable that the specific volume lies between 2 and 3 as between 3 and 4. But the specific density is inversely proportional to the specific volume, so that if the volume is v, the density is $1/v$. Hence the density of this substance must lie between ½ and ¼ (that is, between 4/8 and 2/8) and therefore it is just as likely that it will lie between ½ and 3/8 as between 3/8 and ¼. This, however, is equivalent to saying that it is just as likely that the *specific volume* will lie between 2 and 8/3 as between 8/3 and 4. And this contradicts our first result.

Probability as Relative Frequency

Difficulties of this nature have led to the interpretation of probability as the *relative frequency* with which an event will occur in a class of events. Thus when we say that the probability of a given coin falling heads is ½, we mean that as the number of throws increases indefinitely (in the long run) the ratio between the number of heads and the total number of throws will be about (that is, will not materially differ from) $1/2$. Such a statement is of course an assumption or hypothesis as to the actual course of nature, and therefore needs factual evidence. Such evidence may be rational (in the sense of deduction from previous knowledge as to the constitution of things) or statistical. We may know that pennies are symmetrical and from our knowledge of mechanics infer that the forces making a penny fall head are bound to balance those which make it fall tail. Or we may rely on purely empirical observation as evidence that in the long run neither head nor tail predominates. In the physical sciences, such as meteorology or genetics, and also in practical affairs, such as insurance, we rely on both kinds of evidence. But the statistical evidence is not only indispensable but also apt to be more prominent. We must not, however, completely identify the meaning of a hypothesis with the actual amount of statistical evidence for it available at any given time. A hypothesis as to the nature of things asserts something in regard to all possible phenomena or members of a given class. It can therefore never be completely proved by any number of finite observations. But if we have several hypotheses, the one that agrees best with observable truths that are statistically formulated is naturally preferable.

From this point of view we can understand more clearly the function of the mathematical theory of probability. Suppose we start with the hypothesis that the probability of a male birth is $1/2$. The calculus of probability may then be used to deduce and predict the frequency with which families with 2 male children or families with 2 children of opposite sexes should in the long run occur. Now it may happen that in some particular community *all* the children born during one year are girls. Will this disprove the assumption that the probability of a male birth is $1/2$? Not at all! The calculus shows that on our assumption such an occurrence is *extremely improbable* but not impossible. However, the calculus may also show that such an actual "exceptional" occurrence is in greater conformity with (or would be *less improbable* than) some other assumption. A large number of repetitions of "exceptional" occurrences may thus increase the probability of some other hypotheses, and diminish the probability of our original one. Thus the hypothesis that the probability of a male child is $105/205$ is more constant with actual statistical observations.

The calculus of probabilities thus systematizes our experiences on the simplest available assumptions that will also explain apparent exceptions. Of course no hypothesis concerning the probability of an event can be absolutely refuted by a finite number of observations, since even very large discrepancies from theoretically most probable results are not impossible. But statistical results can show some hypotheses to be more improbable than others.

On this view probability is not concerned with the strength of subjective feelings but is grounded in the nature of classes of events, and objective data are needed to determine their probability. We must note, however, that on this view the probability of a unique event is meaningless. When we appear to be talking of the probability of singular events, we must be understood to be speaking elliptically of that phase of the event which it has in common with other possible events of a certain kind. Hence, when we say that the probability of a head with a *given* coin on a *definite* toss is 1/2, what we must mean is that in a *long series* of such throws heads will turn up about half the time. When we say that in tossing a coin twice, the probability of 2 heads is 1/4, what we must mean is that in *a long series* of sets of 2 throws each, sets which contain 2 heads are approximately 1/4 of all the sets.

An immediate corollary from these remarks is a caution against what is known as the "gambler's fallacy." Suppose we enter a game played with a coin assumed to be "fair," that is, for which the probability of heads is 1/2 and the tosses are independent. Suppose there is a run of 20 heads in succession and we wish to bet on the next throw. What is the probability that the next throw is a head? Many players cannot resist the conclusion that the probability is *less* than 1/2, on the ground, presumably, that heads and tails must "even up" if the coin is fair. But this conclusion is erroneous, and all gambler's systems devised to make winnings secure when playing with *fair* instruments are inevitably fatal to those who employ them. For if the coin is indeed fair, the 20 heads which have already turned up do not affect the results of the 21st throw; when we say the probability of heads on the 21st throw is 1/2, we are talking elliptically of a *large series* of throws. On the other hand, if the coin is not fair, but loaded to favor heads, then clearly the probability of heads on the 21st throw is *greater,* and not less, than 1/2. Laplace tells a story of a man about to become a father. As the date of his wife's delivery approached, he noticed that during the preceding month more girls were born in the community than boys. He therefore placed large bets giving odds favoring the birth of a son.

4. We must note, finally, that no event is probable intrinsically.

but only in terms of its membership in certain classes or series of events. The probability of heads with a coin tossed by hand may be $1/2$; the probability of heads with this same coin shaken in a cup may be altogether different. The event "heads" is here referred to two different classes. And in general the class of events to which a specified event belongs must be explicitly noted in evaluating its probability.

The frequency theory of probability as stated so far has had certain objections urged against it. It does not *seem* capable of interpreting what we mean by the probability of a theory being true, or by the probability of propositions which deal with singular events. We often declare that the heliocentric theory is more probable than the geocentric theory. What does this mean on the frequency theory? And we repeatedly assert propositions like the following: *It is probable that it will rain tonight; It is improbable that Hercules was a historical figure; It is probable that even if Napoleon had been victorious at Waterloo, he would have been unable to remain Emperor of France for much longer.* Such statements are not easily interpreted on the ordinary form of the frequency theory. But these objections are really not fatal, and can be answered by altering the technical expression of the frequency theory.

Probability as the Truth-Frequency of Types of Arguments

This brings us back to the analysis of probable inference we have outlined at the beginning of this chapter—a third interpretation of probability due to Charles Peirce. We have already indicated the objective foundations of probability on this view. We now wish to exhibit the scope of this interpretation.

Suppose a street railway company desires to obtain a city franchise, and decides on bribery as the most effective way of persuading the city fathers to grant it. Great care must be taken, however, for if an alderman genuinely filled with civic virtue were to be approached, the whole game might be spoilt. Should the company approach Alderman A? Now the following facts, believed to be relevant, may be known about Mr. A:

1. He is an alderman, and that means a professional politician.
2. He is a jovial Irishman, ready to see the point of a joke.
3. He is a devout Catholic, and professes high moral principles.
4. He owns real estate, and is suspected of sharp practices in connection with it.
5. He is a member of the local school board, and offers prizes to children for devoted school service.

6. He is not on record as having ever protested against corruption in public office.

Is it probable that he will accept payment for his vote if the bribe is proffered in a proper manner? Let us consider the first item. If that were the sole circumstance known about Mr. A, the truth-frequency theory would interpret the probability of Mr. A's accepting a bribe in the following way. Consider the class of *true* propositions n, obtained from the expression "X is a politician," by giving particular values to the variable X. Consider also the class of propositions n_t, obtained by giving the same values to X in the expression "X is a politician and X is a bribe-taker." Some of the resulting propositions of the second set are true, others are false. Then the limiting value of the ratio $\frac{n_t}{n}$ is defined as the probability that any given individual, say Mr. A, will take bribes, on the evidence that he is a politician. That is, the probability of a proposition being true is the relative frequency with which a class of inferences yield true conclusions from true premises. In general, we do not know the precise numerical value of this ratio. In that case we say that a conclusion is probable on the evidence if the class of such inferences lead to true conclusions *more often than not.*

But what if the evidence for a proposition is more complicated than that we have so far considered? In that case the *analysis* of the argument is more complicated: but the *interpretation* of the probable inference remains the same. If we were to consider the first two items about Mr. A, the class of propositions n would be obtained from the expression "X is a politician, and X is a jovial Irishman," while the class n_t would be obtained from "X is a politician, and X is a jovial Irishman, and X is a bribe-taker." The limiting value of the ratio $\frac{n_t}{n}$ would again define the probability of Mr. A's being a bribe-taker on the evidence that he was a jovial Irish politician. Similar considerations apply if we were to take as evidence all of the six true propositions known about Mr. A.

In most cases, as we have already said, the numerical value of the measure of probability is not known. In those cases we must be content with more or less vague impressions, and sometimes with sheer guesses, as to its magnitude. Very often, too, the evidence may be so complex that a numerical evaluation of the truth-frequency becomes impossible on practical grounds. This, however, is not fatal to this interpretation, since we can reason on indeterminate ratios as well as we can on determinate ones. The great merit of the truth-frequency theory lies (1) in the success with which it can interpret definite numerical probabilities as well as the indeterminate ones, and (2) in its ability to give an objective reading to the probability of propositions dealing with singular events.

1. The truth-frequency theory can take over all the theorems of the calculus of probability, and accept the statistical foundation for probability, simply by making a few verbal changes in the terminology. Instead of talking about *events,* such as "heads," the truth-frequency theory discusses the *proposition This coin will fall head uppermost on the next throw.* Instead of talking about a *class of events,* this theory discusses a *class of inferences.* For it is very clear that the relative frequency with which *This coin falls head uppermost on toss X* is true, or when *This coin is thrown under specific conditions on toss X* is true, must be the same as the relative frequency with which the event "heads" occurs in a series of tosses with the coin. In the same manner, independent, exclusive, and complex events are discussed in terms of independent, exclusive, and compound propositions.

2. The probability of actual singular events is evaluated by the truth-frequency theory in terms of the *kind of evidence* which is supplied for each. And the probative force of evidence depends upon objective matters of fact. To say *It is probable it will rain tonight* means that the truth of propositions stating the present behavior of the barometer, the changes in temperature, the cloudiness of the sky, and so on, is *in fact* accompanied, with a certain relative frequency, by the truth of propositions stating the occurrence of rainfall within a determinate number of hours.[4]

A final caution will help us to avoid frequent confusions on this point. Just as it is meaningless to speak of a body at rest or in mo-

[4] It may be worth repeating that on the view of probability here advanced, the question of the truth of a proposition concerned with a single event (for example, Caesar's crossing the Rubicon) or of a theory (for example, the Copernican one), is equivalent to the question: With what frequency are propositions or theories of a certain class true if there is as much evidence for them as there is for the proposition or theory being considered? Hence on this view a theory highly probable on one set of evidence may cease to be so as the evidence is increased. This, however, follows not from the subjectivity but from the relative character of probability. Indeed, the reason that the psychological interpretation of probability has held the field so long is the mistaken prejudice that all relativity is psychological. It has been, perhaps, a dim recognition of the relative character of probability which has militated against the common form of the frequency theory. For to speak of the probability of an event generally carries the suggestion that the probability is an intrinsic characteristic of the event itself.

tion except in reference to some other body, so it is meaningless to speak of the probability of an event occurring, or a set of propositions being true, except relative to certain evidence or material assumptions. If despite that we all do often speak of some bodies being at rest, it is because we so usually assume our earth as a refference body that it is not generally necessary to mention it. So in philosophy when we speak of all material propositions or theories as only probable, we understand this in reference to the whole body of available relevant knowledge as evidence.

This removes a difficulty often felt in regard to the probability of philosophical theories as to the total world. "Universes," as Peirce puts it, "are not as plentiful as blackberries." But logically the actual world is one of a class of possible ones, and the probability of any theory in respect to it is the relative frequency with which, according to our estimate, theories of the given type are true on the sort of evidence actually available.

The specific difficulties encountered in the study of probable inferences lie in analyzing the great variety of such inferences into their components, in evaluating the probative force of each, and in determining whether the components are independent of one another. That is not a task for an elementary book. We shall, however, have occasion to study more complicated forms of probable inferences in a later chapter.

CHAPTER IX

SOME PROBLEMS OF LOGIC

§ 1. THE PARADOX OF INFERENCE

A richer understanding of the nature of formal logic can be obtained if we consider some critical onslaughts upon it. Our discussions of traditional logic as well as those of generalized or modern logic and mathematics have made clear that in every valid argument the conclusion *necessarily* follows from the premises. At the same time, we have indicated that the conclusion is not merely a *verbal* transformation of the premises: more than a knowledge of language is needed to prove a theorem in geometry or to determine the weight of evidence for a given proposition. As a consequence, many students have been led to serious doubt as to the usefulness or the validity of formal logic.

On the one hand, it has been said, since the conclusion necessarily follows from the premises, the conclusion must be "contained" in the premises. If the conclusion were not "contained" in the premises, it would be quite arbitrary whether we inferred one rather than another, perhaps incompatible, proposition from them; and "validity" would be a meaningless sound. On the other hand, it has been said, the conclusion must be different from the premises, and a valid inference should advance us to something "new" or "novel"; if it did not, an inference would be useless. This "paradox of inference" may be stated in the following form: *If in an inference the conclusion is not contained in the premise, it cannot be valid; and if the conclusion is not different from the premises, it is useless; but the conclusion cannot be contained in the premises and also possess novelty; hence inferences cannot be both valid and useful.*

This criticism of formal logic, although it has been often urged, is based upon several confusions. We must examine what is meant when it is said that the conclusion "is contained" in the premises, and that the conclusion represents something "novel." The questions are closely connected. Let us consider the second one first.

1. It is essential to distinguish the psychological novelty which a conclusion may have from any logical novelty it may be supposed to have. A conclusion may be surprising or unexpected even though it is correctly implied by the premises. Certainly to most men all the consequences of Euclid's assumptions are not present to their minds when they contemplate the assumptions. Even in less complex arguments psychological novelty is very frequently the rule. The following is a frequently cited story of Thackeray's. "An old abbé, talking among a party of intimate friends, happened to say, 'A priest has strange experiences; why, ladies, my first penitent was a murderer.' Upon this, the principal nobleman of the neighborhood enters the room. 'Ah, Abbé, here you are; do you know, ladies, I was the Abbé's first penitent, and I promise you my confession astonished him!' " The reader may add that the conclusion of the syllogism no doubt surprised the ladies. In a puzzle invented by C. L. Dodgson about two clocks, the unexpectedness of the conclusion from premises freely granted is clearly illustrated. "Which is better, a clock that is right only once a year, or a clock that is right twice every day? 'The latter,' you reply, 'unquestionably.' Very good, now attend. I have two clocks: one doesn't go *at all,* and the other loses a minute every day: which would you prefer? 'The losing one,' you answer, 'without doubt.' Now observe; the one which loses a minute a day has to lose twelve hours, or 720 minutes, before it is right, and is therefore right about once every two *years,* whereas the other is evidently right as often as the time it points to comes around, which happens twice a day. So you've contradicted yourself *once.*"

It may therefore be granted as a well-attested fact about human minds that the conclusion of an argument is not in general known to them when they inspect or believe the premises, especially if a long chain of inferences is required to reach the conclusion. And it may well be that if it were otherwise, we would not perform explicit inferences, and that deduction would be unnecessary. *But this has nothing to do with the validity of an inference.* An inference may be valid even though the conclusion is quite familiar, and to teachers of geometry the cogency of the demonstration of the Pythagorean theorem has not disappeared simply because they know exactly what is coming next at each step of the proof. If the reader, unlike Euclid, "immediately sees" that the second part of his Proposition 29—If a straight line falls on two parallel lines, it makes the interior angles on the same side equal to two right angles—asserts precisely the same state of affairs as does his Postulate 5—If a straight line falling on two straight lines makes the interior angles on the same side less than two right angles, the two straight lines, if produced, meet on that side—he will not need an elaborate demonstration, as did Euclid, for the theorem. Nevertheless, the theorem is a necessary consequence of the assumption.

The question of *psychological* novelty is, therefore, not one for logic. On the other hand, by *logical* novelty must be meant the logical independence of what is said to be the "conclusion" from its "premises." And it is clear that if the argument is to be valid, the conclusion cannot, so long as it is dependent upon the premises, possess logical novelty.

2. We must now examine what is meant when it is said that the conclusion "is contained" in the premises. In the first place, "to be contained" is clearly a *spatial* metaphor, and it surely cannot be meant that the conclusion is contained or is present in the premises as a desk is in a room, or even as a chicken is in an egg. In the second place, we have already disposed of the view that the conclusion is psychologically or explicitly present to our minds when we entertain the premises from which we infer it. What meaning, then, can we assign to the assertion that the conclusion is contained in the premises? Simply this: The conclusion in a valid argument is *implied by* the premises. And the paradox disappears when it is once seen that the relation of implication between propositions is of such unique type that only confusion is courted when it is replaced by some analogous relation which has some of its formal properties.

Some further remarks may remove any lingering perplexity in the reader. Propositions imply one another, and our inferences are valid in virtue of such *objective* relations of implication. We may *make* inferences; we do not make, but only *discover*, implications. *Which* of the propositions implied by a set of assumptions we *do* infer, is of course not logically determined. That depends upon our extralogical interests and our intellectual skill.

It is convenient also, in this connection, to distinguish between the *conventional meaning* of a proposition, and the propositions it implies. In one sense, of course, such a distinction is quite gratuitous, since after we discover some of the implied propositions they are taken as constituting part of the *meaning* of the premises. Thus, having discovered that Euclid's axioms imply that the angle sum of a triangle is equal to two right angles, we often regard this theorem as characterizing the full import of the axioms. Nevertheless, although no sharp or final line of division can be drawn between what is the agreed or conventional meaning of a proposition and its logical consequences, such a distinction is recognized in practice. The tyro in geometry understands in some measure what is meant by saying that through a point outside a line only one parallel can be drawn to it, even though he may not know to what other proposition he is committed when he accepts this one. Let us then agree to designate as *conventional meaning* that minimum of meaning which is required if a group of inquirers can be said to address themselves to the *same* proposition. Thus, the conventional meaning of *All men are mortal* may be that the class of men is included in the class of mortals. In such a case, the meaning of the proposition *All immortals are non-men* is not a part of its conventional meaning, but is the meaning of a proposition *implied* by it.

This distinction is useful for the purpose of stating clearly the answer to the paradox. Since only the conventional meaning of the premise is necessarily before the reasoner's mind, the conventional meanings of some of the implied propositions may be absent, so that when these are found to be implied by a set of premises, a feeling of novelty may result. On the other hand, from the point of view of the relations between conventional meanings, the meaning of the implied propositions is always connected with ("contained in") the meaning of the premises.

The thesis that the conclusion of an inference is essentially connected with its premises, so that the latter could not be true if the former were false, has sometimes been understood to preclude the possibility of physical change or physical novelty. An adequate discussion of this question would lead us into metaphysics. But the reader should have no difficulty in rejecting such an interpretation if he remembers that the relations of implication hold not in virtue of the premises being empirically true, but in virtue of the logical relations between premises and conclusion. These relations are applicable to a changing world—indeed, we know of change only in relation to certain relative constants. The question whether there is any constancy in the physical world involves empirical as well as logical considerations. But *if* any proposition in physical science turns out to be true in fact, it reveals certain structural identities which are common characteristics of different or successive states.

§ 2. IS THE SYLLOGISM A *Petitio Principii?*

A somewhat different attack on the usefulness of formal logic has been made since the days of Aristotle. It has been leveled specifically against the syllogism. However, if the charge is well founded, it is fatal to the value of all deductive reasoning. But it is sufficient to examine the specific form of the criticism. John Stuart Mill, who has renewed the ancient charge upon the syllogism, although he believed he was replying to it, has stated it as follows:

"We have now to inquire, whether the syllogistic process, that of reasoning from generals to particulars, is, or is not, a process of inference; a progress from the known to the unknown: a means of coming to a knowledge of something which we did not know before.

"Logicians have been remarkably unanimous in their mode of answering this question. It is universally allowed that a syllogism is vicious if there be anything more in the conclusion than was assumed in the premises. But this is, in fact, to say, that nothing ever was, or can be, proved by syllogism, which was not known, or assumed to be known, before. . . .

"It must be granted that in every syllogism, considered as an argument to prove the conclusion, there is a *petitio principii*. When we say,

All men are mortal,
Socrates is a man,
therefore
Socrates is mortal;

it is unanswerably urged by the adversaries of the syllogistic theory, that the proposition, Socrates is mortal, is presupposed in the more general assumption, All men are mortal: that we cannot be assured of the mortality of all men, unless we are already certain of the mortality of every individual man: that if it be still doubtful whether Socrates, or any other individual we choose to name, be mortal or not, the same degree of uncertainty must hang over the assertion, All men are mortal: that the general principle, instead of being given as evidence of the particular case, cannot itself be taken for true without exception, until every shadow of doubt which could affect any case comprised with it, is dispelled by evidence *aliundè;* and then what remains for the syllogism to prove? That, in short, no reasoning from generals to particulars can, as such, prove anything: since from a general principle we cannot infer any particulars, but those which the principle itself assumes as known."[1]

In order to understand this criticism the reader must distinguish the charge that the premises of a syllogism could not be true unless the conclusion were also true from the charge that we must *know* the truth of the conclusion in order to establish the truth of one or other of the premises. The first charge, the reader will realize, is perfectly harmless. It is not a criticism of the syllogism, but is a statement of the conditions of validity for any deductive inference. It is the second charge which is serious and to which we must address ourselves. It raises the pertinent question of whether any part of the evidence employed to establish the *truth* of a proposition is, in turn, established by evidence of which that *very* proposition is a part. The syllogism begs the question, the criticism runs, because the premises advanced as evidence for the conclusion can be advanced only if the conclusion is known to be true. This charge may therefore be summed up in the following dilemma: If *all* the facts of the premises have been examined, the syllogism is needless; if some have not been, it is a *petitio*. But either all the facts have been examined or some have not. Therefore, the syllogism is either useless or circular. This charge, rightly understood, is not leveled at the validity of deductive inferences as such.

The question, then, is whether the universal premise of a syllogism can ever be asserted as true without examining all of its instances, and if it cannot be asserted as true, whether the syllogism is therefore useless. Now there are undoubtedly some cases when a universal proposition is established by examining all its instances. Thus, when we assert *All the known planets revolve around the sun* the evidence for it is that *Mercury revolves around the sun, Venus revolves around the sun,* and so on for all the known planets. A universal so obtained is called an *enumerative* universal, and may with justice be considered as merely summarizing neatly such a collection of singular propositions. If, then, we were to argue that Jupiter revolves around the sun because all known planets do so and Jupiter is a known planet, the charge that we are arguing in a circle would be justified.

Is it the case, however, that enumerative universals are the typical universals employed in inquiry? Does the proposition *All men are mortal* merely summarize the collection: *The man Adam is mortal, The man Abel is mortal,* and so on, down the line? As we shall see in the sequel, such a view is an absurd interpretation of scientific

[1] *A System of Logic,* 1875, 2 vols., Vol. I, p. 210. The first edition was published in 1843.

method. At this place we can only suggest alternative views. In the first place, a universal may express a resolution to act in certain ways that may have been made independently of any specific occasion to which it nevertheless applies. Thus it may be a rule that *All policemen shall be at least five feet eight inches in height*, and we may know that this rule is and will be enforced without knowing any members of the present or a future police force. If in such a case we infer that Smith is at least five feet eight because he is on the police force, the argument is not circular.

But secondly, and this is the important case, a universal proposition may be advanced as a hypothesis in order to discover the solution of some practical problem or to unify our knowledge. In such a case, the universal is not known to be *certainly* true. Nevertheless, the evidence for it may be more adequate and stronger than for any individual proposition that is a verifiable consequence of it but which is not included in the antecedent evidence for the theory. Thus one consequence of Newtonian physics is that a pair of double stars will revolve around their common center of gravity in elliptic orbits. But Newtonian physics is established with considerable security without first examining double stars; and the evidence for the theory may be greater than that for the existence of such elliptic orbits even if we try to examine double stars directly. The inference that double stars do move on such orbits is therefore not circular The conclusion is, indeed, not certain, since the theory is not certainly true. But only a mistaken idea of science, such as Mill's, which demands absolute certitude in matters of fact, would reject it as useless for that reason. Similarly, we infer the mortality of any living man from the premise *All men are mortal* without basing this premise upon an enumeration of dead men. For we know that *All men have organic bodies* and *All organic bodies disintegrate with time* are supported by evidence of a far-reaching character which itself does not rest in turn upon an examination of the mortality of any one human being.

It is interesting to consider Mill's attempted defense of the syllogism. According to him, when we infer that the Duke of Wellington is mortal (he was alive when Mill was writing) from the premises *All men are mortal* and *The Duke is a man* the major premise is not the real basis for the inference. He says:

"Assuming that the proposition, The Duke of Wellington is mortal, is immediately an inference from the proposition, All men are mortal; whence do we derive our knowledge of that general truth? Of course from observation. Now, all which man can observe are individual cases. From these all general truths must be drawn, and into these they may be again resolved; for a general truth is but an aggregate of particular truths; a comprehensive expression, by which an indefinite number of individual facts are affirmed or denied at once. But a general proposition is not merely a compendious form for recording and preserving in the memory a number of particular facts, all of which have been observed. Generalization is not a process of mere naming, it is also a process of inference. From instances which have been observed, we feel warranted in concluding, that what we found true in those instances, holds in all similar ones, past, present, and future, however numerous they may be. We then, by that valuable contrivance of language which enables us to speak of many as if they were one, record all that we have observed, together with all that we infer from observations, in one concise expression; and have thus only one proposition, instead of an endless number, to remember or to communicate. The results of many observations and inferences, and instructions for making innumerable inferences in unforeseen cases, are compressed into one short sentence. . . .

"If, from our experience of John, Thomas, etc., who once were living, but are now dead, we are entitled to conclude that all human beings are mortal, we might surely without any logical inconsequence have concluded at once from those instances, that the Duke of Wellington is mortal. The mortality of John, Thomas, and others is, after all, the whole evidence we have for the mortality of the Duke of Wellington. Not one iota is added to the proof by interpolating a general proposition. . . . Not only *may* we reason from particular to particular without passing through generals, but we perpetually do so reason." [2]

Thus, on Mill's view, the charge of *petitio* against the syllogism is quashed simply by denying that the premises of the syllogism are the true basis for the conclusion, and by insisting that the conclusion is not part of the data from which the universal premise is inferred. It will not have escaped the reader, however, that Mill's attempt to defend the syllogism culminates in regarding an inference as valid when it is not so without a further premise. Why may we infer from the death of John, Thomas, and the rest that the Duke of Wellington is mortal? Because, says Mill, what was true in those instances holds in all similar ones. If, however, we state

[2] *A System of Logic*, 1875, 2 vols., Vol. I, p. 210. The first edition was published in 1843.

this argument formally, we find, lo and behold! a syllogism: *What is true for the case of John, Thomas, and so on is true for all similar instances; but John, Thomas, and so on died; therefore the Duke, being a similar instance, is mortal.* And the major premise of this syllogism cannot, even on Mill's view, be regarded as simply an aggregate of particular propositions. In many cases, such as the theory of gravitation, universals cannot possibly be conceived as being memoranda for singular propositions. In such cases the absolute indispensability of universal propositions is even more evident than in the simple illustrations given by Mill. But we are trespassing on a discussion which we must reserve until later.

We may therefore summarize the discussion of the syllogism as a *petitio:*

1. The charge that the syllogism is a *petitio principii* is significant only if we are interested in the evidence for the *material* truth of premises and conclusion. Even if this charge were sustained, the syllogism as a mode of valid inference would not be impugned.

2. Those universal premises which express our resolutions, commands, laws, may be asserted without examining all the possible instances to which they may be applied.

3. Universal premises may be asserted as probably true on evidence which does not contain as parts every verifiable consequence of those premises.

And it is simply not true that every universal proposition is a shorthand device for summarizing a collection of previously known singular propositions.

§ 3. THE LAWS OF THOUGHT

Logic has often been defined as the study of the "laws of thought." And in particular three principles—the principle of identity, the principle of contradiction, and the principle of excluded middle—have been taken as the necessary, and sometimes as the sufficient, conditions for valid thinking. We wish to examine these principles, discuss whether they are indeed laws of thought, and indicate finally the nature of logical principles.

The three principles mentioned have been formulated in several ways. One formulation of them, in the order stated above, is *If anything is* A *it is* A; *nothing can be both* A *and not* A; *anything must be either* A *or not* A. It is preferable, however, to consider the following alternative formulation first. The *principle of identity* asserts that: *If any proposition is true, it is true.* The *principle of contradiction* asserts that: *No proposition can be both true and false.* The *principle of excluded middle* states that: *Any proposition must be either true or false.*

It will doubtless strike the reader that in either formulation none of the principles asserts anything about anybody's *thoughts.* In the second formulation, which we shall take as particularly relevant to logic, the so-called laws of thought state something about *propositions.* The principle of contradiction, for example, does not say that we cannot *think* a proposition to be both true and false. If it did assert that, it would be most certainly false, as the fact that men often believe contradictory propositions clearly shows. There is no psychological impossibility, unfortunately, to think confusedly and inconsistently. If, therefore, these principles are representative of logical principles, we must conclude that the subject matter of logic is not human thoughts at all. On the other hand, if the qualification is made that the laws of thought do not refer to human thinking as a process in time, but to the conditions of valid thinking, the reply is much the same. The conditions of valid thinking are themselves not thoughts. In fact, as the reader knows, logic studies the relations between sets of propositions in virtue of which some limitation is placed upon the possible truth or falsity of one set by the possible truth or falsity of another set.

But after our lengthy discussion of many principles of logic, it should also be clear that while these three "laws of thought" state essential logical properties of propositions, they are not an *exhaustive* statement of logical principles. The principle of the syllogism, the principles of tautology, simplification, absorption, and others discussed in Chapter VI have an equal claim with the traditional three to belong to the foundations of logic. It may be thought, perhaps, that all the other logical principles may be derived from these three by a chain of logical steps. This, however, would be a mistake. The so-called laws of thought are not a sufficient basis from which to deduce all the other logical principles.

It is also an error to suppose that any one of the three "laws" may be deduced logically from the others without assuming it in the process of deduction. Into these matters we cannot go in further detail. But the reader will recall our previous discussion of demonstrability. Even if the other principles of logic could be derived from the traditional three, that would not make these more important, or more certain, than any of the others.

Criticisms of the Three "Laws"

The significance of the three "laws of thought" may be made more evident if we consider some criticisms which have been made of them.

1. It has been denied, for example, that the *principle of identity* is universally true, because a proposition may be true at one time and false at another. Thus it is said that *The sun is shining* may be true today but false tomorrow, and perhaps even later on the same day. The objection, however, arises from a confusion. The expression, "The sun is shining," does not fully express the proposition which is judged true or false. There is an implicit time and place reference, which is suppressed (because it is taken for granted) in the expression, but which is essential in the proposition judged. As it stands, therefore, the expression is a *propositional function* (and not a proposition) of the form *The sun is shining at (place) x on (time) y*. If we supply that reference, as in the expression *The sun is shining at New York on January 1, 1932, the proposition cannot be true on one day and false on another.* We must distinguish, therefore, the time and place *in* predication (the time and place reference *in* the proposition) from the time and place *of* predication (the time and place *at* which a position is judged). The truth or falsity of a proposition is independent of the time and place *of* predication, so that it is correct to say of a proposition that "once true, always true, once false, always false."

2. Similarly, the universality of the *principle of contradiction* has been denied on the ground that in some cases two apparently contradictory propositions may both be true. Thus it is said *The floor is wet* and *The floor is not wet* may both be true; also *This penny has a circular shape* and *This penny has an elliptical shape* (where the subject is the same penny) may both be true. This apparent violation of the principle of contradiction may be resolved as was the criticism of the principle of identity. In the first pair of expressions the time *in* predication is not specified; in the second pair, the place *in* predication, the place at which the penny presents the shape, is omitted. If these omissions are supplied, in neither of the pairs are the propositions contradictories.

Another objection has arisen from a consideration of what have been known traditionally as *sophisms*, which play an important rôle in modern logic. Suppose a man asserts, "I am lying." If he is speaking the truth, "I am lying" is true. But in that case, the man is not speaking the truth, so that "I am lying" is false. But in that case he *is* speaking the truth, and "I am lying" is true, and so on, *ad infinitum.* Here seems to be a proposition which is both true and false.

Common sense will easily dispose of this difficulty by recognizing that the man who says, "I am lying," and says nothing else, has not *asserted* anything, and therefore has not committed himself to any proposition. The difficulty arises from confusing a group of words which form a *sentence* with a *proposition,* which alone is true or false. The sentence, "I am lying," would indicate a proposition only if it referred to some *other* assertion of the speaker, which would thus be characterized as a lie. In such a case the paradox obviously vanishes.

The recognition of the fact that the sentence, "I am lying," is not a complete and independent proposition, but can only serve as such if it points to some other proposition, is at the basis of an elaborate and carefully worked-out doctrine known as the *theory of types.* According to the theory of types the assertion, "I am lying," is a proposition only if it has as its own subject matter a collection of propositions which do *not* include the proposition *I am lying.* This proposition is then of different type from those which it is *about,* and cannot, without contradiction, be regarded as asserting anything about propositions which are of the *same* type as itself. In other words, "I am lying" must be interpreted as, "There is a proposition which I am affirming and which is false." But this assertion itself cannot be any one of the propositions referred to. If the speaker subsequently wishes to deny that he was lying, the proposition expressing the denial must be of a *higher* type than *"I am lying."* In this way, propositions can be arranged in hierarchies or types. so that any proposition may be about a proposition of a lower type, but never about a proposition of the same or a higher type. The principle used to avoid such contradictions has been named the *vicious-circle principle,* and has been stated as follows: *Whatever involves* all *of a collection must not be one of the collection.*

3. Finally, the *principle of excluded middle* has been challenged on the ground that another alternative is possible to the truth or falsity of a proposition. Thus, it is claimed that it is not necessary that one of the following two propositions be true: *He is older than his brother* and *He is younger than his brother* for there is the alternative that *He is as old as his brother.* But the objection confuses the contrary of a proposition with its contradictory. The contradictory of *He is older than his brother* is not *He is younger*

than his brother, but *He is not older than his brother.* To this pair of contradictories the principle of excluded middle does apply.

Another objection has been made on the ground that things change, often insensibly, so that the line between the truth and the falsity of a proposition is difficult to draw, even if it is not altogether arbitrary. Thus it is said *He is mature* and *He is not mature* are formal contradictories, and yet we cannot decide which one of them is true. This objection does not deny the principle, since the latter simply states that one of a pair of contradictories must be true, but does not tell us which one. It is true, however, that "maturity" may denote a vaguely defined character, so that it may be difficult to draw a line between maturity and the lack of it. But in such cases, because there is a region of indetermination in the application of our concepts we must either make further distinctions as to what is meant by "maturity," or agree upon some conventional standard, such as age, which will fix the denotation of the concept.

Finally, it has been said that to the two alternatives *true* and *false,* there is a third, the *meaningless.* Thus, according to Mill, "Abracadabra is a second intention," is neither true nor false, but meaningless. To which the proper reply is that the principle of excluded middle is applicable only to propositions, and that a meaningless expression is not a possible object to which the principle may be applied. However, the question of what constitutes a meaningful expression is a large one, and we cannot more than suggest some of the problems. Is, "Wisdom has a low electrical resistance," either true or false—is it a proposition? In what sense may we deny that a number has weight? Such questions lead to a discussion of the categories or types of being and the general conditions of significance.

§ 4. THE BASIS OF LOGICAL PRINCIPLES IN THE NATURE OF THINGS

We turn now to the first formulation of the three so-called laws of thought. This formulation is an obvious counterpart of the propositional formulation. And it expresses, perhaps even more clearly, that their subject matter is certain *general or generic traits of all things whatsoever.* And the same may be said of *all* the principles of logic. From this point of view, logic may be regarded as the study of the most general, the most pervasive characters of both whatever is and whatever may be. From what has already been said, however, the reader will recognize that the principle of identity (If anything is *A*, it is *A*) does not deny the possibility of change and does *not* affirm that if a poker is hot it will always remain hot. It *does* say that anything whatsoever in some definite context and occasion has some determinate character. If a poker is hot here and now, then it is hot, and not, in this one of its aspects, something else. If the coin has a round shape at this time and from this point of view, then the shape cannot also be not round. And if evenness or oddness can be significantly predicated of a number, then such a number must be either odd or even.

The insight that logical principles express the most general nature of things was first clearly expressed by Aristotle. At the same time he recognized that since the general nature of things is the ground for the correctness or incorrectness of reasoning, that general nature is also expressed in the principles of logic or inference. According to him, therefore, logic studies the nature of anything that is; "it investigates being as being." It is differentiated from other sciences because while the other disciplines examine the properties which distinguish one subject matter from another, logic studies those truths which hold for everything that is, and not for some special subdivision of what is apart from the others. As a consequence, logical principles must be formal—they represent the common characters of any subject matter, and they cannot be employed to differentiate one subject matter from another. Instead of regarding the abstractions of logic as a fault, we must regard them as a virtue. For we require to know only the most general characters of a subject matter (that which it has in common with everything else) in order to reason upon it validly. We need not encumber our thought with useless intellectual baggage if we reason intelligently. As principles of being, logical principles are universally applicable. As principles of inference, they must be accepted by all, on pain of stultifying all thought. Logical principles, therefore, are not independent of questions of truth. For when we draw a conclusion from the premises correctly we are tacitly recognizing the truth of the proposition, which is grounded in the general nature of things, that the premises do imply the conclusion. (We must mention in passing that this view of the nature of logic is not accepted by all thinkers.)

It is important, however, to be clear about the sense in which logical principles are principles of being. As has been noted before, it has often been supposed that logical principles are "better known" or "more certain" than any other principles. Whether this is so or not is not the significant fact about the principles. Logical principles are involved in every proof, and in that sense every proof depends upon them *whether we know them explicitly or not,* or *whether we have confidence in them or not.* We have already pointed out that the fundamental assumptions in a system need not be more familiar than the theorems. That we know logical principles at all, is not a consequence or a condition of their expressing pervasive characters of whatever is.

It has also been supposed that we can prove logical principles to be necessarily true by showing that they are involved in every reflective inquiry. This also is an error, if by proof is meant what is ordinarily understood by proof. Logical principles in their full generality cannot be proved, since every such attempted proof must assume some or all of them. That which is required in every proof cannot itself be demonstrated. Nevertheless, logical principles are confirmed and exhibited in every inference that we draw, in every investigation which we successfully bring to a close. They are discovered to hold in every analysis which we undertake. They are inescapable, because any attempt to disregard them reduces our thoughts and words to confusion and gibberish.

It has also been supposed that because logical principles are first principles in the sense that every subject matter verifies them, they are prior to existence and condition it. Now there is no doubt that every significant proposition, if true, limits the subject matter, and prohibits something else to be true. In this sense, and only in this sense, do logical principles condition existence. It is an error, however, to suppose that logical principles are prior to existence in the sense that logical principles were *first in time.* On this point Aristotle himself has said all that needs to be said: "The fact of the being of a man carries with it the truth of the proposition that he is, and the implication is reciprocal: for if a man is, the proposition wherein we allege that he is is true, and conversely, if the proposition wherein we allege that he is is true, then he is. The true proposition, however, is in no way the cause of the being of the man, but the fact of the man's being does seem somehow to be the cause of the truth of the proposition, for the truth or falsity of the proposition depends on the fact of the man's being or not being." [3] The priority of logic lies simply in its expression of the utmost generality possible.

[3] *Categoriae,* Chap. 12, in *Works,* ed. by W. D. Ross, 1928, Vol. I, p. 14b.

BOOK II

APPLIED LOGIC AND SCIENTIFIC METHOD

CHAPTER X

LOGIC AND THE METHOD OF SCIENCE

Formal logic as studied in the first part of this work deals with the possible relations (in regard to truth and falsity) between propositions, no matter what their subject matter. This gives us the *necessary* conditions for valid inference and enables us to eliminate false reasoning, but that is not *sufficient* to establish any material or factual truth in any particular field. Formal logic shows us that any such proposition must be true *if* certain others are so. The categorical assertion that our premises are actually true cannot be a matter of logic alone without making the latter identical with all knowledge. Logic, then, is involved in all reasoned knowledge (which is the original meaning of "science") but is not the whole of it.[1] This enables us to regard all science as applied logic, which was expressed by the Greeks in calling the science of any subject, for example, man, or the earth, the logic of it—anthropo*logy*, or geo*logy*.

The great prestige of the natural sciences, acquired largely by their aid to modern technology and by their successful fight against the ancient mythology that was sanctified by various authorities, has led us to apply the term "science" only to these or to similarly highly developed branches of knowledge and to deny it to ordinary knowledge of affairs, no matter how well founded. Thus no one thinks of a railroad time-table or of a telephone book as science even though the knowledge in it is accurate, verifiable, and organized in a definite order. We reserve the term "science" for knowledge which is general and systematic, that is, in which specific propositions are all deduced from a few general principles. Now we need not enter here into the quarrel which arises because archeologists, historians, descriptive sociologists, and others wish to call their more empirical knowledge science. We shall try to show later

[1] The German *Wissenschaft* is still used to mean both "knowledge" and "science."

that all the logical methods involved in proving the existence of laws are involved in establishing the truth of any historical event. In determining the weight of evidence for any human event we must reason from general propositions in regard to human affairs, though such propositions are generally implicitly rather than explicitly assumed.

If we look at all the sciences not only as they differ among each other but also as each changes and grows in the course of time, we find that the constant and universal feature of science is its general method, which consists in the persistent search for truth, constantly asking: Is it so? To what extent is it so? Why is it so?—that is, What general conditions or considerations determine it to be so? And this can be seen on reflection to be the demand for the best available evidence, the determination of which we call logic. Scientific method is thus the persistent application of logic as the common feature of all reasoned knowledge. From this point of view scientific method is simply the way in which we test impressions, opinions, or surmises by examining the best available evidence for and against them. And thus a critical historian like Thucydides can be more scientific than the more credulous Livy, and a sound philologist like Whitney can be more scientific than the more hastily speculative Max Müller. The various features of scientific method can naturally be seen more clearly in the more developed sciences; but in essence scientific method is simply the pursuit of truth as determined by logical considerations. Before determining this in detail, it is well to distinguish between scientific method and other ways of banishing doubt and arriving at stable beliefs.

Most of our beliefs, we have already indicated, rest on the tacit acceptance of current attitudes or on our own unreflective assumptions. Thus we come to believe that the sun revolves around the earth daily because we see it rise in the east and sink in the west; or we send a testimonial to the makers of a certain toothpaste to the effect that it is an excellent preserver of teeth because we have had no dental trouble since we have used that preparation; or we offer alms to some beggar because we perceive his poverty by his rags and emaciated appearance. But too often and sometimes, alas! too late, we learn that not all "seeing" is "believing." Beliefs so formed do not stand up against a more varied experience. There is too little agreement in opinions so formed and too little security in acting upon them. Most of us then find ourselves challenged to support or change our opinions. And we do so by diverse methods.

The Method of Tenacity

Habit or inertia makes it easier for us to continue to believe a proposition simply because we have always believed it. Hence, we may avoid doubting it by closing our mind to all contradictory evidence. That frequent verbal reiteration may strengthen beliefs which have been challenged is a truth acted upon by all organized sects or parties. If anyone questions the superior virtues of ourselves, our dear ones, our country, race, language, or religion, our first impulse and the one generally followed is to repeat our belief as an act of loyalty and to regard the questioning attitude as ignorant, disloyal, and unworthy of attention. We thus insulate ourselves from opinions or beliefs contrary to those which we have always held. As a defense of this attitude the believer often alleges that he would be unhappy if he were to believe otherwise than he in fact does. But while a change in opinion may require painful effort, the new beliefs may become habitual, and perhaps more satisfying than the old ones.

This method of tenacity cannot always secure the stability of one's beliefs. Not all men believe alike, in part because the climate of opinion varies with historical antecedents, and in part because the personal and social interests which men wish to guard are unlike. The pressure of opinions other than one's own cannot always be so disregarded. The man who tenaciously holds on to his own way occasionally admits that not all those who differ from him are fools. When once the incidence of other views is felt, the method of tenacity is incapable of deciding between conflicting opinions. And since a lack of uniformity in beliefs is itself a powerful source of doubt concerning them, some method other than the method of tenacity is required for achieving stable views.

The Method of Authority

Such a method is sometimes found in the appeal to authority. Instead of simply holding on doggedly to one's beliefs, appeal is made to some highly respected source to substantiate the views held. Most propositions of religion and conduct claim support from some sacred text, tradition, or tribunal whose decision on such questions is vested with finality. Political, economic, and social questions are frequently determined in similar fashion. What one should wear at a funeral, what rule of syntax one should follow in writing, what rights one has in the product of his labor, how one should behave in some social crisis like war—these are problems repeatedly resolved by the authoritative method.

We may distinguish two forms of the appeal to authority. One form is inevitable and reasonable. It is employed whenever we are unable for lack of time or training to settle some problem, such as, What diet or exercise will relieve certain distressing symptoms? or, What was the system of weights which the Egyptians used? We then leave the resolution of the problem to experts, whose authority is acknowledged. But their authority is only relatively final, and we reserve the right to others (also competent to judge), or to ourselves (finding the time to acquire competence), to modify the findings of our expert. The second form of the appeal to authority invests some sources with infallibility and finality and invokes some external force to give sanction to their decisions. On questions of politics, economics, and social conduct, as well as on religious opinions, the method of authority has been used to root out, as heretical or disloyal, divergent opinions. Men have been frightened and punished into conformity in order to prevent alternative views from unsettling our habitual beliefs.

The aim of this method, unanimity and stability of belief, cannot be achieved so long as authorities differ. Buddhists do not accept the authorities of the Christians, just as the latter reject the authority of Mahomet and the Koran. In temporal matters experts frequently disagree and are often found in error. Moreover, authoritative regulation of all beliefs is not feasible practically, and much must be left to be decided in some other way. The method of authority has thus to be supplemented, if not replaced, by some other method for resolving doubt and uncertainty.

The Method of Intuition

A method repeatedly tried in order to guarantee stable beliefs is the appeal to "self-evident" propositions—propositions so "obviously true" that the understanding of their *meaning* will carry with it an indubitable conviction of their *truth*. Very few men in the history of philosophy and that of the sciences have been able to resist at all times the lure of intuitively revealed truths. Thus all the great astronomers, including Copernicus, believed it to be self-evident that the orbits of the planets must be circular, and no mathematician or physicist before Gauss seriously doubted the proposition that two straight lines cannot enclose an area. Other examples of propositions which have been, or still are, believed by some to be self-evident are: that the whole is greater than any one of its parts; that the right to private property is inalienable; that bigamy is a sin; that nothing can happen without an adequate cause.

Unfortunately, it is difficult to find a proposition for which at some time or other "self-evidence" has not been claimed. Propositions regarded as indubitable by many, for example, that the earth is flat, have been shown to be false. It is well known that "self-evidence" is often a function of current fashions and of early training. The fact, therefore, that we feel absolutely certain, or that a given proposition has not before been questioned, is no guarantee against its being proved false. Our intuitions must, then, be tested.

The Method of Science or Reflective Inquiry

None of the methods for settling doubts we have examined so far is free from human caprice and willfulness. As a consequence, the propositions which are held on the basis of those methods are uncertain in the range of their application and in their accuracy. If we wish clarity and accuracy, order and consistency, security and cogency, in our actions and intellectual allegiances we shall have to resort to some method of fixing beliefs whose efficacy in resolving problems is independent of our desires and wills. Such a method, which takes advantage of the objective connections in the world around us, should be found reasonable not because of its appeal to the idiosyncrasies of a selected few individuals, but because it can be tested repeatedly and by all men.

The other methods discussed are all inflexible, that is, none of them can admit that it will lead us into error. Hence none of them can make provision for correcting its own results. What is called *scientific method* differs radically from these by encouraging and developing the utmost possible doubt, so that what is left after such doubt is always supported by the best available evidence. As new evidence or new doubts arise it is the essence of scientific method to incorporate them—to make them an integral part of the body of knowledge so far attained. Its method, then, makes science progressive because it is never too certain about its results.

It is well to distinguish between scientific method and general skepticism. The mere resolution to doubt all things is not necessarily effective. For the propositions most in need of questioning may seem to us unquestionable. We need a technique that will enable us to discover possible alternatives to propositions which we may regard as truisms or necessarily true. In this process formal logic aids us in devising ways of formulating our propositions explicitly and accurately, so that their possible alternatives become clear. When thus faced with alternative hypotheses, logic develops their consequences, so that when these consequences are compared with observable phenomena we have a means of testing which hypothesis is to be eliminated and which is most in harmony with the facts of observation. The chapters that follow are an expansion of this simple statement.

CHAPTER XI

HYPOTHESES AND SCIENTIFIC METHOD

"Those who refuse to go beyond fact rarely get as far as fact. . . . Almost every great step [in the history of science] has been made by the 'anticipation of nature,' that is, by the invention of hypotheses which, though verifiable, often had very little foundation to start with."—T. H. Huxley.

"How odd it is that anyone should not see that all observation must be for or against some view, if it is to be of any service."—Charles Darwin.

§ 1. THE OCCASION AND THE FUNCTION OF INQUIRY

In the second book of his fascinating *History,* Herodotus recounts the sights that met him on his travels to Egypt. The river Nile aroused his attention:

"Now the Nile, when it overflows, floods not only the Delta, but also the tracts of country on both sides the stream which are thought to belong to Libya and Arabia, in some places reaching to the extent of two days' journey from its banks, in some even exceeding that distance, but in others falling short of it.

"Concerning the nature of the river, I was not able to gain any information either from the priests or from others. I was particularly anxious to learn from them why the Nile, at the commencement of the summer solstice, begins to rise, and continues to increase for a hundred days—and why, as soon as that number is past, it forthwith retires and contracts its stream, continuing low during the whole of the winter until the summer solstice comes around again. On none of these points could I obtain any explanation from the inhabitants, though I made every inquiry, wishing to know what was commonly reported—they could neither tell me what special virtue the Nile has which makes it so opposite in its nature to all other streams, nor why, unlike every other river, it gives forth no breezes from its surface.

"Some of the Greeks, however, wishing to get a reputation for cleverness, have offered explanations of the phenomena of the river, for which they have accounted in three different ways. Two of these I do not think it worth while to speak of, further than simply to mention what they are. One pretends that the Etesian winds [the northwest winds blowing from the Mediterranean] cause the rise of the river by preventing the Nile-water from running off into the sea. But in the first place it has often happened, when the Etesian winds did not blow, that the Nile has risen according to its usual wont; and further, if the Etesian winds produced the effect, the other rivers which flow in a direction opposite to those winds ought to present the same phenomena as the Nile, and the more so as they are all smaller streams, and have a weaker current. But these rivers, of which there are many both in Syria and in Libya, are entirely unlike the Nile in this respect.

"The second opinion is even more unscientific than the one just mentioned, and also, if I may so say, more marvellous. It is that the Nile acts so strangely because it flows from the ocean, and that the ocean flows all round the earth.

"The third explanation, which is very much more plausible than either of the others, is positively the furthest from the truth; for there is really nothing in what it says, any more than in the other theories. It is, that the inundation of the Nile is caused by the melting of snows. Now, as the Nile flows out of Libya [Central Africa], through Ethiopia, into Egypt, how is it possible that it can be formed of melted snow, running, as it does, from the hottest regions of the world into cooler countries? Many are the proofs whereby anyone capable of reasoning on the subject may be convinced that it is most unlikely this should be the case. The first and strongest argument is furnished by the winds, which always blow hot from these regions. The second is, that rain and frost are unknown there. Now, whenever snow falls, it must of necessity rain within five days; so that, if there were snow, there must be rain also in those parts. Thirdly, it is certain that the natives of the country are black with the heat, that the kites and the swallows remain there the whole year, and that the cranes, when they fly from the rigors of a Scythian winter, flock thither to pass the cold season. If then, in the country whence the Nile has its source, or in that through which it flows, there fell ever so little snow, it is absolutely impossible that any of these circumstances could take place.

"As for the writer who attributes the phenomenon to the ocean, his account is involved in such obscurity, that it is impossible to disprove it by argument. For my part I know of no river called Ocean, and I think that Homer, or one of the earlier poets, invented the name and introduced it into his poetry." [1]

Herodotus then goes on to state his own explanation of the behavior of the Nile.

Has the reader ever been guilty of believing or saying that the way to find out what the truth is, is to "study the facts" or to "let the facts speak for themselves"? Then let him examine this quotation for the light it may throw on the nature of the circumstances under which contributions to knowledge are made. We have suggested in the introductory chapter of the present Book that unless habitual beliefs are shaken into doubt by alterations in our familiar environment or by our curiosity, we either do no thinking at all, or our thinking, such as it is, has a routine character. We wish now to reinforce this suggestion and indicate its importance in understanding the nature of reflective or scientific method.

This excerpt from Herodotus illustrates clearly the Greek zest for scientific knowledge and speculation. But it also illustrates the great difference between the habit of simple acceptance of apparently stray, disconnected information, and the attitude that searches for some order in facts which are only superficially isolated. The observable inundation of the Nile was to many a brute fact, unconnected with other familiar but isolated facts. For Herodotus, however, the behavior of the Nile was not simply a brute fact. It presented a *problem* that could be resolved only by finding some general *connection* between the periodic inundation of the Nile and *other* facts.

It is an utterly superficial view, therefore, that the truth is to be found by "studying the facts." It is superficial because no inquiry can even get under way until and unless *some difficulty is felt* in a practical or theoretical situation. It is the difficulty, or problem, which guides our search for some *order among the facts,* in terms of which the difficulty is to be removed. We could not possibly discover the *reasons* for the inundation of the Nile unless we first recognized in the inundation a *problem* demanding solution.

If some problem is the occasion for inquiry, the *solution* of the problem is the goal and function of the inquiry. What constitutes a satisfactory solution of a problem, and in particular of the problem: Why does the Nile overflow its banks? The sort of answer for

[1] *History,* tr. by George Rawlinson, 1859, 4 vols., Vol. II, pp. 24-29.

which Herodotus was looking was the discovery of a connection between the fact of the Nile's behavior and *other* facts; in virtue of that connection, apparently isolated facts would be seen to be *ordered* facts. And in general, scientific investigations must begin with some problem, and aim at an order connecting what at first sight may seem unrelated facts. But the ability to perceive in some brute experience the occasion for a problem, and especially a problem *whose solution has a bearing on the solution of other problems,* is not a common talent among men. For no rule can be given by means of which men can learn to ask significant questions. It is a mark of scientific genius to be sensitive to difficulties where less gifted people pass by untroubled with doubt.

§ 2. THE FORMULATION OF RELEVANT HYPOTHESIS

How does such a search for an order among facts proceed? The reader must note in the first place that a problem cannot even be *stated* unless we are somewhat familiar with the subject matter in which we discover the problem. The Greeks found a problem in the behavior of the Nile because, among other reasons, they were acquainted with the behavior of other rivers, and because the behavior of these other rivers was known to them to be connected with such things as wind, snowfall, and evaporation.

In order to state some obscurely felt difficulty in the form of a determinate problem, we must be able to *pick out,* on the basis of *previous knowledge,* certain elements in the subject matter as *significant.* Thus Herodotus noted the *distance covered* by the overflowing waters, the *time* at which the inundation *begins,* the *time* at which the overflow reaches its *maximum,* and the absence of *breezes* at the river's surface. It was in terms of such distinguishable and repeatable elements in the total situation known as "the inundation of the Nile" that Herodotus stated his difficulty. But his attention was drawn to these elements, rather than to others, because he was familiar with certain *theories* dealing with the behavior of rivers. It was his familiarity with such theories which made him look to facts like the winds, snowfall, or evaporation rather than to other facts in order to find a connection between them and the Nile's behavior.

We cannot take a single step forward in any inquiry unless we begin with a *suggested* explanation or solution of the difficulty which originated it. Such tentative explanations are suggested to us by something in the subject matter and by our previous knowledge.

When they are formulated as propositions, they are called *hypotheses.*

The function of a hypothesis is to *direct* our search for the order among facts. The suggestions formulated in the hypothesis *may* be solutions to the problem. *Whether* they are, is the task of the inquiry. No one of the suggestions need necessarily lead to our goal. And frequently some of the suggestions are incompatible with one another, so that they cannot all be solutions to the same problem.

We shall discuss below the formal conditions a satisfactory hypothesis must fulfill. The reader should note at this point that Herodotus examined three hypotheses (besides his own) for solving the problem of the Nile's periodic inundation. He accepted his own, after rejecting the other three. As a matter of fact, all four explanations are false. Nevertheless, the procedure he followed in rejecting some hypotheses and accepting others is still a model of scientific method.

How important hypotheses are in directing inquiry will be seen clearly if we reflect once more on the frequent advice: "Let the facts speak for themselves." For what *are* the facts, and *which* facts should we study? Herodotus could have observed the rise and retreat of the Nile until the end of time without finding in that particular repeated fact the sort of connections he was looking for—the relations of the inundation to the rainfall in Central Africa, for example. His problem could receive a solution only with the discovery of an invariable connection between the overflow of the Nile and some other fact. But *what* other fact? The number of other facts is endless, and an undirected observation of the Nile may never reveal either the other facts or their mode of connection. Facts must be *selected* for study on the basis of a hypothesis.

In directing an inquiry, a hypothesis must of necessity regard some facts as *significant* and others as not. It would have been humanly impossible for Herodotus to examine the relations of the Nile to *every other* class of events. Such a task, however, would have been regarded by him as preposterous. For most of these other facts, such as the number of prayers offered by the Egyptians every day, or the number of travelers visiting Naucratis each season, were judged by him to be *irrelevant.*

What is meant by saying that some hypotheses express "relevant" connection of facts, and others do not? The melting of snows is a relevant fact for understanding the Nile's behavior, Herodotus might have explained, because *on the basis of previous knowledge* melting snow can be regarded as related more or less constantly and in some determinate manner with the volume of rivers. But the number of visitors in Naucratis each season is not relevant to the Nile's behavior, because no such relation is known to exist between changes in the visiting population of a city and variations in the volume of rivers. A hypothesis is believed to be relevant to a problem if it expresses determinate modes of connections between a set of facts, including the fact investigated; it is irrelevant otherwise.

No rules can be stated for "hitting upon" relevant hypotheses. A hypothesis may often be believed to be relevant which subsequent inquiry shows to be not so. Or we may believe that certain facts are irrelevant to a problem although subsequent inquiry may reveal the contrary. *In the absence of knowledge concerning a subject matter, we can make no well-founded judgments of relevance.* It follows that the valuable suggestions for solving a problem can be made only by those who are familiar with the kinds of connections which the subject matter under investigation is capable of exhibiting. Thus the explanation of the Nile's periodic overflow as due to heavy rainfall would not be very likely to occur to anyone not already familiar with the relation between rain and swollen rivers. The hypotheses which occur to an investigator are therefore a function, in part at least, of his previous knowledge.

§ 3. THE DEDUCTIVE DEVELOPMENT OF HYPOTHESES

Let us now reëxamine the procedure of Herodotus in terms of the distinctions already familiar.

The search for an explanation of the Nile's behavior was a search for a *general rule* which asserts a *universal* connection between facts of that kind and other facts of different kind. The task of Herodotus was to show that the general rule which was suggested to him in the form of a hypothesis *did truly and in fact* apply to the specific problem at hand. How did he perform it? The argument which Herodotus employed to reject the first theory may be stated as follows: The defender of the theory offers the following argument:

If the Etesian winds blow, the Nile rises (*general rule*).
The Nile rises for one hundred days beginning with the summer solstice (*observed fact*).
∴ The Etesian winds blow, beginning with the summer solstice (*inferred event*).

The inference is, of course, invalid as a conclusive proof. But its proponent may claim that the reasoning is a *presumptive probable inference,* so that the conclusion is probable on the evidence. Herodotus shows that this is not the case. He points out that we can find an occasion when the Nile rises (*observed case*) and the Etesian winds do not blow. Such a case is obviously not explained by our general rule. He therefore concludes that the hypothesis of the winds will not *always* account for the inundation of the river. But he is not content with this, for the defender of the theory may perhaps be satisfied with an explanation of the overflow which is not invariable. Herodotus showed further that the logical consequences of the Etesian wind theory were *contrary* to the known facts. In order to do this, he had therefore to point out some of the other consequences of that theory by discovering what it *implied.* His argument continues:

If the blowing of the Etesian winds produced inundations, other rivers should behave as the Nile does (*elaborated rule*).
These other rivers do not overflow their banks (*observed fact*).
∴ The blowing of the Etesian winds does not invariably produce inundations.

This inference is a valid mixed hypothetical syllogism. Herodotus has therefore shown that the Etesian-wind theory cannot be regarded as a satisfactory explanation of the problem.

In this rejection of the first theory, Herodotus was compelled to elaborate it deductively. The importance of this step can be seen even more clearly by considering his rejection of the third theory. This may be stated as follows: If there are periodic melting snows in the interior of Africa, then the Nile will inundate periodically. Herodotus rejects this explanation not because he can *actually observe* the absence of snow in Central Africa, but because he can observe what he believes to be the consequences of Central Africa's being a warm country. And since he rejects the possibility of snowfall in warm places, he also rejects the theory of melting snows as the cause of the Nile's behavior. Let us restate part of his argument:

If hot winds blow from a region, then that region itself is hot (*general rule*).
Hot winds blow from the interior of Africa (*observed fact*).
∴ The interior of Africa is hot (*inferred fact*).

If snow falls in a region, then that region cannot have a hot climate (*rule*).
The interior of Africa *is* hot (*inferred fact from the previous inference*).
∴ Snow does not fall in the interior of Africa (*inferred fact*).

From this analysis we may conclude that the deductive elaboration of a hypothesis must follow its formulation. For we can discover the full meaning of a hypothesis, whether it is relevant and whether it offers a satisfactory solution of the problem, only by discovering what it *implies*. It is worth noting that Herodotus rejected the second theory simply on the ground that it was obscurely stated, so that it was impossible to find out what it did imply.

We are therefore already in the position to appreciate how important the technique of deduction is for scientific method. In the chapter on mathematics we have seen how a complex set of assumptions may be explored for their implications. The techniques we have discussed there are relevant for the deductive elaboration of any theory. Without writing a textbook on some special science one cannot illustrate the full scope of those methods in a particular subject matter. But by attending to a few more relatively simple examples the reader can appreciate the indispensability for scientific procedure of developing a hypothesis deductively.

Galileo's study on falling bodies is one of the most far-reaching in modern times. He had shown that if we neglect the resistance of air, the velocity with which bodies fall to the ground does not depend on their weight. It was known that bodies pick up speed as they approach the ground. But it was not known what the relation is between the velocity, the space traveled, and the time required for the fall. Of what general law could the fall of a body be regarded as an instance?

Galileo considered two hypotheses. According to the first, the increase in the velocity of a freely falling body is proportional to the *space* traversed. But Galileo argued (mistakenly, as we now know) that one consequence of this assumption is that a body should travel *instantaneously* through a portion of its path. He believed this was impossible, and therefore rejected the proposed law of the increase in velocity.

Galileo next considered the hypothesis that the change in velocity of a freely falling body during an interval of time is proportional to that interval. This assumption may be expressed in modern notation as: $v = at$, where v represents the velocity, a the velocity acquired in one second, and t the number of seconds the body has fallen. It may also be expressed by saying that the acceleration of a falling body (defined as the change in velocity during any unit interval of time) is constant.

But the assumption that the acceleration is constant could not be put to the test *directly*. Galileo was compelled to strengthen his argument by *deducing other consequences* from the acceleration hypothesis, and showing that these consequences were capable of direct verification. The argument was strengthened because these consequences had not previously been known to be true. For example, he deduced from the hypothesis $v = at$, the proposition: The distances freely falling bodies traverse are proportional to the square of the time of their fall.

Instances of this rule can be established experimentally. Thus a body which falls for two seconds travels four times as far as a body which falls only one second; and a body falling three seconds travels nine times as far as a body falling one second. This, therefore, strengthens the evidence for the hypothesis that bodies fall so that their acceleration is constant.

In a similar fashion, Galileo deduced other propositions from the acceleration hypothesis, all of which he could verify with much precision. In this way the evidence for that hypothesis was increased. *But it was possible to increase it only after exploring its directly verifiable implications.*

Nevertheless, the evidence for the acceleration hypothesis always remains only *probable*. The hypothesis is only probable on the evidence because it is always logically possible to find some other hypothesis from which all the verified propositions are consequences. Nevertheless, it shows itself the best available so long as it enables us to infer and discover an ever greater variety of true propositions. A comprehensive theory is established as true with a high probability by showing that various *samplings* from its logical consequences are empirically true.

Let us now summarize the general features of Galileo's procedure. We find that he *selected* some *portion* of his experiences for study. His experiments from the Tower of Pisa resolved some of his doubts. But the resolution of these doubts only raised others. If the behavior of freely falling bodies did not depend upon their weight, upon what did it depend? The ancients, as well as his own contemporaries, had already isolated some properties of bodies as *irrelevant* to their behavior in falling. The temperature, the smell, the color, the shapes of the bodies, were tacitly assumed to be irrelevant qualities. The ancients also regarded the distance and the duration of fall as unimportant. But this assumption Galileo refused to make. And he ventured to formulate hypotheses in which these properties of bodies were the determining factors of their behavior.

This selection of the relevant factors was in part based on his previous knowledge. Galileo, like the ancients, neglected the color and smell of bodies because general experience seemed to indicate that their color or smell could vary without corresponding changes in their behavior when falling. In part, however, the selection was based on a tentative guess that properties heretofore regarded as unimportant were in fact relevant. Galileo had already made successful researches in physics, in which the quantitative relations exclusively studied by the mathematics of his day played a fundamental rôle. He was also well read in ancient philosophy, and had an unbounded confidence that the "Book of Nature" was written in geometric characters. It was not, therefore, with an *unbiased* mind, it was not with a mind empty of strong convictions and interesting suggestions, that Galileo tried to solve for himself the problems of motion. It was a conviction with him that the only relevant factors in the study of motion were velocity, time, distance, and certain constant proportions.

We may thus distinguish two sets of ideas which Galileo employed in studying the motions of bodies. One set, by far the larger, consisted of his mathematical, physical, and philosophical convictions, which determined his choice of subjects and their relevant properties. The other set consisted of the *special* hypotheses he devised for discovering the relations between the relevant factors. The first set was a relatively stable collection of beliefs and prejudices. It is very likely Galileo would have held on to these, even if neither of his two hypotheses on falling bodies had been confirmed by experiment. The second set, especially at the stage of scientific development in Galileo's time, was a more unsettled collection of suggestions and beliefs. Thus Galileo might easily have sacrificed his very simple equations between velocity, time, distance, and acceleration for somewhat more complex ones if his experiments had demanded the latter.

It is these special assumptions which become formulated consciously as hypotheses or theories. And it is to a more careful study of the conditions which such hypotheses must meet that we now turn.

§ 4. THE FORMAL CONDITIONS FOR HYPOTHESES

1. In the first place, a hypothesis must be formulated in such a manner that deductions can be made from it and that consequently a decision can be reached as to whether it does or does not explain the facts considered. This condition may be discussed from two points of view.

a. It is often the case—indeed the most valuable hypotheses of science are of this nature—that a hypothesis cannot be directly verified. We cannot establish directly by any simple observation that two bodies attract each other inversely as the square of their distances. The hypothesis must therefore be stated so that by means of the well-established techniques of logic and mathematics its implications can be clearly traced, and then subjected to experimental confirmation. Thus the hypothesis that the sun and the planet Mars attract each other proportionally to their masses, but inversely as the square of their distances, cannot be directly confirmed by observation. But one set of consequences from this hypothesis, that the orbit of Mars is an ellipse with the sun at the focus, and that therefore, given certain initial conditions, Mars should be observable at different points of the ellipse on stated occasions, is capable of being verified.

b. Unless each of the constituent terms of a hypothesis denotes a determinate experimental procedure, it is impossible to put the hypothesis to an experimental test. The hypothesis that the universe is shrinking in such a fashion that all lengths contract in the same ratio is empirically meaningless if it can have no consequences which are verifiable. In the same way the hypothesis that belief in a Providence is a stronger force making for righteous living than concern for one's fellow man can have no verifiable consequences unless we can assign an experimental process for measuring the relative strength of the "forces" involved.

2. A second, very obvious, condition which a hypothesis must satisfy is that it should provide the answer to the problem which generated the inquiry. Thus the theory that freely falling bodies fall with constant accelerations accounts for the known behavior of bodies near the surface of the earth.

Nevertheless, it would be a gross error to suppose that false hypotheses—that is, those whose logical consequences are not all in agreement with observation—are always useless. A false hypothesis may direct our attention to unsuspected facts or relations between facts, and so increase the evidence for other theories. The history of human inquiry is replete with hypotheses that have been rejected as false but which have had a useful purpose. The phlogiston theory in chemistry, the theory of caloric, or specific heat substance, the corpuscular theory of light, the one-fluid theory of electricity, the contract theory of the state, the associationist theory of psychology—these are a few examples of such useful hypotheses. A more obvious illustration is the following: The ancient Babylonians entertained many false notions about the magical properties of the number seven. Nevertheless, because of their belief that the heavenly bodies visible to the naked eye which move among the fixed stars had to be seven in number, they were led to look for and find the rarely seen planet Mercury. "Wrong hypotheses, rightly worked," the English logician De Morgan remarked, "have produced more useful results than unguided observation." [2]

3. A very important further condition must be imposed upon hypotheses. As we have seen, Galileo's theory of acceleration enabled him not only to account for what he already knew when he formulated it, but also to *predict* that observation would reveal certain propositions to be true whose truth was not known or even suspected at the time the prediction was made. He was able to show, for example, that if the acceleration of a freely falling body was constant, then the path of projectiles fired from a gun inclined to the horizon would have to be a parabola. A hypothesis becomes *verified,* but of course *not proved* beyond every doubt, through the successful predictions it makes.

Let us change the illustration to make the point clearer. Let us imagine a very large bag which contains an enormous number of slips of paper. Each piece of paper, moreover, has a numeral written upon it. Suppose now we draw from the bag without replacing it one slip of paper at a time, and record the numeral we find written on each. The first numeral we draw, we continue to imagine, is 3, the second is 9. We are now offered a fortune if we can state what the five successive numerals beginning with the hundredth drawing will be.

What reply shall we give to the implied question? We may say, perhaps, that one answer is as good as another, because we suspect that the order in which the numerals appear is completely random. We may, however, entertain the hypothesis that the numeral we obtain on one drawing is *not* unrelated to the numeral we obtain

[2] *A Budget of Paradoxes* (Open Court Edition), Vol. I, p. 87.

on another drawing. We may then look for an *order* in which the numerals appear. On the *general* hypothesis that there is such an order we may then offer a *special* hypothesis to account for the sequence of the numerals. For it is clear that we can *try* to formulate such a law of sequence, even if in fact the numerals do not appear in any determinate sequence. The supposition we may make at this time, that the numerals appear in an ordered array, need not prevent us at some subsequent time from affirming, on the basis of better evidence, that they do not.

Let us accept the general hypothesis of order. The problem then is to find the *particular* order. Now the particular law or formula that we may entertain will be largely determined by our previous knowledge and our familiarity with mathematical series. On the basis of such familiarity it may appear plausible that the numeral drawn is connected with the *number of the drawing*. Other modes of connection may, of course, be entertained; the numerals drawn may be supposed connected with the *time* at which they are drawn, for example. Let us, however, accept the suggestion that the numeral is a function of the number of the drawing. Several formulae expressing this mode of connection will occur to everyone familiar with algebra. Thus we may offer as the law of the series the formula $y_1 = 3^n$, where n is the number of the drawing and y_1 the numeral drawn. When $n = 1$, y_1 is 3; and when $n = 2$, y_1 is 9. This hypothesis, therefore, completely accounts for the known facts.

But we know several other hypotheses which will also completely account for the known facts. $y_2 = 6n - 3$; $y_3 = \frac{3}{2}(n^2 + n)$; $y_4 = 2n^2 + 1$; $y_5 = \frac{n^3}{3} + \frac{11n}{3} - 1$. are four other formulae which will do so. And it is easy to show that an endless number of different expressions can be found which will perform the same function. If we reject this multitude of hypotheses without even a cursory examination, it is because we think we have some *relevant knowledge* for considering only these five.

But are all these five formulae equally "good"? If the discovery of an order determining the numerals *already drawn* were the only condition imposed upon a hypothesis, there would indeed be no reason for preferring one formula to another. However, we desire that our laws or formulae should be truly *universal:* they must express the *invariable* relations in which the numerals stand to one another. Hence the hypothesis to be preferred is the one which can *predict* what will happen, and from which we can infer what has *already* happened, even if we did not know what has happened when the hypothesis was formulated. Accordingly, we can calculate that if any one of these five formulae is universally applicable to the series of drawings, then on the third drawing from the bag we should obtain the following numerals: 27 if the first is true; 15 if the second is true; 18 if the third is true; 19 if the fourth is true; and 19 if the fifth is true.

It is extremely important to state the hypothesis and its consequences *before* any attempt at verification. For in the first place, until the hypothesis is stated we do not know what it is we are trying to verify. And in the second place, if we deliberately choose the hypothesis so that it will in fact be confirmed by a set of instances, we have no guarantee that it will be confirmed by other instances. In such a case we have not guarded against the fallacy of selection, and the "verification" is not a test or check upon the hypothesis so chosen. The logical function of prediction is to permit a genuine verification of our hypotheses by indicating, prior to the actual process of verification, instances which may verify them.

If, therefore, in our illustration the third numeral to be drawn should happen to be 19, the first three formulae would be eliminated. The remaining two would have faced the challenge of a larger body of experience. Nevertheless, we cannot be sure that these two formulae are the *only* two which could have expressed the order of the sequence of the numerals.

It becomes evident that a function of verification is to supply satisfactory evidence for eliminating some or all of the hypotheses we are considering. We are supposing we have been left with the two formulae: y_4 and y_5. Both have been imagined to be successful in predicting the third numeral. However, what we have said of what is necessary for a hypothesis to predict successfully applies not only to the third drawing, but to all subsequent drawings. If a hypothesis expresses a universal connection it must maintain itself and not be eliminated in the face of *every possible* attempt at verification. But since, as in our illustration, it is often the case that more than one hypothesis is left in the field after a finite number of verifications, we cannot affirm one such hypothesis to the *exclusion* of the others. We can *try,* however, by repeating the process, to eliminate all the relevant alternatives to some *one* hypothesis. This is an ideal which guides our inquiry, but it can rarely, if ever, be realized. And we are fortunate indeed if the hypotheses we had initially regarded as relevant are not all eliminated in the development of the inquiry.

To say that a hypothesis must be so formulated that its material consequences can be discovered means, then, that a hypothesis must be *capable of verification*. At the time a hypothesis is developed, it may be impossible to verify it actually because of practical or technical difficulties. The logical consequences of a hypothesis may be such that much time may have to elapse between the time of drawing the inference and the time of the predicted consequence. Thus a total eclipse of the sun was required for testing one of the consequences of the theory of relativity. But while a hypothesis is frequently incapable of immediate verification, and while it can never be *demonstrated* if it asserts a truly universal connection, it must be *verifiable*. Its consequences, as we have already observed, must be stated in terms of *determinate* empirical operations.

It follows that unless a hypothesis is explicitly or implicitly *differentiating* in the order it specifies, it cannot be regarded as adequate. A hypothesis must be *capable of being refuted* if it specifies one order of connection rather than another.

Consider the proposition *All men are mortal,* which is a hypothesis to account for the behavior of men. Is this a satisfactory formulation? If we should find a man who is two hundred years old, need this instance cast any doubt on the universality of the mortality theory of men? Certainly a defender of the theory would not have to think so. But what if we found a gentleman as old as one of the Struldbrugs? The defender of the theory could still maintain that his hypothesis is perfectly compatible with such an instance. We may reflect, however, that the hypothesis is so stated that *no matter how aged a man we could produce,* the hypothesis would not be refuted. The hypothesis, to be satisfactory, must be modified so that an experimental determination is possible between it and any contrary alternative.

A hypothesis, if it has verifiable consequences, cannot pretend to explain *no matter what* may happen: the consequences which are capable of being observed if the hypothesis is true cannot all be the same as the verifiable consequences of a contrary hypothesis. In our example, the hypothesis receives the proper modification if it is stated in the form *All men die before they reach the two-hundredth anniversary of their birth*. In this form, a five-hundred-year-old gentleman would definitely refute the hypothesis.

Many theories which have a wide popular appeal fail to meet the condition we have specified. Thus the theory that whatever happens is the work of Providence, or the will of the unconscious self, is unsatisfactory from the point of view we are now considering. For that theory is *not* verified, if *after* the "happening" we can interpret the event as the work of Providence or of the unconscious self. In fact, the theory is so poorly formulated that we cannot state what its logical consequences are, and therefore what should be the nature of some future event. The theory does not enable us to predict. It is not verifiable. It does not differentiate between itself and any apparently contrary theory, such as that whatever happens is fortuitous.

4. One further condition for satisfactory hypotheses remains to be considered. In our artificial illustration we found that after the third drawing, two hypotheses still remained in the field. How are we to decide between them? The answer seems not to be difficult in this case. Since the formula y_4 will yield a different numeral for $n = 4$ than will the formula y_5, the fourth drawing will enable us to verify one of them and eliminate the other, or perhaps eliminate both. But what if we should be dealing with two hypotheses of which all the consequences we can actually verify are the same?

We must distinguish two cases in which this may happen. Suppose, as the first case, that two investigators wish to determine the nature of a closed curved line they find traced on a piece of ground. One says it is a curve such that the distances from points on it to a certain fixed point are all equal. The other says the curve is such that the area inclosed by it is the largest one that can be inclosed by a curve of that length. It can be shown, however, that all the logical consequences of the first hypothesis are the same as those of the second. Indeed, the two hypotheses are not different logically. If the two investigators should quarrel about their respective theories, they would be quarreling either about words or about their esthetic preferences for the different formulations of what is essentially the same theory.

It may happen, however, that two theories are not logically equivalent although the consequences in which they differ are incapable of being tested experimentally. Such a situation may arise when our methods of observation are not sensitive enough to distinguish between the logically distinct consequences. For example, the Newtonian theory of gravitation asserts that two bodies attract each other inversely as the "second power" of their distances; an alternative theory may assert that the attraction is inversely proportional to the 2.00000008 power of their distances. We cannot detect experimentally the difference between the two theories. What further condition must be imposed so that we mav be able to decide in such cases between rival hypotheses?

The answer we shall examine is that the *simpler* one of two hypotheses is the more satisfactory. We may cite as a familiar example the heliocentric theory formulated by Copernicus to account for the apparent motions of the sun, moon, and planets. The geocentric theory of Ptolemy had been formulated for the same purpose. Both theories enable us to account for these motions, and in the sixteenth century, apart from the question of the phases of Venus, neither theory permitted a prediction which could not be made by the other theory. Indeed, it has been shown that for many applications the two theories are mathematically equivalent. Moreover, the theory of Ptolemy had the advantage that it did not go counter to the testimony of the senses: men could "see" the sun rise in the east and sink in the west; the heliocentric view, from the point of view of "common sense," is a very sophisticated explanation. Nevertheless, Copernicus and many of his contemporaries found the heliocentric theory "simpler" than the ancient theory of Ptolemy, and therefore to be preferred. What are we to understand by this? We must try to analyze what is meant by "simplicity."

a. "Simplicity" is often confused with "familiarity." Those not trained in physics and mathematics doubtless find a geocentric theory of the heavens simpler than a heliocentric theory, since in the latter case we must revise habitual interpretations of what it is we are supposed to see with our eyes. The theory that the earth is flat is simpler than the theory that it is round, because the untutored man finds it difficult to conceive of people at the antipodes walking on the surface of the earth without falling off. But "simplicity" so understood can be no guide for choosing between rival hypotheses. A new and therefore unfamiliar hypothesis would never be chosen for its simplicity. What is simple to one person is not so to another. To say that Einstein's theory of relativity is simpler (in this sense) than Newton's physics is clearly absurd.

b. One hypothesis is said to be simpler than another if the number of independent types of elements in the first is smaller than in the second. Plane geometry may be said to be simpler than solid geometry, not merely because most people find the first easier to master than the second, but also because configurations in three independent dimensions are studied in the latter and only two are studied in the former. Plane projective geometry is simpler in this sense than plane metric geometry, because only those transformations are studied in the first which leave invariant the colinearity of points and the concurrence of lines, while in the second type of geometry there is added the study of transformations which leave invariant the congruence of segments, angles, and areas. So also theories of physics are simpler than theories of biology, and these latter simpler than the theories of the social sciences.

A theory of human behavior which postulates a single unlearned impulse, for example, sex desire, or self-preservation, is often believed to be simpler in this sense than a theory which assumes several independent unlearned impulses. But this belief is mistaken, because in theories of the first type it is necessary to introduce special assumptions or qualifications of the single impulse in order to account for the observed variety of types of human behavior. Unless, therefore, *all* the assumptions of a hypothesis are explicitly stated, together with the relations between them, it is impossible to say whether it is in fact simpler than another hypothesis.

c. We are thus led to recognize another sense of simplicity. Two hypotheses may be both capable of introducing order into a certain domain. But one theory may be able to show that various facts in the domain are related on the basis of the *systematic implications of its assumptions.* The second theory, however, may be able to formulate an order only on the basis of special assumptions formulated *ad hoc* which are unconnected in any systematic fashion. The first theory is then said to be simpler than the second. Simplicity in this sense is the *simplicity of system.* A hypothesis simple in this sense is characterized by *generality.* One theory will therefore be said to be more simple or general than another if the first can, while the second cannot, exhibit the connections it is investigating as *special instances* of the relations it takes as fundamental.

The heliocentric theory, especially as it was developed by Newton, is systematically simpler than the theory of Ptolemy. We can account for the succession of day and night and of the seasons, for solar and lunar eclipses, for the phases of the moon and of the interior planets, for the behavior of gyroscopes, for the flattening of the earth at the poles, for the precession of the equinoxes, and for many other events, in terms of the fundamental ideas of the heliocentric theory. While a Ptolemaic astronomy can also account for these things, *special* assumptions have to be made in order to explain some of them, and such assumptions are not systematically related to the type of relation taken as fundamental.

Systematic simplicity is the kind sought in the advanced stages of scientific inquiry. Unless we remember this, the changes that are taking place in science must seem to us arbitrary. For changes in theory are frequently made for the sole purpose of finding some more general theory which will explain what was heretofore explained by two different and unconnected theories. And when it is said we should prefer the simpler one of two theories, it is systematic simplicity which must be understood. As we shall see presently, it is not easy at an advanced stage in a science to find a satisfactory hypothesis in order to explain some difficulty. For not every hypothesis will do. The explanation demanded is one in terms of a theory *analogous* in certain ways to theories already recognized in other domains. Such a demand is clearly reasonable, because if it is satisfied we are one step nearer to the ideal of a coherent *system* of explanations for an extensive domain of facts. In this sense, Einstein's general theory of relativity, although its mathematics is more difficult than that of the Newtonian theory of gravitation, is simpler than the latter. Unlike the latter, it does not introduce forces *ad hoc*.

It must be said, however, that it is difficult to differentiate between the relative systematic simplicity of two theories at an advanced stage of science. Is the Schrödinger wave theory more or less simple than the Heisenberg matrix theory of the atom? Here we must allow for an incalculable esthetic element in the choice between rival theories. But while there is an element of arbitrariness in thus choosing between very general theories, the arbitrariness is limited, for the theory chosen is still subject to the other formal conditions we have examined.

§ 5. FACTS, HYPOTHESES, AND CRUCIAL EXPERIMENTS

Observation

A hypothesis, we have said, must be verifiable, and verification takes place through experiment or sense observation. Observation, however, is not so simple a matter as is sometimes believed. A study of what is involved in making observations will enable us to offer the *coup de grâce* to the utterly misleading view that knowledge can be advanced by merely collecting facts.

1. Even apparently random observation requires the use of hypothesis to interpret what it is we are sensing. We can claim, indeed, that we "see" the fixed stars, the earth eclipsing the moon, bees gathering nectar for honey, or a storm approaching. But we shall be less ready to maintain that we simply and literally *see* these things, unaided by any theory, if we remember how comparatively recent in human history are these explanations of *what* it is we see.

Unless we identify observation with an immediate, ineffable experience, we must employ hypotheses even in observation. For the objects of our seeing, hearing, and so on, acquire meaning for us only when we link up what is directly given in experience with what is not. This brilliant white spot of light against the deep-blue background—it has an incommunicable quality; but it also *means* a star many light-years away. In significant observation we *interpret* what is immediately given in sense. We *classify* objects of perception (calling this a "tree," that a "star") in virtue of noted similarities between things, similarities which are believed to be significant because of the theories we hold. Thus, a whale is classified as a mammal, and not as a fish, in spite of certain superficial resemblances between whales and fish.

2. Observation may be erroneous. The contradictory testimony of witnesses claiming to have "seen" the same occurrence is a familiar theme of applied psychology. Every day in our courts of law men swear in good faith to having seen things which on cross-examination they admit they were not in a position to observe. This is satirized in Anatole France's *Penguin Island* in the replies given by the villagers of Alca when they were asked for the color of the dragon who had brought destruction in the darkness of the night before. They answered:

"Red.

"Green.

"Blue.

"Yellow.

"His head is bright green, his wings are brilliant orange tinged with pink, his limbs are silver grey, his hind-quarters and his tail are striped with brown and pink bands, his belly bright yellow spotted with black.

"His color? He has no color.

"His is the color of a dragon." [3]

No wonder, after hearing this testimony, the Elders remained uncertain as to what should be done! But if uninterpreted sense experience were observation, how could error ever arise?

3. The hypothesis which *directs* observation also determines in large measure what factors in the subject matter are noted. For this reason, unless the conditions under which an observation is made are known the observation is very unreliable, if not worthless. Changes are most satisfactorily studied when only a single factor is

[3] Translation by A. W. Evans, Bk. II, Chap. VI.

varied at a time. Of what value, then, is an observation that a certain liquid boils at 80° C. if we do not also observe its density and the atmospheric pressure? But, clearly, only some theory will lead us to observe all the relevant factors; only a theory will indicate whether atmospheric pressure is a single factor, or whether it may be distinguished into several others, as force is into magnitude and direction.

4. All but primitive observations are carried on with the aid of specially devised instruments. The nature and limitations of such instruments must be known. Their readings must be "corrected" and interpreted in the light of a comprehensive theoretical system.

These points are made in a striking manner by the French physicist Pierre Duhem. "Enter a laboratory; approach the table crowded with an assortment of apparatus, an electric cell, copper wire covered with silk, small cups of mercury, spools of wire, an iron bar carrying a mirror; an experimenter is plugging into small openings the metal end of a pin whose head is ebony; the iron oscillates, and by means of a mirror which is attached to it, throws upon a celluloid scale a luminous band; the forward and backward motion of this luminous spot enables the physicist to observe minutely the oscillations of the iron bar. But ask him what he is doing. Will he answer, 'I am studying the oscillations of an iron bar which carries a mirror?' No, he will answer that he is measuring the electric resistance of the spools. If you are astonished, if you ask him what his words mean, what relation they have with the phenomena he has been observing and which you have noted at the same time as he, he will answer that your question requires a long explanation, and that you should take a course in electricity." [4]

Is it not imperative, therefore, that the sharp distinction frequently made between fact and hypothesis be overhauled? Facts, we have seen, are not obtained by simply using our organs of sense. What, then, are facts? Are they, as is sometimes asserted, hypotheses for which evidence is considerable? But in that case does this evidence consist only of *other* hypotheses for which the evidence is considerable, and so on *ad infinitum?*

Facts

We must, obviously, distinguish between the different senses of "fact." It denotes at least four distinct things.

1. We sometimes mean by "facts" certain discriminated elements in sense perception. That which is denoted by the expressions

[4] *La théorie physique,* p. 218.

"This band of color lies between those two bands," "The end of this pointer coincides with that mark on the scale," are facts in this sense. But we must note that no inquiry *begins* with facts so defined. Such sensory elements are *analytically sought out by us,* for the purpose of finding reliable signs which will enable us to test the inferences we make. All observation appeals ultimately to certain *isolable* elements in sense experience. We search for such elements because concerning them universal agreement among all people is obtainable.

2. "Fact" sometimes denotes the propositions which *interpret* what is given to us in sense experience. *This is a mirror, That sound is the dinner bell, This piece of gold is malleable,* are facts in this sense. All inquiry must take for granted a host of propositions of this sort, although we may be led to reject some of them as false as the inquiry progresses.

3. "Fact" also denotes propositions which truly assert an invariable sequence or conjunction of characters. *All gold is malleable, Water solidifies at zero degree Centigrade, Opium is a soporific,* are facts in this sense, while, *Woman is fickle,* is not a fact, or at least is a disputed fact. What is *believed* to be a fact in this (or even in the second) sense depends clearly upon the evidence we have been able to accumulate; ultimately, upon facts in the first sense noted, together with certain assumed universal connections between them. Hence, whether a proposition shall be called a fact or a hypothesis depends upon the state of our evidence. The proposition *The earth is round* at one time had no known evidence in its favor; later, it was employed as a hypothesis to *order* a host of directly observable events; it is now regarded as a fact because to doubt it would be to throw into confusion other portions of our knowledge.

4. Finally, "fact" denotes those things existing in space or time, together with the relations between them, in virtue of which a proposition is true. Facts in this sense are neither true nor false, they simply *are:* they can be apprehended by us in part through the senses; they may have a career in time, may push each other, destroy each other, grow, disappear; or they may be untouched by change. Facts in this fourth sense are distinct from the hypotheses we make about them. A hypothesis is true, and is a fact in the second or third sense, when it does state *what* the fact in this fourth sense is.

Consequently, the distinction between fact and hypothesis is never sharp when by "fact" is understood a proposition which may indeed be true, but for which the evidence can never be complete. It is the function of a hypothesis to reach the facts in our fourth sense. However, at any stage of our knowledge this function is only partially fulfilled. Nevertheless, as Joseph Priestley remarked: "Very lame and imperfect theories are sufficient to suggest useful experiments which serve to correct those theories, and give birth to others more perfect. These, then, occasion farther experiments, which bring us still nearer to the truth; and in this method of *approximation,* we must be content to proceed, and we ought to think ourselves happy, if, in this slow method, we make any real progress." [5]

Crucial Experiments

In the light of these remarks on the distinction between fact and hypothesis, we must reconsider, and qualify, our previous discussion of the verification of hypotheses. It is a common belief that a *single crucial experiment* may often decide between two rival theories. For if one theory implies an experimentally certifiable proposition which contradicts a proposition implied by a second theory, by carrying out the experiment, the argument runs, we can definitely eliminate one of the theories.

Consider two hypotheses: H_1, the hypothesis that light consists of very small particles traveling with enormous speeds, and H_2, the hypothesis that light is a form of wave motion. Both hypotheses explain a class of events *E,* for example, the rectilinear propagation of light, the reflection of light, the refraction of light. But H_1 implies the proposition p_1 that the velocity of light in water is *greater* than in air; while H_2 implies the proposition p_2 that the velocity of light in water is *less* than in air. Now p_1 and p_2 cannot both be true. Here, apparently, is an ideal case for performing a crucial experiment. If p_2 should be confirmed by experiment, p_1 would be refuted, and we could then argue, *and argue validly,* that the hypothesis H_1 cannot be true. By 1850 experimental technique in physical optics had become very refined, and Foucault was able to show that light travels faster in air than in water. According to the doctrine of crucial experiments, the corpuscular hypothesis of light should have been banished to limbo once for all.

Unfortunately, matters are not so simple: contemporary physics has revived Newton's corpuscular hypothesis in order to explain certain optical effects. How can this be? What is wrong with the apparently impeccable logic of the doctrine of crucial experiments?

[5] *The History . . . of Discoveries relating to Vision, Light, and Colours,* 1772, p. 181.

The answer is simple, but calls our attention once more to the intimate way in which observation and theory are interrelated. In order to deduce the proposition p_1 from H_1, and in order that we may be able to perform the experiment of Foucault, many *other* assumptions, K, must be made about the nature of light and the instruments we employ in measuring its velocity. Consequently, it is not the hypothesis H_1 alone which is being put to the test by the experiment—it is H_1 and K. The logic of the crucial experiment therefore is as follows: If H_1 and K, then p_1; but p_1 is false; therefore either H_1 is false or K (in part or completely) is false. Now if we have good grounds for believing that K is not false, H_1 is refuted by the experiment. Nevertheless the experiment really tests both H_1 *and* K. If in the interest of the coherence of our knowledge it is found necessary to revise the assumptions contained in K, the crucial experiment must be reinterpreted, and it need not then decide against H_1.

Every experiment, therefore, tests not an isolated hypothesis, but the *whole body* of relevant knowledge logically involved. If the experiment is claimed to refute an isolated hypothesis, this is because the rest of the assumptions we have made are believed to be well founded. But this belief may be mistaken.

This point is important enough to deserve another illustration. Let us suppose we wish to discover whether our "space" is Euclidean, that is, whether the angle sum of a physical triangle is equal to two right angles. We select as vertices of such a triangle three fixed stars, and as its sides the paths of rays traveling from vertex to vertex. By making a series of measurements we can *calculate* the magnitude of the angles of this triangle and so obtain the angle sum. Suppose the sum is less than two right angles. *Must* we conclude that Euclidean geometry is not true? Not at all! There are at least three alternatives open to us:

1. We may explain the discrepancy between the theoretical and "observed" values of the angle sum on the hypothesis of errors in measurement.

2. We may conclude that Euclidean geometry is not physically true.

3. We may conclude that the "lines" joining the vertices of the triangle with each other and with our measuring instruments are not "really" straight lines; that is, Euclidean geometry is physically true, but light does not travel in Euclidean straight lines in stellar space.

If we accept the second alternative, we do so on the assumption that light is propagated rectilinearly, an assumption which, although supported by much evidence, is nevertheless not indubitable. If we accept the third alternative, it may be because we have some independent evidence for denying the rectilinear propagation of light; or it may be because a greater coherence or system is introduced into the body of our physical knowledge as a consequence of this denial.

"Crucial experiments," we must conclude, are crucial against a hypothesis only if there is a relatively stable set of assumptions which we do not wish to abandon. But no guarantees can be given, for reasons we have stated, that some portion of such assumptions will never be surrendered.

§ 6. THE RÔLE OF ANALOGY IN THE FORMATION OF HYPOTHESES

The reader of this chapter, noticing that it is nearing its end, may perhaps finally lose his patience. "You have told me what a hypothesis means, how central a position it occupies in all inquiry, and what the requirements for a hypothesis are. For all this I thank you. But why don't you tell me how I am to discover a satisfactory hypothesis—what rules I should follow?"

In a succeeding chapter we shall consider several suggested rules for making discoveries. Meanwhile, we must perhaps try the reader's patience still further, first, by quoting what a great wit replied to a similar question, and second, by considering critically a piece of advice that is sometimes given as an aid in discovering hypotheses. The wit is De Morgan. "A hypothesis must have been started," he wrote, "not by rule, but by that sagacity of which no description can be given, precisely because the very owners of it do not act under laws perceptible to themselves. The inventor of hypothesis, if pressed to explain his method, must answer as did Zerah Colburn [a Vermont calculating boy of the early nineteen-hundreds] when asked for his mode of instantaneous calculation. When the poor boy had been bothered for some time in this manner, he cried out in a huff, 'God put it into my head, and I can't put it into yours.' "[6]

The advice is that analogies or resemblances should be noted between the facts we are trying to explain and other facts whose explanation we already know. But which analogies, we are tempted

[6] *A Budget of Paradoxes*, Vol. I, p. 86.

to ask? We can always find *some* resemblances, although not all of them are significant. What we have already said about relevance is applicable here. Nevertheless, it is true enough that if previously established knowledge can be used in new settings, analogies must be noted and exploited.

It is a mistake, however, to suppose that we always explicitly notice precise analogies and then rationally develop their consequences. We generally begin with an unanalyzed feeling of vague resemblance, which is discovered to involve an explicit analogy in structure or function *only by a careful inquiry*. We do not *start* by noting the structural identity in the bend of a human arm and the bend of a pipe, and then go on to characterize the latter as an "elbow." Nor do we notice first the slant of the eyes and thinness of the lips of Orientals, and then conclude that they look alike. Usually it is rather the other way.

Moreover, considerations of analogy are not always on hand when we wish to formulate a satisfactory hypothesis. For though a hypothesis is generally satisfactory only when it does have certain *structural analogies* to other well-established theories, it is not easy to formulate hypotheses which meet this condition. When we study the behavior of gases, we wish to find a theory analogous to those *already* established to account for the behavior of matter in motion. This is not an easy task, as the history of the kinetic theory of gases shows. The analogy of a hypothesis to others is therefore a *condition we impose* upon it, in the interest of the systematic simplicity of all our knowledge, before such analogy can aid in any discovery. And when we succeed in formulating a hypothesis analogous to others, this is an *achievement,* and the starting-point of further inquiry.

CHAPTER XII

CLASSIFICATION AND DEFINITION

§ 1. THE SIGNIFICANCE OF CLASSIFICATION

We have been calling the reader's attention to the fact that the process of classifying things really involves, or is a part of, the formation of hypotheses as to the nature of things. It is well to consider this in some detail.

There is a general feeling, shared by many philosophers, that things belong to "natural" classes, that it is by the nature of things that fishes, for instance, belong to the class of vertebrates, just as vertebrates "naturally" belong to the class of animals. Those who hold this view sometimes regard other classifications as "artificial." Thus a division of animals into those that live in the air, on land, and in water would be regarded as artificial. This distinction involves a truth which is confusedly apprehended. Strictly speaking, the last division, or any division of animals according to some actual trait arbitrarily chosen, is perfectly natural. For in every classification, we pick out some one trait which all the members of the class in fact possess, and therefore we may call it natural. All classification, however, may also be said to be artificial, in the sense that we select the traits upon the basis of which the classification is performed. For this reason controversies as to what is the proper classification of the various sciences are interminable, since the various sciences may be classified in different ways, according to the objectives of such classification.

Various classifications, however, may differ greatly in their logical or scientific utility, in the sense that the various traits selected as a basis of classification differ widely in their fruitfulness as principles of organizing our knowledge. Thus the old classification of living things into animals that live on land, birds that live in the air, and fish that live in water gives us very little basis for systematizing all that we know and can find out about these creatures. The habits and the structure of the porpoise or the whale have many more significant features in common with the hippopotamus or the horse than with the mackerel or the pickerel. The fact that the first two animals named have mammary glands and suckle their young, while all species of fish deposit their eggs to be fertilized, makes a difference which is fundamental for the understanding of the whole life cycle. In the same way, the fact that some animals have a vertebral column, or, to be more exact, a central nervous cord, is the key which enables us to see the significance of the various structures and enables us to understand the plan of their organization and functioning. Some traits, then, have a higher logical value than others in enabling us to attain systematic knowledge or science.

When, therefore, it is said that the business of science is first to gather the facts and then to classify them, we do not have a clear or adequate account of the situation. Some classification is involved in determining what facts we should gather; but this is not all. The most important thing is to pick out that trait in the objects studied which will be the most significant clue to their nature.

Obviously, there can be no *a priori* rules as to how we may hit upon such significant traits. Generally it depends upon genius, except that, other things being equal, we can say that he who has more knowledge is more likely to reject irrelevant or insignificant traits.

Formal logic, however, may aid us by defining the objects or traits considered so that our reasonings about them may be accurate, and may permit of being put into systematic deductive form.

§ 2. THE PURPOSE AND THE NATURE OF DEFINITION

The language of everyday conversation is notoriously vague, and the language of even technical treatises is not always very much better. Everyone is familiar with the difficulty of deciding whether certain micro-organisms are "plants" or "animals," whether certain books are or are not "obscene," whether a certain symphony is or is not the work of a "genius," whether a given society is or not a "democracy," whether we do or do not have certain "rights." Such words are vague, because their denotation shades off imperceptibly into the denotation of other words. Many of the fatuities of actual thinking take place because the inescapable vagueness of most words makes a careful check upon one's thoughts well-nigh impossible. The vagueness of ordinary words is one of the principal reasons why technical vocabularies must be constructed in the special sciences.

To the vagueness of words their ambiguity must be added as a serious danger to accurate thinking. Serious blunders in reflective thinking occur because the meaning that a word has in some context is replaced, without the fact being noticed, by an allied but different meaning. A famous instance of how the ambiguity of words may invalidate a reasoned discourse, is found in Mill's *Utilitarianism.* Mill is trying to prove "that happiness is desirable, and the only thing desirable, as an end." He argues as follows: "What ought to be required of this doctrine—what conditions is it requisite that the doctrine should fulfil—to make good its claim to be believed? The only proof capable of being given that an object is visible, is that people actually see it. The only proof that a sound is audible, is that people hear it: and so of the other sources of our experience. In like manner, I apprehend, the sole evidence it is possible to produce that anything is desirable, is that people do actually desire it." [1] Now to say that a thing is "desirable" may mean either that it should be the object of desire, or that it is in fact the object of desire. These two meanings are different. But in order that Mill may prove his thesis that happiness is the only end, "desirable" must be taken in the first sense; all his argument shows, however, is that happiness is "desirable" in the second sense.

Ambiguity arising from the grammatical structure of sentences, rather than from the ambiguity of its constituent words, was a common feature of the deliverances of the ancient oracles. Thus a celebrated response of an oracle was, "Pyrrhus the Romans shall, I say, subdue."

Much of the best effort of human thought must go, therefore, to delimit the vagueness of words and eliminate their ambiguity. Vagueness can be reduced, but never completely eliminated. Ambiguity also can with care be successfully overcome. Thus the specific meaning of an ambiguous word may be determined from the context in which it is found on a specific occasion. For example, as we have noted before, when Christ declares, "Blessed are they that mourn: for they shall be comforted," it is clear from the context that the "mourners" meant are those who "hunger and thirst after righteousness."

But such a method of clarifying the meaning of a word is not always possible or even desirable. A much more deliberately devised

[1] Chap. IV.

process must be employed, and a standard or formal rule for defining symbols must be adopted. Let us examine it.

The reader is no doubt familiar with the famous scene in Molière's *Le Bourgeois Gentilhomme* between Monsieur Jourdain and the Teacher of Philosophy. We reproduce it somewhat abridged:

Teach. What do you wish to learn?

M. Jour. Everything I can, for I am intensely anxious to be learned; it troubles me that my father and mother did not see to it that I was thoroughly grounded in all knowledge when I was young.

Teach. An admirable sentiment: *Nam sine doctrina vita est quasi mortis imago.* Doubtless you know Latin and understand that?

M. Jour. Yes, but proceed as though I did not know it: explain to me what it means.

Teach. It means, "Without knowledge, life is little more than the reflection of death."

M. Jour. That Latin is right. . . . I must tell you something. I am in love with a person of high estate, and I would like you to help me to write something to her in a billet-doux, which I propose to let fall at her feet.

Teach. Very good.

M. Jour. Something very gallant.

Teach. Certainly. Do you wish to write in verse?

M. Jour. No, no, no verses.

Teach. You only want prose?

M. Jour. No. I do not want either prose or verse.

Teach. It must really be either one or the other.

M. Jour. Why?

Teach. Because, monsieur, one can only express oneself in prose or verse.

M. Jour. Is there nothing but prose or verse?

Teach. No, monsieur: all that is not prose is verse; and all that is not verse is prose.

M. Jour. And what is it when one speaks?

Teach. Prose.

M. Jour. What? when I say, "Nicole, bring me my slippers, and give me my nightcap," is that prose?

Teach. Yes, monsieur.

M. Jour. Upon my word! I have spoken prose for more than forty years without knowing anything about it; I am infinitely obliged to you for having taught me this.[2]

We shall compare the above "lesson" with the following (also abridged) scene from Plato's dialogue *Euthyphro*. Socrates meets Euthyphro, who is on his way to the Athenian court in order to accuse his father of murder. Socrates is surprised, and asks Euthyphro whether it is pious to behave thus to one's father. Euthyphro thereupon claims adequate knowledge about the nature of piety.

"*Soc*. . . . What is piety, and what is impiety?

"*Euth*. Piety is doing as I am doing; that is to say, prosecuting anyone who is guilty of murder, sacrilege, or of any other similar crime and not to prosecute them is impiety.

"*Soc*. But . . . I would rather hear from you a more precise answer, which you have not as yet given, my friend, to the question, What is 'piety'? When asked, you only replied, Doing as you, charging your father with murder.

"*Euth*. And what I said was true, Socrates.

"*Soc*. No doubt, Euthyphro; but you would admit that there are many other pious acts?

"*Euth*. There are.

"*Soc*. Remember that I did not ask you to give me two or three examples of piety, but to explain the general idea which makes all pious things to be pious. Do you not recollect that there was one idea which made the impious impious, and the pious pious? Tell me what is the nature of this idea, and then I shall have a standard to which I may look.

"*Euth*. I will tell you, if you like.

"*Soc*. I should very much like.

"*Euth*. Piety, then, is that which is dear to the gods, and impiety is that which is not dear to them.

"*Soc*. Very good, Euthyphro; you have now given me just the sort of answer which I wanted. But whether what you say is true or not I cannot as yet tell, although I make no doubt that you will prove the truth of your words."[3]

Nominal Definition

We have before us now several attempts at the definition of verbal symbols. There are important differences between some of them, which we must note. To M. Jourdain, who knew no Latin,

[2] Act II, Scene VI.

[3] Plato, *op. cit.*, Vol. II, pp. 79-81.

the explanation of the Latin sentence consisted in a *translation.* He was informed of the meaning of a set of symbols with which he had previously been totally unfamiliar by being told that they were equivalent to a set of symbols with which he had been familiar. Ordinarily, we regard translations as true or false. Thus if the words *"sine pecunia"* were used for "without knowledge" those who know Latin would call it a false translation. If, however, there were no reference to the fact that the new words were part of the language historically called Latin, the question of truth or falsity would not be involved. There would simply be a substitution of a new set of words or symbols for old familiar ones, as is the case in the creation of cryptograms, private codes, and artificial languages, as well as in the invention of technical terms in the various sciences. Thus the word "sociology" was invented by Auguste Comte as a name for the study of human relations in organized group life, and other writers have chosen to follow him. But the word might have been introduced to denote the study of legal or business partnerships, the phenomena of clubbing together, or the way things in general are associated together. That, unlike many other proposed new terms, this one has been generally adopted, and that its denotation has been confined to human relations but not restricted to any special form of them, are results of choice, to which we may agree or not as we please without thereby asserting anything true or false. This is also the case when in mathematics we introduce symbols like + for "plus" after the latter had become used as equivalent to "added to." Careful writers since Aristotle have been aware of this and have often used the imperative form to define a new word, for example, Let the process of grasping meanings be called "apperception."

A *nominal definition,* then, is an agreement or resolution concerning the use of verbal symbols. A new symbol called the *definiendum* is to be used for an already known group of words or symbols (the *definiens*). The definiendum is thus to have no meaning other than the definiens. In the *Principia Mathematica* by Whitehead and Russell a definition of this type is written by putting the definiendum to the left and the definiens to the right with the sign of equality between them and the letters "Df." to the right of the definiens. Thus *implication,* symbolized by $\supset$, is defined thus: $p \supset q = p' \vee q$. Df. Or, in words, "p implies q" is equivalent by definition to "not p or q." In algebra the same procedure could be followed. Exponents could be introduced as follows: $a^2 = a \;.\; a$. Df.

A nominal definition, then, is a resolution and not anything true or false—though of course the assertion that anyone has or has not consistently lived up to his resolution may be true or false. And since that which is neither true nor false cannot be a proposition, nominal definitions cannot be real premises of any argument. There are no implications of truth or falsity in words themselves.

But while nominal definitions do not extend our real knowledge, they aid in scientific inquiry in the following ways:

1. In the first place, we economize space, time, and attention or mental energy if we use a new and simple symbol for a group of old familiar ones. Thus if we continued to use ordinary words and did not introduce such technical terms of higher mathematics and physics as "differential coefficient," "energy," "entropy," and the like, our expressions would become so long and involved that we could not readily grasp the complex relations indicated by these terms. Thus, it is easier to read Newton's *Principia* translated into the technical language of the modern calculus than in the more familiar language of geometry in which Newton wrote.

2. The translation of the familiar into unfamiliar terms tends to clarify our ideas by depriving our symbols of accidental or irrelevant associations. Familiar or ordinary words have strong emotional associations and carry penumbras of suggested meanings which obstruct the process of rigorous deduction.

Definition by Denotation

Another way in which the meaning of words is clarified is by exhibiting a part of their *denotation*. Thus the word "prose" was explained to M. Jourdain by giving him examples to which it can be correctly applied. For psychological reasons this method may have something to recommend it. Such a method, however, does not yield a "definition" in any usual sense of that word. We may understand what a word means when we know what it symbolizes, that is, to what it may be applied; but we do not thereby define its meaning.

Euthyphro's attempt to define "piety" in this way was naturally unsatisfactory to Socrates. That which is offered as an instance of piety may also be an instance of something else. Unless we have some sense of the connotation of the term, how can we be at all sure that we can recognize in the example what it is an instance of? It is partly for this reason that Socrates rejected Euthyphro's first attempt.

Real Definitions

Euthyphro grasped the nature of a satisfactory definition in his second trial. We must examine his attempted definition of "piety," for it introduces us to *real,* as contrasted with *verbal definition.*

Both Socrates and his friend knew, in a rough way, what "piety" was. They understood, that is, to what sort of acts the term could be applied correctly. But in seeking for a definition of "piety," Socrates was searching for an *analysis* of that which the term represented. Consequently, he was pleased with the *sort* of answer Euthyphro gave, although, as the dialogue shows, he reiected it as false. Euthyphro's definition may be put in the form:

Piety. = . that which is dear to the gods. Df.

Like a nominal definition, this *real* definition defines the *word* "piety" by means of an equivalent group of words. But, and this is the important point, the definiens is an *analysis* of the idea, form, type, or universal *symbolized* by "piety." Both the definiens and the definiendum refer to the same thing or character. They each possess a meaning independently of the process of definition which equates them. The definiens, however, indicates the *structure* of that to which both refer.

A *real definition,* therefore, is a genuine proposition, which may be either true or false. Since the definiendum and the definiens must symbolize the same universal, and since the definiens must express the structure of that universal, a real definition can be true only if the two sides of the definition are equivalent in meaning and the right-hand side represents a correct analysis of it.

We may give another illustration of real definitions. Everyone may be supposed to be familiar with the meaning of "similar figures." Such figures resemble one another in a way most people untutored in geometry would find it hard to state, but which they can identify in a crude way. The following is a real definition of *similarity:*

Figure A is similar to figure A'. = .The ratio of the distance between any two points P, Q, on A and the distance between the corresponding points P', Q', on A', is constant. *Df.*

This is a true definition of what is ordinarily meant by simliar figures, because the right-hand side means precisely what the left-hand side does, and at the same time the right-hand side offers an analysis of the structure of that which both sides symbolize.

We may now survey some of the purposes of definitions.

Psychological Motives for Definitions

There is, in the first place, the desire to learn the meaning of new words. This may be satisfied by expressing that meaning in more familiar words. In the second place, there is a desire to find a conveniently short expression for one that is long and cumbersome. Thus instead of using the phrase "the son of my mother's sister" we introduce the shorter one "my cousin." In the third place, we wish to make the meaning of a word better known to us by resolving that meaning into its constituent elements. This requires a real definition. All these motives are *psychological.*

The reader may have noted that the definiens is generally a longer expression than the definiendum, not only in nominal definitions, where it is to be expected, but in real definitions also. This fact is intimately connected with the psychological purpose of definitions. Since the definiens contains a larger number of symbols than the definiendum, it brings to the mind a larger number of ideas also. These ideas, however, are structurally related, so that they limit one another and at the same time are equivalent as a whole to the meaning of the definiendum. Thus in the definition of similarity above, the right-hand side contains the symbols "ratio," "distance," "corresponding points," "constant"; the notions which they represent are familiar, and they are so organized that the fringe of vagueness each one may have does not affect the sense of the complex whole.

This same psychological phenomenon is more clearly observable if, as is sometimes done, the meaning of a word is clarified by means of a series of synonyms. Thus "to be honest" means "to be candid, equitable, frank, genuine, ingenuous, straightforward, trustworthy, upright." No one of the so-called synonyms has precisely the same meaning as "honest." But the intensions of the synonyms overlap, so that they mutually delimit each other. The part of the intensions which is common to all may then convey, more or less precisely, the meaning of the word required.

Logical Purpose of Definitions

But these psychological reasons for making definitions must not be confused with the *logical* function that definitions have. Logically, definitions aim to lay bare the principal features or structure of a concept, partly in order to make it definite, to delimit it from other concepts, and partly in order to make possible a systematic exploration of the subject matter with which it deals. A real definition may always serve as the premise, or part of the premise, of a logical inquiry concerning a subject matter. Thus from the definition of similar figures, together with other premises, we can deduce the theorem that the volumes of any two similar figures are to each other as the cubes of any two corresponding distances. Aristotle saw this clearly when he declared that "the basic premises of demonstrations are definitions."[4]

Unfortunately the terminology concerning these matters has undergone much change, so that any attempt to bring together traditional and modern opinions must seem confusing. In the technique of modern mathematics, as we have already seen, all real definitions are implicit. No explicit definitions except nominal ones are required. However, what Aristotle called "undemonstrable definitions" which reveal the essence of a subject matter, appear in modern logical techniques as axioms or primitive propositions. Such axioms define the subject matter *implicitly,* as one which satisfies or verifies the axioms. For example, the nature of electricity is defined by Maxwell's equations, the nature of gravitation by Newton's laws. It is perhaps unnecessary to remind the reader, however, that while in a given system the real definitions or axioms may be logically prior to all the theorems, these axioms are not first in the order of the development of our knowledge, nor are they more evident or certain than any of the theorems which they imply.

We have drawn a sharp distinction between verbal and real definitions. In practice, however, the distinction is never so sharp, and even in definitions which seem altogether verbal there is generally some reference to the analysis of what the words stand for. Words are so fundamentally symbolic that it would be strange if it were otherwise. Moreover, the emotional associations and overtones of words may often prevent a clear apprehension of the issues at stake. This is particularly true in the social sciences. Words like "democracy," "liberty," "duty," have a powerful emotive function; they are frequently used as battle cries, as appeals to emotions, and as substitutes for thought. Many of the disputes about the true nature of property, of religion, of law, which undoubtedly arise from a conflict of emotional attitudes, would assuredly disappear if the precisely defined equivalents were substituted for these words.

[4] *Posterior Analytics,* in *Works,* ed. by W. D. Ross, Vol. I, 1928, p. 90b.

However, issues other than emotional ones may also be involved. Religion, for example, has sometimes been defined in terms of some dogma, sometimes in terms of a social organization and ritual, and sometimes in terms of emotional experiences. The resulting conflicts over the meaning or essence of religion have been regarded, perhaps not without some justice, as conflicts over words. But this is only a half-truth. For the disputants frequently have their eye on a concrete phenomenon which presents all these aspects. The quarrels over the right definition of religion are attempts to locate the fundamental features of a social phenomenon. For if those features are taken as the definition of religion, it is possible to deduce many important consequences from it. Thus if belief in some doctrine is the essence of religion, other things follow than if some type of emotional experience is taken as defining religion: in the one case there is an emphasis upon intellectual discipline and conformity, in the other, an emphasis upon esthetic elements and a neglect of theology.

The age-long dispute about the nature of law involves similar issues. Is "law" to be construed as a command, as a principle certified by reason, or as an agreement? The controversy is not simply about words. It is concerned with making one rather than another aspect of law central, so that the appropriate consequences may be drawn from it. A schoolroom illustration is the question, "Is a bat a bird?" The two parties to the dispute concerning the answer may agree that a bird is a warm-blooded vertebrate having its fore limbs modified as wings, and yet not agree as to whether a bat is a bird. Why? Because one party to the dispute may believe there is a closer affinity of the bat to rodents than to birds, and may wish to regard those common features of rodents as central in the bat.

We may now summarize our discussion of real definitions. A real definition involves two sets of expressions, each with a meaning of its own and these meanings are equivalent if the definition is true. In a true definition, the definiens may be substituted for the definiendum without any alteration of sense. The definiens must be easier to understand even though a longer or more complicated expression than the definiendum if the psychological as well as the logical function of the definition is to be fulfilled.

Are there any general rules which are of help in the formulation of definitions? We shall reserve the reply until after we have examined the traditional discussion of definition.

§ 3. THE PREDICABLES

Aristotle's discussion of definition is central to his entire theory of science, and is itself based upon his analysis of the possible ways in which a predicate may be related to the subject. His inquiry grew out of his reflections upon the method and results of the speculations of Socrates and Plato. His writings upon the syllogism cannot really be understood without reference to his analysis of the possible kinds of propositions, each kind depending upon the nature of the relation between subject and predicate. This analysis, called the *theory of the predicables,* was in turn closely connected with fundamental metaphysical doctrines, especially with the doctrine of fixed natural kinds or types. Into these important matters we cannot go except for a brief discussion of the predicables.

Aristotle obtains an exhaustive enumeration of the possible relations between predicate and subject, in the following way: Every predicate must be either convertible with its subject or not; that is, if *A* is *B,* then either *B* is so related to *A* that if anything is *B* it is *A,* or this is not the case. If it is convertible (Aristotle also calls it commensurable), it either signifies its essence, in which case it is the *definition;* or it is a *property.* If the predicate is not convertible with the subject, either it is contained in the definition of the subject, in which case it is the *genus* or the *differentia,* or it is not contained in the definition, in which case it is an *accident.* A predicate must therefore stand to the subject in some one of these five possible relations: it must be either definition, property, genus, differentia, or accident. We must now explain the significance of each of these distinctions. But the reader must understand that the subject term is taken by Aristotle to represent a form, type, or universal, and not a singular, concrete thing. The predicables indicate the possible ways in which universals are related to one another. The concrete individual as such is not a subject matter for science, according to Aristotle; only in so far as the individual embodies a type or form is a science of individuals possible. It is never of Socrates as an individual, but only of Socrates as "man" that we may have scientific (or systematic) knowledge. Aristotle's discussion of the predicables, therefore, stressed the intensional aspect of terms. But it is possible to give an extensional interpretation of them also, and traditionally this has usually been done.

Definition

"A 'definition,'" according to Aristotle, "is a phrase signifying a thing's essence." [5] By the essence of a thing he understood the set of fundamental attributes which are the necessary and sufficient conditions for any concrete thing to be a thing of that type. It approximates to what we have called the conventional intension of a term. Thus the essence or definition of a circle is that it is a plane figure every point of which is equidistant from a fixed point. The predicate (a plane figure every point of which is equidistant from a fixed point) is convertible or commensurate with the subject: it may be predicated of everything that is a circle, and everything to which it can be applied is a circle. The predicate is the essence, because it tells *what* a circle is, so that all the "peculiarities" of the circle necessarily follow from it.

Genus

The definition contains two terms as components, the *genus* and the *differentia*. "A 'genus' is what is predicated in the category of essence of a number of things exhibiting differences in kind." [6] Thus the genus of "circle" is "plane figure." The circle, on the other hand, is a *species* of plane figure. But "plane figure" is also the genus of "triangle," "ellipse," "hyperbola," and so on. These different species exhibit differences in kind, but they all belong to the same genus.

Differentia

The *differentia* is that part of the essence which distinguishes the species from the other species in the same genus. The differentia of "circle" is "having all its points equidistant from a fixed point"; the differentia of "triangle" is "being bounded by three straight lines."

The distinction between genus and differentia was absolute for Aristotle, and was connected with his metaphysical views.[7] But from a purely logical or formal point of view, the distinction is absolute only within a specific context. For consider the definition, "Man is a rational animal." According to Aristotle, the genus is "animal," the differentia is "rational." But formally we may regard with equal right "rational" as the genus and "animal" as the differentia. This

[5] *Op. cit.*, p. 101b.

[6] *Ibid.*, p. 102a.

[7] *Cf. ibid.*, 122b.

will be clear if we express the definition explicitly as a logical conjunction of two attributes. Thus, X is a man: = : X is rational and X is an animal. It doesn't make any logical difference which conjunctive is regarded as the more important. The logical function of the differentia is to limit or qualify the genus. And this function is performed by either term in the definition with respect to the other. A definition may, therefore, be regarded as *the logical product of two terms.* This interpretation is particularly adapted to an extensional emphasis upon the predicables.

The relation of a genus to its species is clearly illustrated by the device known as the Tree of Porphyry. The following is the traditional illustration, and has evoked from Bentham the characterization of "the matchless beauty of the Tree of Porphyry."

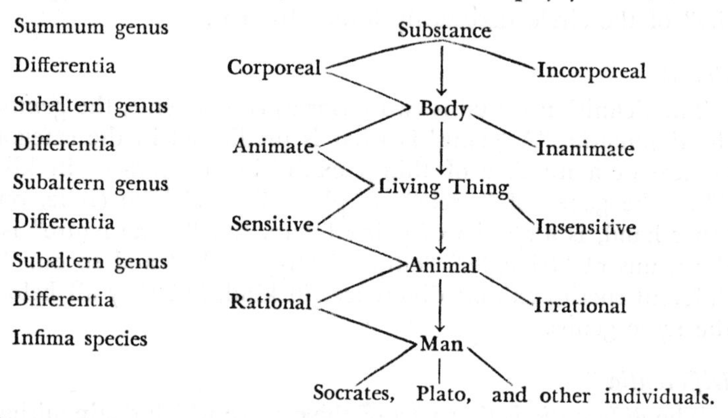

The reader will note, however, that the relation between the genus "animal," say, to its species "man," is different from the relation of the species man to its individual members. The first is a relation between a *class* and its *subclass,* the second a relation between a *class* and its *members.* Porphyry, who considerably modified Aristotle's theory of the predicables, also confused it irreparably.

Property

"A 'property' is a predicate which does not indicate the essence of a thing, but yet belongs to that thing alone, and is predicated convertibly of it. Thus it is a property of man to be capable of learning grammar: for if A be a man, then he is capable of learning grammar, and if he be capable of learning grammar, he is a man." [8] Thus, a property of a circle is that it has the maximum area with a given perimeter; another property is that the product of the segments of the chords passing through a fixed point is constant. The property is an attribute which follows necessarily from the definition.

The distinction between essence and property was regarded by Aristotle as absolute, for a subject has, according to him, only one essence. From a purely logical point of view, however, the distinction is absolute only relatively to a given system. Thus if we define a circle as the locus of points equidistant from a fixed point, we can formally deduce the property that its area is maximum with a given perimeter. On the other hand, if the circle is defined as the plane figure having a maximum area with a given circumference, it follows necessarily that all its points are equidistant from a fixed point. The rôles of definition and property are therefore interchangeable. Which character of a subject is taken as the definition turns upon extralogical considerations. Hence, while the distinction between essence and property is perfectly sound, it is absolute only within a given system. We have already seen, in connection with the discussion of the nature of mathematics, that there are no intrinsically undemonstrable propositions or intrinsically undefinable terms. The points we made there are relevant here. We have also suggested above that the "undemonstrable definitions" of Aristotle are the axioms of modern mathematical technique. The reader will therefore have no difficulty in interpreting the "properties" which flow from the definition as none other than the theorems of a system which are implied by the axioms. Unfortunately, in the example above we have quoted from him, Aristotle does not show how the property of being capable of learning grammar follows from the definition of man.

Accident

Finally, "an 'accident' is (1) something which, though it is none of the foregoing,–i.e., neither a definition nor a property nor a genus–yet belongs to the thing: (2) something which may possibly either belong or not belong to any one and the self-same thing, as (e.g.) the 'sitting posture' may belong or not belong to the self-same thing." [9] To have a triangle inscribed in it is, therefore, an accident of the circle. From a purely logical point of view, an accident is a proposition not formally derivable from the definition. So stated,

[8] *Cf. ibid.*, 102ª.

[9] *Ibid.*, 102ᵇ.

it is perhaps unnecessary to warn the reader once more that an accidental predicate is not to be predicated of a concrete individual, but only of an individual as representing a *kind.* Thus, snub-nosedness is an accident not of Socrates as an *individual;* but of Socrates as a *man.* Man, the *type,* need not be, although it may be, conjoined with snub-nosedness. Snub-nosedness is an accident, because it is not a necessary consequence of being a man.

Such, in brief, is the Aristotelian theory of predicables. In terms of the doctrine, therefore, the condition which satisfactory definitions must satisfy is that they be stated in terms of genus and differentia.

§ 4. RULES FOR DEFINITIONS

It is convenient, however, to discuss the rules for satisfactory definitions without restricting ourselves to the Aristotelian analysis. The following rules are the substance of those usually given:

1. A definition must give the essence of that which is to be defined. The definiens must be equivalent to the definiendum—it must be applicable to everything of which the definiendum can be predicated, and applicable to nothing else.
2. A definition must not be circular; it must not, directly or indirectly, contain the subject to be defined.
3. A definition must not be in the negative where it can be in positive terms.
4. A definition should not be expressed in obscure or figurative language.

We shall comment briefly upon each of these precepts.

1. The first rule expresses in different words the substance of our discussion in the previous sections of this chapter. When the traditional doctrine of the predicables is made the basis for discussion, this rule may be replaced by the injunction that a definition must be *per genus et differentiam.* Real definitions are definitions of words, and at the same time are analyses of the universal symbolized by both the definiens and the definiendum.

We have already called the reader's attention to the fact that in modern treatments of mathematics real definitions are *implicit,* the subject being defined in terms of the axioms which it must satisfy. It frequently happens, therefore, that several undefined terms must be defined on the basis of their relations to one another, and not in isolation from one another. Thus in Hilbert's study of the foundations of geometry, points, lines, and planes are taken as the "undefined" elements. But they are implicitly defined by the axioms. These axioms state the relations which must hold between points taken by themselves, lines taken by themselves, planes taken by themselves, and also the relations between points and lines, points and planes, and so on. But whether explicit or implicit, the definition should be so selected that the attributes known to belong to the things defined must be formally derivable from the definition.

Since, therefore, the logical aim of definitions is to state those features of a thing from which its other features follow, it is not always possible to satisfy the psychological motives behind the need for definitions. When the psychological objectives of definitions are emphasized, it is often said that the definiens should contain more familiar ideas than the definiendum. But if the logical goal of definitions is in the foreground, it may be advantageous to neglect this advice and to use less familiar notions in the definiens than in the definiendum. The undefined terms (and of necessity there must be undefined terms in every system) should be so selected as to give scope for a deductive treatment of the subject matter. Such undefined terms cannot be made meaningful by further definition, but only by some carefully selected process of exhibiting that which they denote. In some instances the undefined terms may be invested with significance by a direct process of exhibition. In others, however, the denotation of such terms cannot be exhibited. This is generally true of terms defined implicitly through the axioms. Such terms, although they function as undefined elements in the system, are *virtually* defined by the system itself. Thus in electrical theory a hypothetical electrical fluid may be an undefined term. Its meaning becomes known to us, however, in virtue of the fact that many of the properties of such a fluid with which the theory endows it can be directly exhibited.

2. If the term to be defined, or some synonym, appears in the definiens, no logical advance has been made in the analysis of the concept for which it stands, although it may be that the psychological purpose of the definition is satisfied. Thus if "courage" is defined through its synonym "bravery," the meaning of "courage" may have become clearer to us because we are more familiar with the meaning of "bravery." But the net effect of the definition is verbal, and the *structure* of "courage" (what it signifies, not the *word*) has not been analyzed. Such tautological definitions some-

times escape detection. The present rule is violated if the sun is defined as "the star which shines by day": for "day" itself is defined in terms of the shining of the sun.

A definition may seem to violate this rule when in fact it does not do so. A famous example is Russell's definition of "number." According to him, "A number is anything which is the number of some class." Here "number" is defined in terms of "the number of some class." The definition does not violate the present rule, for the definiendum is "number," or "number in general," while the definiens contain the term "number of some class." Definitions of this type are frequent in mathematics. Thus the series $u_0 + u_1 + u_2 + \ldots u_n + \ldots$ is defined to be *convergent,* if the sequence of successive terms $S_0 = u_0$, $S_1 = u_0 + u_1$, $S_2 = u_0 + u_1 + u_2 + \ldots$, $S_n = u_0 + u_1 + \ldots + u_n + \ldots$ is convergent.

3. It is obviously preferable to define a thing in terms of what it *is,* rather than in terms of what it is *not.* For in general, to state what a thing is not does not sufficiently delimit it from other things. Thus to define a watch as a timepiece which is not a clock will be unsatisfactory if there are other timepieces besides watches and clocks. However, it is easy to overemphasize this rule, for in some cases an adequate definition can be given this way. Thus to define a scalene triangle as one which is neither equilateral nor isosceles delimits perfectly scalene triangles from all others, provided it is stated in what system of geometry the triangle is to be included. In some cases negative definitions are inescapable. Thus the definition of an orphan as a child who has not parents must of necessity be in negative terms, for the state of orphanhood is a denial of the state of having parents. Other instances, like "independence," "parallel," "bankrupt," or "insolvent" will readily occur to the reader. Moreover, whether a definition is considered negative or positive often depends upon linguistic conventions. Some languages may possess a positive term for an idea which must be expressed negatively in another language. Finally, a definition may have the appearance of being negative simply because one of the terms in it is negative in form. Thus to define a drunkard as a man who is intemperate in drink is not to violate the present rule: intemperance itself is defined in terms of an excessive imbibing of alcoholic liquids.

4. The chief danger from definitions expressed in figurative language is that the metaphors which are employed may suggest meanings that they are not intended to convey. Thus to define a

king as the "captain of the ship of state" may be misleading because it may suggest that a king can guide the destinies of a nation by following a charted path. The injunction that the definiens should not be obscure expresses the psychological motives for definitions. Samuel Johnson's definition of a net as a "reticulated fabric, decussated at regular intervals, with interstices at the intersections" is a classic example of a definition which violates this psychological requirement.

However, the occurrence in the definiens of terms *unfamiliar* to most readers does not make the definition obscure. In physics, the definition of "the action of a system of particles," is given as "the sum for all the particles of the mean momentum for equal distances multiplied by the distance traversed by each particle." This definition is by no means obscure to the competent student of analytical dynamics, whatever it may appear to be to the untrained.

§ 5. DIVISION AND CLASSIFICATION

According to the traditional account, definition consists in the analysis of a species in terms of its genus and differentia. But a genus may be differentiated into other species as well. Thus the genus "plane figure" may be differentiated not only into the species "triangle," but into the species "quadrilateral," "conic section," and so on. The exhibition of the various species in the same genus is called *logical division,* or more simply, *division.* The genus with which the process of division starts is called the *summum genus.* Now the species obtained by a division may be capable of further division. The species with which a division ends is called the *infima species,* while the species intermediary between the summum genus and the infima species are called the *subaltern genera.*

The process of division, from an extensional point of view, is the breaking-up of a class into its constituent subclasses. Division is therefore related to definition, because it marks off the limits of the extension of a class denoted by a term. If, however, division is looked at from the point of view not of its constituent species, but of its individual members, the process is allied to *classification.* While division breaks up a genus into species, classification is the grouping of individuals into classes, and these classes into wider ones.

A number of rules have been stated for satisfactory logical division. They are also applicable to classification. They are:

1. A division must be exhaustive.
2. The constitutent species of the genus must exclude one another.
3. A division must proceed at every stage upon one principle, the *fundamentum divisionis.*

Thus if we divide rational numbers into odd integers and even integers, the first rule is violated, for we have omitted the fractions. The purpose of the first rule is to take account of every species in the genus. We violate the second rule when we divide the genus "quadrilateral" into "rhomboids," "parallelograms," "rectangles," since if anything is a rectangle it is also a parallelogram. The principle upon which a division is made is called the *fundamentum divisionis.* In dividing the genus "professor" into "mathematicians," "physicists," and so on, the *fundamentum divisionis* is the subject matter which they profess; if we divide it into "dull lecturers," "brilliant lecturers," and so on, the principle is their rhetorical ability. A division which conforms to the third rule will necessarily conform to the second. But the converse is not true. Thus, the division of the genus "number" into "odd," "even," "fractional," yields exclusive species, although the principle of the division is not single.

However, these rules, although unexceptional from a formal point of view, are of little help in practice. They express an ideal rather than state a method. Moreover, the ideal is inadequate for a well-developed science; it is more suitable to sciences in their infancy.

Until we have explored a subject matter thoroughly we cannot achieve either a satisfactory definition, a satisfactory division, or a satisfactory classification. In the first place, we can never be sure, in any existential subject matter, that the division or classification is exhaustive. A hitherto unknown and unpredictable aspect of the subject matter may suddenly turn up and undo, or at least call for a serious revision of, our efforts at system. Nor can we be certain that the subaltern genera are in fact exclusive. Indeed, this warning is a corollary from the proposition that the division cannot be known for certain to be exhaustive, for a hitherto unfamiliar subclass may turn up which possesses the common characters of several of the already recognized species.

In the second place, the process of scientific classification is much more groping and less formal than the rules would suggest. Even before science was deliberately pursued, everyday experiences compelled the recognition of *kinds of things* in which certain groupings of qualities occurred more or less invariably. Thus unreflective experience takes cognizance of trees, earth, animals, and so on, on the basis of obvious similarities between instances of these types. With growth of knowledge, however, features that are less obvious may be taken as the basis for classification or division. Thus although the porpoise is like a fish in many ways, it is classified in modern biology as a mammal because it suckles its young. The basis of classification depends on the discovery of some significant traits, significant in the sense that on the basis of the traits the subject matter can be organized into a system. Such traits, however, are only slowly discovered, and cannot be determined on formal grounds alone.

All sciences in their early days are classificatory, and almost any arbitrary scheme of grouping objects may be tentatively adopted in the interest of mastery of the subject matter. The classification of genera in modern biology still does not conform to the third of the rules above. Anthropology has not yet grown out of the classificatory stage, and until recently chemistry too was content to classify its subject matter in terms of elements, compounds, and reactions. Today, however, chemistry is organized on the basis of physical principles, which show more clearly than the older scheme the structure of its subject matter and the interrelation of chemistry and other sciences.

An exhaustive and exclusive division can always be obtained by dividing a genus in terms of a differentia and its negative. Aristotle obtained an exhaustive set of possible relations between a subject and predicate by this method. It is called *dichotomous division.* It can be represented as follows:

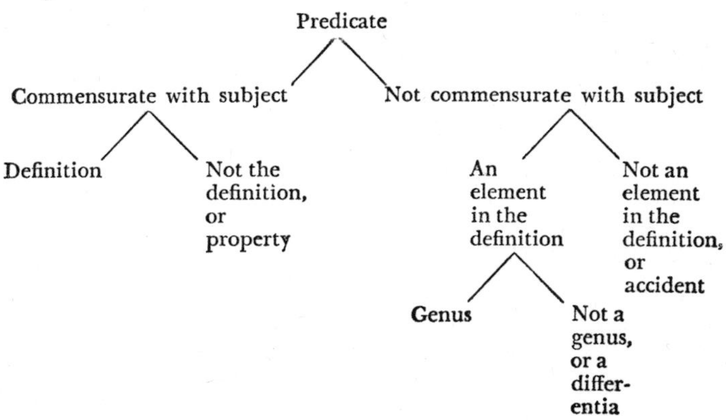

Nevertheless, although dichotomy insures exhaustiveness and exclusiveness of the species, it is not much of an advance over ordinary division. The practical difficulty of finding significant principles of division still remains. And in dichotomous division we cannot be sure that all the subclasses have members. Moreover, the method is somewhat clumsy, and modern symbolic logic has shown how dichotomous division can be effected in an almost mechanical manner. Thus suppose we wish to classify the population of the United States on the bases of sex, of being over thirty years of age, and of being in good or exceptional health. Let 1 represent, as usual, the universe of discourse; a those of male sex, a' those of female sex; b those over thirty, b' those thirty or under thirty; c those in good or exceptional health, c' those in poor health. Then the population of the United States is divided into eight groups as follows:

$$\begin{aligned} 1 &= (a + a') = (a + a')\ \ (b + b') = (a + a')\ \ (b + b')\ (c + c') \\ &= abc + abc' + ab'c + ab'c' + a'bc + a'bc' + a'b'c + a'b'c' \end{aligned}$$

The symbol abc will then represent the males over thirty and in good health; $a'bc'$ will represent the females over thirty who are in poor health, and so on.

CHAPTER XIII

THE METHODS OF EXPERIMENTAL INQUIRY

§ 1. TYPES OF INVARIANT RELATIONS

The search for order among facts is a difficult task. Few succeed in it. But it has always been the hope of some thinkers that easily learned rules might be found according to which anyone undertaking such a task may be assured that success will crown his efforts. And some writers on scientific method have proudly believed they had actually found such rules. Francis Bacon was one of them. "Our method of discovering the Sciences," he wrote, "is such as to leave little to the sharpness and strength of men's wits, but to bring all wits and intellects nearly to a level. For as in drawing a straight line, or describing an accurate circle by the unassisted hand, much depends on its steadiness and practice, but if a rule or a pair of compasses be applied, little or nothing depends upon them, so exactly is it with our method." [1] The methods which Bacon recommended for discovering the causes of things are popularly believed to express the nature of scientific method. They were elaborated by John Stuart Mill and formulated by him as the *methods of experimental inquiry.*

Before examining these methods, let us note some preliminary difficulties. In Chapter XI we tried to indicate that isolated facts do not constitute a science, and that the object of science is to find the order among facts. But what kind of order? It is commonly believed that science is interested exclusively in *causal* order. The analysis of the meaning of "causality" is a most difficult task. We cannot enter upon it here, for it is not a task for the logician. But we must observe that various kinds of order are sometimes confounded as identical with the causal order. We may then be pre-

[1] *Novum Organum,* Bk. I, LXI.

pared to state the general character of the *sort* of order for which the sciences are in search.

1. There is a type of order so familiar that it generally escapes notice. All of us recognize certain things as water, other things as wood, still others as steel, and so on. Why do we assign special names such as "water," "wood," and "steel"?

We apply the name "water" to something which is a liquid when above a definite temperature, a vapor when above another temperature. This "substance" is generally translucent, odorless, colorless; it has a constant density, and is practically incompressible; it quenches fire and thirst. "Water" denotes a *constant conjunction of properties,* and a name is given to it in order to distinguish it from other such conjunctions or "things." So also for "wood" and "steel."

The vague concept of "thing" denotes, therefore, a very elementary but fundamental type of order. It denotes a certain invariable conjunction or association of properties that is different from other conjunctions. Such a type of order would probably never be called "causal." But the discovery of this kind of order is fundamental for the discovery of any other kind of physical order. Different kinds of things have been recognized in the most primitive stages of man's history. But the process of classifying and cataloguing our experiences is not complete, and perhaps never will be.

2. A type of order frequently recognized is one involving a temporal span or direction. "Iron rusts in moist air," expresses one instance of such an order. It is this kind of order that is generally regarded as causal.

The "common-sense" notion of cause is an interpretation of nonhuman behavior in terms peculiarly adopted to human behavior. Thus, "John broke the window," is supposed to express a causal relation, because there is an agent "John" who *produced* the breaking of the window. So in the proposition above, the moist air is said to be the cause, of which the rusting iron is the effect. And the moist air is said to be the cause because it is believed to *produce* the rusting. In the popular mind, all *changes* require *causes* to explain them, and when found are interpreted as agents producing the change.

It is very difficult to make clear what is understood by causes "producing" their effects. When we reflect on the matter, all that we seem capable of discovering in alleged instances of causality, is an instance of an *invariable relation* between two or more *processes.* It is not the mere *existence* of John which is the cause of the broken window. What is important is the constant relation which holds between a certain kind of behavior of John and a certain other kind of behavior of the glass. One of the characteristics of the causal relation as generally understood is that it is asymmetrical and temporal in nature.

But even "common sense" soon recognizes that the apparent invariability of alleged causal relations is often specious. Iron does not *always* rust in moist air, and a window is not *always* broken when a brick is hurled at it. Even "common sense" discovers that other factors must be present in these situations besides those already noted. Hence it is not moist air alone that is the cause of rusting. Those other factors are then sought for which seem to be necessary for the occurrence of the effects. In this way there is a gradual transition from the crude and approximate uniformities observed in everyday experience, to the more completely analyzed invariant relations of the developed sciences.

3. Many uniformities are expressible by numerical equations. Ohm's law in electricity states that the current is equal to the potential difference divided by the resistance. The principle of the lever states that equilibrium is obtained when the two weights vary inversely to their distances from the fulcrum.

Invariable relations of this type no longer assert a sequence in time, and they are probably never regarded as illustrations of causal order. It is true that in making experiments upon an electric circuit we may alter the current first and subsequently note the change in the potential difference. But what Ohm's law states is not the order in which we make observations; it states that the measurable elements observed stand to each other in the specified invariable relations.

4. A fourth type of order is illustrated in such comprehensive theories as the theory of gravitation, or the kinetic theory of matter. In such theories, not all the elements between which the invariant relations are asserted to hold are directly observable. Nor are all the relations which are asserted to hold between the elements capable of direct experimental confirmation. Thus the atoms, their motions and collisions, the invariance of their average energies, cannot be verified directly.

The function of such comprehensive theories, as we have already seen, is that they enable us to show that many numerical as well as qualitative laws which are in fact experimentally confirmable, are not isolated from one another. Such laws can often be shown to be the necessary consequences of the more abstract and inclusive order asserted in the theory. Thus the numerical relations between the temperature, the volume, and the pressure of gases; the numerical laws connecting the density and the specific heat of gases; the relations between melting-point, pressure, and volume of solids—these are all derivable by logical methods from the assumptions of the kinetic theory of matter.

If we examine these four general types of order, we discover that a generic feature of all of them is the assertion of some kind of invariable relation between various kinds of elements. The relation in some cases may involve a temporal asymmetry; the cause is popularly said to precede the effect in time. In other cases the temporal reference is missing. It is the *invariability* which seems to be significant, both theoretically and practically.

By the *cause* of some *effect* we shall understand, therefore, some appropriate factor invariably related to the effect. If A *has diphtheria at time t* is an effect, we shall understand by its cause a certain change *C,* such that the following holds. If *C* takes place, then *A* will have diphtheria at time *t,* and if *C* does not take place, *A* will not have diphtheria at time *t;* and this is true for all values of *A, C,* and *t,* where *A* is an individual of a certain type, *C* an event of a certain type, and *t* the time.

The search for "causes" may therefore be understood as a search for some invariable order between various sorts of elements or factors. The *specific* nature of this order will vary with the nature of the subject matter and the purpose of the inquiry. Moreover, the *specific* nature of the elements between which the order is sought will also differ for different inquiries. In some cases we already know the invariable order and some of the elements, and then our search is for one or more further elements. Thus, finding a person dead from wounds and knowing the conditions under which such a death results, we look for the murderer. In other cases, we may know the elements and search for an invariable order between them. Thus we may note hot water being poured into a glass and also note that the glass cracks; we may then look for the structure of the relations connecting two such processes. In still other cases, we may notice some change and then look for other as yet unknown changes connected with it in some ways not yet known. Thus we may note the aurora borealis, and then search for the circumstances with which it is connected in some fashion or other.

The kind of elements or changes for which we look depends on the structure of the order in which we are interested. The answer to the question, "Who killed the Archduke Ferdinand at Sarajevo?"

must be of the form: "The person or persons *A, B, C* and so on, are the assassins of the Archduke." On the other hand, the question, "What killed the Archduke?" must be answered according to the kind of specific order for which we are in search, and according to the purpose of the inquiry. One answer may be, "A certain revolver was the cause of his death." Other possible answers are: "Certain social and political conditions were the cause of his death"; "The cessation of the oxygen supply to the cells of his body was the cause of his death." In other words, the kind of order, as well as the kinds of elements we look for, is determined by the nature of the problem which generates the inquiry. What is an adequate answer to one question will not, in general, be adequate to another.

In the light of the great variety in the kinds of specific orders and factors which may be the objects of an inquiry, it may seem preposterous to believe that any general rules can be stated which will enable us to find satisfactory answers to all possible problems. We shall not prejudge the matter, however, and shall examine at length the experimental methods formulated by Mill.

§ 2. THE EXPERIMENTAL METHODS IN GENERAL

The "experimental methods," according to their author, have a twofold function to fulfill. They are, in the first place, *methods of discovering* causal connections. Mill believed that by employing the methods the order in which facts stand to one another can be found. Against his critics he maintained that all inferences from experience are made by means of these methods. If "no discoveries were ever made by the . . . methods . . . none were ever made by observation and experiment; for assuredly if any were, it was by processes reducible to one or other of those methods." [2] It is these methods which supply the first generalizations upon which all subsequent construction of hypotheses depends.[3]

But secondly, the methods have a *demonstrative* function as well. Mill conceived the business of logic to be concerned with proof. Inductive logic, according to him, should supply "rules and models (such as the Syllogism and its rules are for ratiocination) to which if inductive arguments conform, those arguments are conclusive, and not otherwise." [4] The methods are therefore to be *tests* of any

[2] *A System of Logic,* Vol. I, p. 500.
[3] *Ibid.,* p. 501.
[4] *Ibid.,* p. 500.

experimental procedure. Just as the evidence for a proposition is conclusive if the relations between the propositions offered in evidence and the one in question conform to the conditions for necessary inference, so an "inductive argument" is valid, it is claimed, if it conforms to the "experimental methods." The conclusion of investigations into matters of fact could therefore be *absolutely certain.*

We shall be able to evaluate these two claims more clearly if we state what the general nature of the methods is, and if we recall what are the conditions under which an inquiry can be carried on. In the first place, some selected portion of our experience is taken for further study because of its problematic character. The problem must then be formulated in terms of the situation which provokes the inquiry, and an analysis of the situation must be made into a certain number of factors, present or absent, which are believed to be relevant to the solution of the problem. Now the order for which we are in search is expressible, as we have seen, in the form: C is invariably connected with E. And this means that no factor can be regarded as a cause if it is present while the effect is absent, or if it is absent while the effect is present, or if it varies in some manner while the effect does not vary in some corresponding manner. The function of experiment is to determine with regard to each of the factors entertained as a possible cause, whether it is invariably related to the effect. If C and E are two factors or processes, there are four possible conjunctions: we may find either CE, $C\bar{E}$, $\bar{C}E$, or $\bar{C}\bar{E}$, where $\bar{C}$ and $\bar{E}$ denote the absence of these factors. To show that C is invariably connected with E we must try to show that the second and third alternatives do not occur.

When the problematic situation is complex, and contains distinguishable factors as components, we can establish an invariable relation between the effect and some of the possibly causal factors only by showing that such factors do or do not meet this *formal* condition for invariable connection. It is necessary, therefore, to vary the supposedly relevant factors *one at a time,* and consequently to analyze the situation into factors that are relatively independent of one another.

The function of experiment, as we shall see, is *eliminative.* And the methods of experimental inquiry, as we shall also see, have precisely that function.

§ 3. THE METHOD OF AGREEMENT

The Method of Agreement as a Canon of Discovery

Whether we are searching for the cause of some event or for the effect, we begin with a situation which can be identified as *one of a type*. Suppose that one morning we find the flowers in the gardens of a suburban town to have withered over night. How shall we proceed to find the cause?

The first experimental canon instructs us as follows: *If two or more instances of the phenomenon under investigation have only one circumstance in common, the circumstance in which alone all the instances agree, is the cause (or effect) of the given phenomenon.*[5] The phenomenon we are investigating is the withering of flowers; the withered flowers in the several gardens are the instances. We must, accordingly, examine the instances for the common circumstances or factors. We note that the gardens differ in many ways: in the quality of the soil, in the kinds of flowers they grow, in their size, in their location, in the character of their gardeners. We note also that the temperature had fallen very sharply during the night. We conclude on the basis of the canon that the drop in temperature is the cause of the withering of the flowers. Why are we justified in drawing this conclusion? Why do we not say that the quality of the soil is the cause? Because, it may be said, the factors which are absent when the phenomenon is present cannot be invariably connected with the phenomenon: the quality of the soil is *not* the same in each instance of the withered flowers. Hence the invariable circumstance of the drop in temperature must be the cause. And indeed sharp changes in temperature are known to be fatal to flowering plants. The canon seems to be a successful method, therefore, both for discovering and for "proving" causes.

Unfortunately, in this illustration we knew what the cause of the withering of the flowers was *prior* to the application of the method. It is not surprising that we have been able to discover the cause. We must employ the canon to discover the cause of some phenomenon for which we do not happen to know the cause antecedently.

Now baldness in men is a phenomenon for which the cause is not known. If the canon is an effective instrument of discovery, no student of logic need suffer long from a naked scalp. In accordance with the canon, we find two or more bald men and search for a

[5] *A System of Logic,* Vol. I, p. 451.

single common circumstance. But immediately we run into great difficulties. The method requires that the men should differ in all respects except one. We will have rare luck indeed (or we are perhaps rather unimaginative in noting common circumstances) if we succeed in obtaining a collection of men who satisfy this condition. (If we are not too particular about the nature of the "common" circumstance, we may perhaps hold that the common circumstance to all bald men is their being organic bodies.)

Let us waive this point. We then meet a more serious obstacle. How shall we go about identifying the common circumstance or circumstances? If one specimen has blue eyes, must we examine all the others for color of eyes? If one of the bald men confesses to having been brought up on cod liver oil, must we discover whether the others were brought up the same way? But the number of such circumstances to which we might be directed is without limit. Date of birth, books read, food eaten, character of ancestry, character of friends, nature of employment, are some examples of circumstances to which we might pay attention. If, therefore, the *common* circumstance can be found only by examining all the instances for *every* circumstance which some one or other of the bald group may possess, we can never find *all* the common factors in this way. We can carry on the search for a common factor only if we disregard most of the circumstances which we may find as not *relevant* to the phenomenon of baldness. We must, in other words, start the investigation with some hypothesis about the possible cause of baldness. The hypothesis which selects some circumstances as possibly relevant and others as not is constructed on the basis of previous knowledge of similar subject matter. This hypothesis is not supplied by the canon. Without some hypothesis on the nature of relevant factors the canon is helpless to guide us to our goal.

We have been pretending that the circumstances or factors which are present in an instance are distinct from one another, and that each comes labeled, as it were, with a tag saying, "I am a circumstance." But it simply is not true that an instance of a phenomenon is presented to us as a unique set of sharply defined factors, immediately recognizable by us as such, so that each factor can be examined and varied independently of any other. Now the method of agreement requires a comparison of circumstances in two or more instances. Unless, therefore, an analysis of an instance into its factors could be made *prior* to the use of the method, the method would be altogether useless.

How do we divide an instance into its factors, and is every analysis of instances into factors equally valid? Consider the following experiment. In two or more test tubes of unequal sizes and each filled with a liquid of a different color, a precipitate is formed. We wish to determine the cause of the precipitation and we find that each instance can be analyzed into the following circumstances: (1) A test tube of a certain size (2) contains a liquid with a specific color (3) to which sulphuric acid has been added; (4) precipitate is formed. Factors 1 and 2 are eliminated by the canon, since the instances vary with respect to these; the canon fixes upon the addition of the acid as the cause of the phenomenon (4). But this is not the *only* way in which the instances could have been analyzed. We may have tried the following division into circumstances: (1′) A test tube of a certain size (2′) contains a liquid (3′); sulphuric acid is added to different-colored liquids; (4′) precipitate is formed. On the basis of this analysis, the method would conclude that the factor (2′) is the cause of the precipitation. This conclusion is in fact false. It is false because the second method of analyzing the instances is not a "proper" one.

Let us consider another illustration. We wish to know the cause of headaches. We find that it was preceded in some cases by eyestrain, in some cases by indigestion, and in some cases by the hardening or other disturbances of certain blood vessels. If we take this canon literally, none of these are common circumstances and therefore to be considered as causes. This, however, would be an error, due to an inadequate analysis of what is a headache and of what factors in all these mentioned circumstances are relevant to the different kinds of headaches. A greater refinement of the causes must be accompanied by an equal refinement of the effects. If we ask what is the cause of disease, we have grouped a large number of phenomena under one rubric and the diverse causes of the different kinds of disease must be similarly grouped. This point will be discussed more fully under the heading of The Plurality of Causes.

It follows that not every analysis into factors is equally valid. It is not valid because *in the light of our knowledge* we must not separate, in such experiments as the one above, the volume and place occupied by a liquid from what has been added to it. Now the method of agreement cannot inform us which is the proper analysis. It cannot discover for us how to divide instances into factors such that invariable relations can be found to hold between some of them. The method cannot possibly function unless, once more, assumptions about relevant factors are made.

The Method of Agreement as a Canon of Proof

Let us consider next whether the method of agreement is one of proof, even if we must recognize that it is not one of discovery. Does it follow that because a search for the cause of a phenomenon conforms to the conditions stated by the method, the conclusion of the search is thereby *demonstrated* to be true? That it does not so follow can be easily seen. The cause of a phenomenon must be invariably related with the phenomenon. But we cannot examine more than a *limited* number of instances of any alleged invariable relation. Even if we could be absolutely certain that the circumstance claimed as a cause is the *single* common circumstance, can we be certain that it is *invariably* (for an *unlimited* number of instances) connected with the phenomenon? Thus we may find in a very large number of instances of typhoid that the activity of microörganisms is the one common factor that is present. It does not follow that this factor is *always* present in the still unobserved instances of typhoid. Not every actual conjunction of circumstances is *indefinitely* repeated.

The reader may perhaps believe that the inference from the observed conjunction of factors to an invariable conjunction is legitimate in virtue of the "uniformity of nature." We shall not disturb the reader's faith in this familiar doctrine at this point. But, as we shall see presently, such a faith has no evidential value in demonstrating the existence of invariable connections.

Not only will the method not serve to prove the presence of a causal relation; it may, on the contrary, lead us to affirm some factor to be the cause when it is not. We have seen this already in connection with the problem of analyzing instances into circumstances. We can see this in another way. Suppose a professor of hygiene finds that he had a splitting headache on three successive nights. He recollects that on Monday he read for ten hours and then took a walk; on Tuesday, he found the dinner delicacies irresistible, ate too much, and then sought repentance by taking a walk; on Wednesday, he slept during the day and then sought refreshment in a walk. If he were to employ the method of agreement, he might conclude that walking was the cause of his headache. But this is quite contrary to fact, since the walks he took (we happen to know on other grounds) have nothing to do with bringing on the headaches. A false conclusion was drawn by using the method, since the instances to which it was applied were not analyzed properly into the right circumstances.

This illustration suggests another familiar doctrine. The method of agreement does not provide a "water-tight" proof, it is claimed, because there is such a thing as a plurality of causes. The same phenomenon is not always produced by the same cause, Mill believed: "There are often several independent modes in which the same phenomenon could have originated." [6] A house may be destroyed by fire, or by an earthquake, or by cannon fire. Consequently, this method cannot find *the* cause. It was such a reflection which compelled Mill to recognize an imperfection in the canon of agreement and which led him to supplement it with the canon of difference. We shall return to the doctrine of plurality of causes presently. Whether the doctrine is tenable or not, its formulation and adoption by Mill show the need for some criterion of the correct analysis of instances into circumstances. It is a criterion not supplied by the canon.

Employing the canon does not guarantee that *all* the necessary conditions for the occurrence of the phenomenon will be found. Why is the mercury column in barometers generally around thirty inches high? If we employ this method we may conclude that since there is a vacuum at the top of each column, the existence of the vacuum is the cause of the observed height of the mercury. This is a mistake, since we know that the occurrence of a vacuum is not a sufficient condition for the height of the column. The atmospheric pressure, the temperature of the room, are other conditions which are indispensable in order to explain the height of the mercury. The method of agreement may, therefore, overlook certain general conditions which must obtain. It may fix our attention only upon certain obvious, even if necessary, features of the instances.

The Value of the Method of Agreement

The method of agreement is therefore useless as a method of discovery and fallacious as a canon of proof. Has it then no value? It has a limited value, if stated negatively: *Nothing can be the cause of a phenomenon which is not a common circumstance in all the instances of the phenomenon.* Thus stated, it is clearly a method of eliminating proposed causes which do not meet the essential requirements of a cause. A circumstance that is not common to all instances of a phenomenon cannot, by definition, be causally related to it.

A search for causes begins with some assumptions about the pos-

[6] *A System of Logic,* Vol. I, p. 505.

sibly relevant factors. Thus in studying baldness we may begin as follows: Baldness is due to congenital, hereditary factors, or to the characters of the diet, or to the nature of the headwear, or to some previous disease. The method of agreement helps to *eliminate* some or all of the suggested alternatives. We may discover that the character of the food eaten by bald men is not a common feature; and according to the principle of *tollendo ponens* we can conclude therefore that only the three other alternatives remain to be examined, that is, baldness is either congenital, or it is due to the nature of the hats worn, or it is due to some previous disease. We may proceed in this fashion until we have eliminated all the suggested alternatives, or found one or more which cannot be eliminated.

Unless, however, we have been fortunate enough to include the circumstance which is in fact the cause of the phenomenon in the enumeration of alternatives, the method of agreement can never identify the cause. Its function is to help *eliminate irrelevant circumstances.*

§ 4. THE METHOD OF DIFFERENCE

The Method of Difference as a Canon of Discovery

The method of agreement was recognized as faulty by Mill because we cannot be certain that the phenomenon investigated has only one cause. It was believed to be useful in those cases where we could not *alter* the circumstances at will. Hence, it was regarded primarily a method of *observation* rather than *experiment.* It was believed, however, that the shortcomings of this method can be overcome by the use of a second canon, the *method of difference.*

This second method requires two instances which resemble each other in every other respect, but differ in the presence or absence of the phenomenon investigated. Its full statement is: *"If an instance in which the phenomenon under investigation occurs, and an instance in which it does not occur, have every circumstance in common save one, that one occurring in the former; the circumstances in which alone the two instances differ, is the effect, or the cause, or an indispensable part of the cause, of the phenomenon."* [7]

Let us see whether this canon is effective in discovering causes. Suppose the reader buys two fountain pens of like make, fills them with the same kind of ink, places them in his pocket, and takes a

[7] *A System of Logic,* Vol. I, p. 452.

long walk before he sits down to write. When he does, he discovers that one of the pens leaks. What is the cause of this? It seems as if the conditions for applying the canon are all present. The pens are alike, but one leaks and the other does not. And if the reader employs the canon he may "discover" that the rubber sack in one of the pens has lost its elasticity and is extremely porous; the other pen has not this defect. The condition of the rubber, the reader may conclude, is the cause of the leak.

But is the matter as simple as that? If we take the canon seriously, we must conclude that the method cannot be applied in this inquiry, since it requires that the two pens be *exactly* alike in all circumstances except those mentioned. But the two pens do differ in very many ways: one was made before the other, or by a different workman; the shapes of the pens have minute differences; chemical analysis reveals other differences; the pens were not placed in exactly the same position in the reader's pocket, nor were they warmed equally by his body.

If anyone should object that the pens need not be *exactly* alike, but only alike in *relevant* factors, we must reply that it is precisely such judgments of relevance which are required before the canon may be used; and that the canon does not supply this vital information. If, however, the objector should declare that the two pens can be shown to be alike by examining *all* the circumstances, we would be forced to reply that an exhaustive examination of the circumstances is impossible; and that if it were possible, the canon would be unnecessary to discover the factor which is present when the phenomenon is present and absent when it is absent.

This canon, like the previous one, requires, therefore, the antecedent formulation of a hypothesis concerning the possible relevant factors. The canon cannot tell us what factors should be selected for study from the innumerable circumstances present. And the canon requires that the circumstances shall have been properly analyzed and separated. We must conclude that it is not a method of discovery.

The Method of Difference as a Canon of Proof

Is it a method of proof or demonstration? No more than the method of agreement! Whatever value the canon may seem to have depends on the assumption that differences are noted with respect to the presence or absence of a *single* factor. But can the canon assure us that the factor is not *complex*?

Suppose a man is psychically and socially maladjusted. He suffers from uncomfortable dreams. He goes to a psychoanalyst, who persuades him to disclose intimate details of his autobiography, and in particular of his sex life. The man "recovers" from the maladjustment and the dreams cease. The canon seems to be applicable: the "single difference" in the events of the man's life during this period is the expression of his hidden sex desires. Can we validly conclude that the man's talking freely on sex is the cause of his recovery? Certainly not. The change in the man's life may in fact be due simply to his finding in the analyst a *sympathetic audience* on any subject whatever, or it may be due to the cessation of certain organic disturbances unknown to the patient or the analyst.

Can this method demonstrate an *invariable* connection by an examination of *two* instances? Suppose the reader spends a sleepless night, but on the following night he rests peacefully. The reader may be able to convince himself that the single "significant" difference in his behavior on the two days preceding these nights is that he had drunk coffee on the first day but not on the second. Can the reader validly conclude that drinking coffee is (for him) the invariable cause of sleeplessness, other things being equal? It may be true that his sleeplessness on the first night was in fact due to the coffee. Nevertheless, it may be that it is not drinking *coffee as such* which produces the undesirable result. The insomnia may be due to a *drug* that the coffee contained. The reader's sleeplessness on that particular night was, by hypothesis, due to drinking that particular coffee. But it does not follow that in general coffee-drinking is followed by a restless night. The application of the canon does not, therefore, uniformly lead to the detection of the factors in terms of which an invariable relation can be expressed, and it may lead to an affirmation of an invariant relation where none in fact exists. The canon does not safeguard us against the fallacy known as *post hoc, ergo propter hoc:* sleeplessness may *follow* drinking coffee, but sleeplessness may not occur *because* coffee was drunk.

The statement of the canon clearly recognizes that the factor noted by it may be only a *part* of the cause. This is a very important qualification. The invariable relations for which the sciences seek are such that if a determinate set of circumstances are present, some other circumstance will always accompany these. The discovery of a *partial* set of circumstances is often not enough. Now the method of difference cannot guarantee that the sufficient conditions for a phenomenon have been found. We cannot infer that rain is a *sufficient* condition for the rich harvest in one part of the state on the ground that there was a drought in another part, even though the quality of the seeds planted and of the soil and the quantity of sunshine were in all significant ways the same. For it isn't the *rain alone,* but the rain *together with* the soil, seeds, and sunshine, which provides the adequate conditions for a bumper crop. The canon may therefore very easily direct our attention to extremely partial and even superficial factors in the complete situation. On the basis of the canon we might argue that since conditions in Europe in January, 1914, were the same as in July, and since the only relevant difference was the murder of the Archduke Ferdinand, the assassination was the cause of the World War. Without denying the importance of this event in explaining just when the war did take place, no serious student of affairs would hesitate to point out the complicated national, diplomatic, and socio-economic factors which were part of the conditions required to explain the occurrence of the war.

The Value of the Method of Difference

The method of difference cannot, therefore, be regarded either as a method of discovery or as a method of proof. But, like the method of agreement, it has a limited value when stated negatively: *Nothing can be the cause of a phenomenon if the phenomenon does not take place when the supposed cause does.* Thus stated, it is clearly a method of eliminating one or more proposed causes which do not meet the essential requirement of a cause. A circumstance that is present whether the phenomenon is present or not cannot, by definition, be causally related to it. Thus if we are studying rheumatism, we may entertain the hypothesis that it is caused by excessive starch in the diet, or by lack of exercise, or by a focal infection in the teeth. Provided that these alternatives represent an adequate analysis and separation of the circumstances, we may be able to eliminate the diet theory of rheumatism if we can show that large quantities of starch can be consumed without being followed by the effect. Proceeding as we did with the canon of agreement, we may, in accordance with the principle of *tollendo ponens,* be able to eliminate all but one alternative. Again, the method of difference is helpless if we have not the sagacity to include in the alternatives considered for further study the factor which is in fact the cause.

§ 5. THE JOINT METHOD OF AGREEMENT AND DIFFERENCE

The two methods so far considered require conditions for their application which we can never find realized. The first method requires instances which are unlike in every respect except one; the second requires instances which are alike in every respect except one. When the phenomenon is dependent upon a complex set of conditions, it is difficult to separate the factors involved and vary them one at a time. Mill therefore proposed a combination of the two preceding methods. Its formulation is: "*If two or more instances in which the phenomenon occurs have only one circumstance in common, while two or more instances in which it does not occur have nothing in common save the absence of that circumstance; the circumstance in which alone the two sets of instances differ, is the effect, or the cause, or an indispensable part of the cause, of the phenomenon.*" [8]

The statement of the canon, however, is really absurd. According to it we require two *sets* of instances. In one set the phenomenon occurs, and the instances *taken together* must have a single common circumstance, although taken two at a time they may agree in more than one circumstance. In the second set the phenomenon does not occur, and the instances must be so chosen that they have *nothing* in common, taken together, save the absence of the phenomenon. But if we follow these instructions we may include *anything we wish* in the second set, since the *absence* of some one character need be the only identical feature in all of them! Suppose we wish to discover the conditions which make for divorce. According to the method, we must in the first place examine a number of divorced couples, and in the second place, examine a number of cases where divorce has not occurred, for example, among flowers, children, mountains, bachelors, and so on. We could not possibly use *these* negative instances to determine the cause of divorce. We must therefore modify the formulation of the canon. The negative instances must be all of a type in which the phenomenon is *capable* of being present when the adequate conditions are supplied.

As a method of discovery and proof, this canon combines all the defects of the first two canons; its virtues are the virtues of either. However, it does formulate certain aspects of methods employed in making comparisons between large groups. If we were to try the method of difference alone for finding the cause of divorce, we

[8] *A System of Logic,* Vol. I, p. 458.

would require two couples, one divorced, the other not, that are alike in *every* way except one. This is hardly feasible. If, however, we were to examine a large number of married pairs, we might be able to show that some of the circumstances which are common to all of them are not *significant* for their continuing in the married state, provided we could also show that divorced pairs show the *same* common features. We might be unable to identify the cause of divorce by this method. Nevertheless, by examining several large groups we might be able to show some relation between the relative frequency of divorce and such factors as differences in the age, education, health, and so on of the parties to a marriage. Such statistical information may be all that can be obtained. The knowledge of the relative frequency of divorce for individuals differing considerably in age, for example, will be useless for determining whether divorce will terminate the marriage of some particular married pair. It may be very useful in ascertaining how often we may expect divorces in a very large group.

§ 6. THE METHOD OF CONCOMITANT VARIATION

The elimination of irrelevant circumstances, which we have seen to be the function of the preceding canons, cannot be performed by them in all cases. For it is sometimes impossible to exclude or isolate the cause completely. If we wish to find the cause of the rise and fall of the tides of rivers and seas, we cannot use the canon of difference, since we cannot find an instance in which such a body of water does not show the phenomenon of tidal behavior. We cannot show with the method of difference that the sun and moon are the cause of tides, since we cannot eliminate the action of these bodies in any instances. And we cannot use the canon of agreement, because we cannot remove from the instances of tidal behavior such ineradicable common circumstances as the presence of the fixed stars.

In such cases, however, we may notice or introduce variations in the degree or magnitude of the effect, and find a corresponding variation in some circumstance, without thereby *completely* eliminating either the effect or the supposed cause. The method of concomitant variation has been formulated by Mill to deal with such phenomena. Its statement is: "*Whatever phenomenon varies in any manne. whenever another phenomenon varies in some particular manner, is either a cause or an effect of that phenomenon, or is connected with it through some fact of causation."* [9]

This canon can be employed, therefore, only if degrees or magnitudes of effects and causes can be distinguished. The previous canons are *qualitative* methods, since their use requires simply the determination of the presence or absence of some character or quality. The present canon is *quantitative,* and requires the aid of measurement and statistical technique.

The Canon of Concomitant Variation as a Method of Discovery

An examination of the statement of the canon of concomitant variation must make us suspicious of its efficacy as a method of discovery. It declares that if a phenomenon varies in any manner *whenever* another phenomenon varies in some manner, a causal relation is present. Now if the concomitant variation is actually invariable, and this seems to be required by the word "whenever," a causal relation is indeed present. But if in order to employ the canon we must know antecedently that the mode of variation is invariable, of what use is the canon? We do not, in that case, need the canon to discover for us the cause. And the canon by its own admission is perfectly helpless in finding the rule of variation or in demonstrating that a supposed mode of variation is invariable.

This suspicion is strengthened if we try to use the canon. Suppose we notice that the temperature in a region varies in some determinate manner during several months. What is the cause of this variation? We look for some circumstance present during these months which also undergoes some variation. But which circumstances shall we examine? Certainly not *all* the circumstances, not even all the varying circumstances. The formulation of hypotheses and judgments of relevance are required before this canon can be employed.

The complicated causal dependencies between several variables which the natural sciences study cannot possibly be unraveled unless hypotheses based on knowledge of mathematical relationships are formulated concerning them. Even such a relatively simple rule of variation as the inverse-square law of gravitational attraction cannot be obtained by merely observing the behavior of planets.

The Method of Concomitant Variation as a Canon of Proof

The mere presence of a concomitant variation of temperature and some other factor is not sufficient to establish a causal connec-

[9] *A System of Logic,* Vol. I, p. 464.

tion. Suppose the changes in the daily temperature in New York City during one year could be shown to vary with the daily death rate in China in that period. Such correlations would be regarded by most competent judges as fortuitous, because they have some prior knowledge concerning the relevant factors in the production of temperature changes. Even very high correlations, especially in the social sciences, do not necessarily signify an invariable connection. For the phenomena between which such correlations can be established may be in fact unrelated in any way which would warrant our believing them to be invariably connected. A little statistical skill and patience make it possible to find any number of high correlations between otherwise unrelated factors. We do not discover causal connections by first surveying all possible correlations between different variables. On the contrary, we suspect an invariable connection, and then use correlations as corroborative evidence.

Moreover, the correlations obtained on the basis of an examination of a finite number of pairs of variables are unreliable, because we cannot be sure that the rule of variation remains the same outside of the actually observed limits of variation. We may, by good fortune, come to the study of gases by selecting a gas like helium at a high temperature. We may then observe that if the temperature remains constant, the pressure varies inversely as the volume of the gas. We may observe this rule of variation for certain intervals of temperature, and then extrapolate the rule for every value of the temperature, or even for any gas at any constant temperature. But if we do so, we are sure to blunder, since it is now known that Boyle's law is true only for a few gases under ideal conditions. Indeed, a rule of variation which has been found to hold within certain intervals may become altogether inaccurate outside those limits, not only because the rule of variation is different, but also because circumstances negligible within those limits cannot be neglected outside that interval. The period of a pendulum is proportional to the square root of its length if the arc of the swing is small. When the arc of the swing is increased, the period (theoretically and approximately) is still related in this way to the length; nevertheless, the factor of air resistance must now be considered, so that the period can no longer be rendered by that simple formula.

The Value of the Method of Concomitant Variation

The method of concomitant variation cannot therefore be accepted as a method of either discovery or proof. Its value lies partly in suggesting lines of inquiry for causal relations and in helping to corroborate hypotheses of causal connection. Its chief value, however, is to help eliminate irrelevant circumstances. For nothing will be regarded as the cause of a phenomenon if when the phenomenon varies that thing does not, or when the phenomenon does not, that thing does. Consequently, the method will help eliminate those factors suggested by the hypothesis guiding the inquiry which do not conform to this condition. Mill's statement of the canon asserts that if *C* varies whenever *E* varies, *C* and *E* are causally related. We have seen that this claims too much. All that can be affirmed is that *C* and *E* are *not* causally related if *C* and *E* do not vary concomitantly. And even in this modified form the method will not save us from error if the circumstances denoted by *C* and *E* are not properly analyzed.

§ 7. THE METHOD OF RESIDUES

The remaining method of "discovery and proof," the *method of residues,* expresses more clearly than the others the eliminative function of all the canons. Its statement is: *"Subduct from any phenomenon such part as is known by previous inductions to be the effect of certain antecedents, and the residue of the phenomenon is the effect of the remaining antecedents."* [10]

The method very clearly depends upon our making use of some already known causal connections in order to isolate the influence of some other known or assumed cause by means of a *strictly deductive argument.*

A favorite illustration for this method is the discovery of the planet Neptune by Adams and Le Verrier. The motions of the planet Uranus had been studied by the help of Newton's theories. Its orbit was plotted on the assumption that the sun and the planets within the orbit of Uranus were the only bodies which determined its motion. But the calculated positions of Uranus were not in agreement with the observed positions. On the assumption that these differences could be explained by the gravitational action of a planet outside the orbit of Uranus, the position of such a hypothetical planet (behaving according to the usual principles of celestial mechanics) was calculated from the perturbations in the motion of Uranus. And in fact the planet Neptune was discovered in the vicinity of the place calculated for it. This achievement is

[10] *A System of Logic,* Vol. I, p. 460.

therefore credited to the method of residues.

But the argument used for locating Neptune is easily seen to be strictly deductive. We must accept, in the first place, the universality of Newton's theory of gravitation. We must assume, in the second place, that the motion of Uranus is determined by the known bodies within its orbit and a single unknown body outside its orbit. The position of this unknown body can then be calculated if we also know how much of the observed behavior of Uranus is due to the influence of the interior planets. Now the canon of residues itself did not pick out the cause of the discrepancies in the observed behavior of Uranus. A hypothesis concerning the possible source of this discrepancy had to be explicitly introduced. The canon simply expresses the fact that, on the assumptions made, the interior masses were to be eliminated as the causes of the observed discrepancies. It does not suggest *where* the source of the residual phenomena is to be located. It does not demonstrate that the suspected source of such residual phenomena is causally related to them.

In this illustration one further condition for the applicability of this method must be noted. We can calculate the position of the planet Neptune only if we know the law according to which forces of attraction can be compounded. These forces are supposed to act "independently" of one another. This means that if one of the interior planets should fly off from the solar system, the magnitude of acceleration that each of the remaining bodies would contribute to the behavior of Uranus could still be calculated from their known positions and masses. Wherever the forces studied are not independent in this sense (where, in other words, the effect of two forces cannot be calculated from a knowledge of each in isolation) the method of residues cannot be employed.

§ 8. SUMMARY STATEMENT OF THE VALUE OF THE EXPERIMENTAL METHODS

We will now summarize this long discussion of the experimental canons. Every investigation of the cause of a phenomenon P must start with a hypothesis. Suppose H_1, H_2, . . . H_n are a set of alternative hypotheses concerning the possible determining conditions of P. The H's therefore express our sense of what is relevant in any occurrence of P. No observation, experiment, or reasoning can proceed without an explicit or implicit acceptance of the following:

Proposition 1. *Either* H_1, *or* H_2 *or . . .* H_n *is the causal law of* P. The function of the experimental canons is to eliminate some or all of these alternatives. We try to show that in the instances where *P* is present, H_1 does not hold; or that H_1 is true in those cases where *P* is absent as well as where *P* is present; or that a variation occurs in *P* without a correlated change in some factor denoted by H_1. If we are successful in showing any one of these things (and if H_1 represents a proper analysis of circumstances), H_1 is then eliminated as a causal law of *P*. Thus experiment may establish the following:

Proposition 2. H_1 *is not the causal law of P.*

We may then conclude from propositions 1 and 2:

Proposition 3. H_2 *or* H_3 *or, . . .* H_n *is the causal law of P.*

The same procedure can now be undertaken for H_2, and so on. And we may be successful in eliminating all the alternatives but H_n. Provided that H_n cannot be eliminated, we may conclude that H_n is the causal law of *P,* on the assumption that the *nH*'s are the only possible causal laws.

But it is clear that this procedure is efficacious in finding causal laws only if the following be true.

a. Proposition 1 must be based upon a proper analysis of the circumstances attendant upon *P*. The *H*'s must express the relevant relations of *P* to certain other factors.

b. The *n* alternatives *H* must include the true causal law of *P*. If we have not been fortunate enough to include the true causal law, all the alternatives may be eliminated and the cause of *P* not be ascertained. But no directions can be given how to include the true law in an enumeration of possible laws. The difficult step in extending our knowledge consists, therefore, in finding propositions of the form *If* H *then* P where *H* is a suitable hypothesis or theory from which the phenomenon *P* can be shown to be a consequence.

c. Proposition 3 is obtained by strictly necessary reasoning from propositions 1 and 2.

d. The concluding propositions are not *demonstrated* to be true unless propositions 1 and 2 are in fact true. But we can rarely, if ever, be certain that proposition 1 is an exhaustive statement of all possible causal laws for *P*.

The canons of experimental inquiry are not therefore capable of *demonstrating* any causal laws.

The experimental methods are neither methods of proof nor methods of discovery. The canons which formulate them state in a more explicit manner what it is we generally *understand* by a causal or invariant relation. They *define* what we mean by the relation of cause and effect, but do not *find* cases of such a relation. The hope of discovering a method that will "leave little to the sharpness and strength of men's wits" is one which finds no support from a careful study of the procedure of the sciences.

But while the methods we have discussed have the defects pointed out, they are of undoubted value in the process of attaining truth. For in eliminating false hypotheses, they narrow the field within which true ones may be found. And even where these methods may fail to eliminate all irrelevant circumstances, they enable us with some degree of approximation to so establish the conditions for the occurrence of a phenomenon, that we can say one hypothesis is logically preferable to its rivals.

§ 9. THE DOCTRINE OF THE UNIFORMITY OF NATURE

The claim that the experimental methods are capable of demonstrating with complete certainty universal, invariable connections rests on a belief that "nature is uniform." Induction, according to Mill, consists in inferring from a finite number of observed instances of a phenomenon, that it occurs in *all* instances of a certain class which resemble the observed instance in certain ways. But according to Mill, the very statement of what induction is requires an assumption concerning the order of the universe. The assumption is that "there are such things in nature as parallel cases, that what happens once, will, under a sufficient degree of similarity of circumstances, happen again." [11]

This assumption may be expressed in various ways: that nature is uniform, that the universe is governed by general laws, that the same cause will under similar circumstances be accompanied by the same effect. In *some* form, however, so the claim runs, it is required for induction. Every induction may be thrown into the form of a syllogism, and the principle of the uniformity of nature will then appear as the "ultimate major premise of all inductions." [12]

Mill puts the matter as follows: "The induction, 'John, Peter, etc., are mortal, therefore all mankind are mortal,' may . . . be thrown into a syllogism by prefixing as a major premise (what is at any rate a necessary condition of the validity of the argument) namely, that what is true of John, Peter, etc., is true of all mankind. But how came we by this major premise? It is not self-evident;

[11] *A System of Logic,* Vol. I, p. 354. [12] *Ibid.,* p. 356.

nay, in all cases of unwarranted generalization, it is not true. How, then, is it arrived at? Necessarily either by induction or ratiocination; and if by induction, the process, like all other inductive arguments, may be thrown into the form of a syllogism. This previous syllogism it is, therefore, necessary to construct. There is, in the long run, only one possible construction. The real proof that what is true of John, Peter, etc., is true of all mankind, can only be that a different supposition would be inconsistent with the uniformity which we know to exist in the course of nature. Whether there would be this inconsistency or not, may be a matter of long and delicate inquiry; but unless there would, we have no sufficient ground for the major of the inductive syllogisms. It hence appears, that if we throw the whole course of any inductive argument into a series of syllogisms, we shall arrive by more or fewer steps at an ultimate syllogism, which will have for its major premise the principle, or axiom, of the uniformity of the course of nature." [13]

We shall not discuss whether the principle of the uniformity of nature is true or whether some such principle is required for making inductive inferences. We wish simply to determine whether the principle *if it were true* would in fact help to demonstrate the existence of some particular instance of a supposed causal relation. We must carefully note the following.

1. The principle is stated in an extremely vague form—"what happens once, will, under a sufficient degree of similarity of circumstances, happen again." But what is a sufficient degree of similarity? The principle does not tell us. In any particular investigation we must rely on other criteria, if there are any, to determine what are the circumstances material to the occurrence of a phenomenon.

2. In the second place, the minor premise of an inductive syllogism, even according to Mill, is a *particular* proposition. Therefore even if we employ a universal major premise, such as the principle of uniformity of nature, the premises are insufficient to *demonstrate* a universal conclusion.

3. Finally, the principle does not affirm that *every* pair of phenomena are invariably related. It simply states that *some* pairs are so connected. To appeal to the doctrine in a particular investigation is therefore useless. If we suspect that tight-fitting hats are the cause of baldness, we employ the canons to eliminate as many circumstances other than tight hats as we can. But no finite number

[13] *A System of Logic,* Vol. I, pp. 357-58.

of observed cases of tight-fitting hats followed by baldness can demonstrate a law which is to hold for an *indefinite* number of cases. The principle of uniformity of nature does not help us. It does not say *which* of the innumerable casual connections between phenomena are invariable; it merely asserts that *some* are. But the task of the particular inquiry is to show that a *designated* pair of phenomena are in causal relation.

§ 10. THE PLURALITY OF CAUSES

The method of agreement is often found to be faulty because we cannot be sure—so it is claimed—that the effect studied may not have more than one cause. It is for this reason that the method of difference was regarded by Mill as a superior experimental procedure. The doctrine of plurality of causes is stated by him as follows: "It is not true . . . that one effect must be connected with only one cause, or assemblage of conditions; that each phenomenon can be produced only in one way. There are often several independent modes in which the same phenomenon could have originated. One fact may be the consequent in several invariable sequences; it may follow, with equal uniformity, any one of several antecedents, or collection of antecedents. Many causes may produce mechanical motion: many causes may produce some kinds of sensations: many causes may produce death. A given effect may really be produced by a certain cause, and yet be perfectly capable of being produced without it." [14]

This doctrine may be given a logical version. The fallacy of affirming the consequent in mixed hypothetical syllogisms may be interpreted as an illustration of the doctrine of plurality of causes. Thus, given *If a number expressed in ordinary algorism has a 5 in the unit place, it is divisible by 5,* we cannot validly infer that a certain number terminates with a 5 in the unit place because it is divisible by 5; the number *may* terminate with a zero. It seems, therefore, that Mill is in the right concerning plurality of causes, and his doctrine is capable of being stated in a more general form and in purely logical terms.

Let us first consider the less general form of the doctrine, as stated by Mill. Suppose a house burns down. What is the cause of this event? Perhaps the house was destroyed because of an overturned kerosene lamp, or because of defective electric wiring, or because of

[14] *Ibid.*, p. 505.

a faulty chimney. The reader may be tempted to retort that the plurality is only apparent. "If the alleged causes of the fire were examined more carefully," he may perhaps say, "a circumstance common to all of them would be found. For example, the occurrence of a rapid oxidation in some part of the house is such a common circumstance. And that common feature of the many alleged causes is *the* cause of the event."

Such an analysis is not very satisfactory. If the reader were investigating that fire for an insurance company and submitted such an analysis, he would not retain his post for long. "The occurrence of a rapid oxidation," the company would doubtless declare, "is an explanation of all fires. It was not your job to discover the most general conditions under which fires occur, for we knew that all the time; it was your job to find the special conditions under which this one occurred."

This hypothetical reply of the insurance company not only indicates the inadequacy of one type of criticism of the doctrine of plurality of causes; it also suggests a more complete reply to the doctrine. For if the doctrine were true, how could we ever be able to infer the cause from an examination of the ruins of a destroyed house? There can be no doubt that we frequently infer the true cause of an effect. Fire insurance companies do so continually. So also the medical examiner is able to establish the real cause of a person's death in spite of the alleged plurality of causes of death.

The more satisfactory reply to the doctrine of plurality of causes is this: When a plurality of causes is asserted for an effect, the *effect* is not analyzed very carefully. Instances which have significant differences are taken to illustrate the *same effect*. These differences escape the untrained eye, although they are noticed by the expert. Thus the way in which a house burns down when an overturned lamp is the cause, is not the same as when defective wiring is the cause. The doctrine of plurality of causes is plausible only if we analyze the causes into a larger number of distinct types than we do the effect. The doctrine overlooks many differentiating factors present in several instances of a so-called effect, and by viewing these instances under their more generic features regards them as instances of the *same* effect. For many purposes it is perhaps convenient to retain this lack of symmetry in the analysis of causes and effects. But it does not follow from this fact of convenience that the usual illustrations of plurality of causes really prove the absence of a one-to-one correspondence between cause and effect.

Let us now turn to the doctrine in its more general or logical form. Must we deny that the fallacy of affirming the consequent is a fallacy? Not if we recognize elementary distinctions, and recall some of our discussions in the chapter on mathematics. Affirming the consequent is a fallacy because the same consequent may follow from more than one antecedent. But, we may ask, if a proposition follows from two distinct sets of premises, does it follow from them in virtue of their being different from each other, or in virtue of their containing something in common?

If the reader remembers our discussion of logical systems, he must acknowledge that the second alternative expresses the true state of affairs. We showed in Chapter VII, § 3, that two systems may be incompatible with each other taken in their entirety although they may have many theorems in common. We explained this by suggesting that the two systems contain a common subsystem. The common theorems of the two systems follow strictly from the axioms of this common subsystem, and not from the axioms of the two systems as such.

Let us state this in different words. A set of premises which is a sufficient condition for a given proposition may contain conditions besides those which are necessary. With care and ingenuity, those portions of the premises which are not required for the conclusion can be eliminated. In this way, we can discover all the conditions necessary for the conclusion. And when the antecedent in an implicative proposition contains the necessary and sufficient conditions for the consequent, it is no longer a fallacy to argue from the affirmation of the consequent to the affirmation of the antecedent.

The fallacy of affirming the consequent is therefore indeed a fallacy, since we do not in general know that the antecedent states the necessary and sufficient conditions for the consequent. For most purposes, science is satisfied with the sufficient conditions for the propositions it wishes to establish. But its goal, which may never be reached, is to find the conditions which are both necessary and sufficient.

The distinction between sufficient conditions and those which are both necessary and sufficient throws further light on the limitations of so-called crucial experiments. Suppose p, a verifiable proposition, follows from theory T_1 but not from T_2. Then a slight modification of T_2, one which leaves its main outlines unaffected, may perhaps be made so that p is implied by the revised T_2 as well as by T_1. Both theories, in such a case, would logically contain the necessary and sufficient conditions for p, although they may contain much else besides. The verification of p, therefore, will not compel us to abandon T_2 if we can continue to use it with scientific profit after slightly altering it. We may conclude with a pertinent remark of Bertrand Russell: "a hypothesis which accounts with a minute exactitude for all known relevant facts must not be regarded as certainly true, since it is probably only some highly abstract aspect of the hypothesis that is logically necessary in the deductions which we make from it to observable phenomena." [15]

[15] *The Scientific Outlook,* 1931, p. 67.

CHAPTER XIV

PROBABILITY AND INDUCTION

§ 1. WHAT IS INDUCTIVE REASONING?

Modern science is often contrasted with the science of antiquity as being "inductive," while the latter was "deductive." According to this view, deductive and inductive reasoning are antithetical modes of inference. Deductive logic is then believed to be concerned with the conditions under which particular or instantial propositions are inferable from universal premises. Inductive logic, on the other hand, is conceived as dealing with those inferences which enable us to derive universal conclusions from particular or instantial premises.

Part of this characterization, as we have already seen, is certainly wrong. The essence of deduction is not the derivation of particular conclusions from universal propositions, but the derivation of conclusions which are *necessarily* involved in the premises. For no conclusion of a deductive inference can be instantial unless at least one of the premises is instantial. The theory of gas engines, a set of universal propositions, can give us no information about the automobile we actually possess unless the instantial proposition is added to the premises that this actual automobile has a gas engine.

But how about induction? Is there a distinct type of inference which proceeds from instantial to universal propositions? Some distinctions should be noted before a determinate answer is given.

1. One of the senses in which Aristotle employed "induction" was to denote the mental process through which a universal character or relation is discriminated and identified in an actual case or event. Our earliest experiences are vague and our attention is directed to certain pervasive qualities in which differences are not recognized. To the infant the world is very likely a "buzzing, blooming confusion," just as to the untrained eye all the trees in a forest

are just trees, or to the untrained ear a symphony is just sound. We pay attention to certain abstract or universal features, like "trees" or "sound," but very little order or structure is recognized within the qualitative whole to which we react. Nevertheless, by examining several cases of qualitative wholes we learn to apprehend a formal pattern in them. Let us imagine Boyle studying the behavior of a gas at a constant temperature. He may write the numerical measures of its volume at different pressures in a parallel column, as follows:

Pressure	*Volume*
1	12
2	6
3	4
4	3

An examination of these few numbers may enable him to recognize in these instances the law that the product of pressure and volume is constant.

Aristotle describes this process of discovering a general rule in a special case of it in a famous passage:

"Though sense-perception is innate in all animals, in some the sense-impression comes to persist, in others it does not. So animals in which this persistence does not come to be have either no knowledge at all outside the act of perceiving, or no knowledge of objects of which no impression persists; animals in which it does come into being have perception and can continue to retain the sense-impression in the soul: and when such persistence is frequently repeated a further distinction at once arises between those which out of the persistence of such sense-impressions develop a power of systematizing them and those which do not. So out of sense-perception comes to be what we call memory, and out of frequently repeated memories of the same thing develops experience; for a number of memories constitute a single experience. From experience again—i.e., from the universal now stabilized in its entirety within the soul, the one beside the many which is a single identity within them all—originate the skill of the craftsman and the knowledge of the man of science. . . .

"We conclude that these states of knowledge are neither innate in a determinate form, nor developed from other higher states of knowledge, but from sense-perception. It is like a rout in battle stopped by first one man making a stand and then another, until the original formation has been restored. . . . Thus it is clear that we must get to know the primary premises by induction; for the method by which even sense-perception implants the universal is inductive."[1]

This process is an important stage in our getting knowledge. Induction, so understood, has been called by W. E. Johnson *intuitive induction.* Nevertheless, this process cannot be called an *inference* by any stretch of the term. It is not a *type of argument* analyzable into a premise and a conclusion. It is a perception of relations and not subject to any rules of validity, and represents the gropings and tentative guessings of a mind aiming at knowledge. Intuitive induction is therefore not antithetical to deduction, because it is not a type of inference at all; and the discovery of the implications of a set of premises requires very much the same sort of guessing and groping. *There can be no logic or method of intuitive induction.*

2. Aristotle, and others after him, have employed "induction" in another sense. Suppose we wish to establish that *All Presidents of the United States have been Protestants.* We may offer as evidence the propositions *Washington, Adams, Jefferson, and so on were Protestants* and *Washington, Adams, Jefferson, and so on were Presidents of the United States.* The evidence is not conclusive unless we know that the converse of the second proposition is also true: unless we know, that is, that *All the Presidents of the United States are Washington, Adams, Jefferson, and so on.* In that case, the argument may be presented as follows: *Washington, and so on, were Protestants; all the Presidents of the United States are Washington, and so on; therefore all the Presidents of the United States have been Protestants.*

Induction, in this sense, means establishing a universal proposition by an exhaustive enumeration of *all* the instances which are subsumable under it. It has been called *perfect* or *complete induction.* Perfect induction is not antithetical to deduction. As we have just seen, *perfect induction is an example of a deductive argument.* The conclusion has been established by strict syllogistic reasoning.

It is evident that a perfect induction is possible only when all the instances of the universal proposition are already known to conform to it. But if general propositions could be employed only if they were the conclusions of a perfect induction, they would be utterly worthless for inferring anything about *unexamined instances.* They could serve simply as mnemonic devices to remind us of the

[1] *Analytica Posteriora,* in *op. cit.,* p. 99b.

host of examined instances which they summarize. Moreover, the *legitimate* application of such universal propositions would always require a circular argument. Thus, suppose we concluded that *Woodrow Wilson was a Protestant* because *All Presidents of the United States have been Protestants* and *Woodrow Wilson was a President of the United States.* The argument is valid. If, however, we examine the evidence for the premise *All Presidents of the United States have been Protestants,* and if this proposition is established by perfect induction, we find that *Wilson was a Protestant* is one of the premises for the proposition in question. Consequently, a proposition must be included among the premises of the argument which serves to establish that very proposition.

3. We are rarely in the position to establish a general proposition by perfect induction, since the number of instances subsumable under it is either too large or inaccessible in space and time. There are classes with an indefinite number of possible members. The real problem in science, and so it has been conceived by logicians from Aristotle down, is to discover the basis for a generalization when the instances *examined* are not *all* the possible instances. Is there any opposition between induction and deduction when induction is understood in this way?

Suppose that we suspect a connection between the color of people's hair and bad temper, perhaps as a result of an unfortunate encounter with a red-haired professor. We find that *A, B, C, D,* who are red-haired, are ill-tempered. We conclude that all red-haired individuals have bad tempers. Here seems to be an inductive inference which establishes a universal proposition on the basis of an examination of some only of its instances. But is this conclusion adequately established? Obviously not, unless we know the truth of the additional proposition, that *Whatever is true of A, B, C, D, is true of all red-haired people.* In that case, however, we may state the argument in a *deductive* form. The reasoning is in fact syllogistic:

1. Whatever is true of *A, B, C, D,* is true of all red-haired people.
2. Ill tempers characterize *A, B, C, D.*
3. ∴ Bad tempers characterize all red-haired people.

When, therefore, we state all the premises of such an inductive argument, we find that not only is there no opposition between induction and deduction, but also that the argument is an example of necessary reasoning. Therefore in none of the senses in which "induction" may be understood is induction a mode of reasoning antithetical to deduction.

Here the reader may object that the foregoing account misses the essence of induction, which is concerned with establishing the *material truth* of universal propositions. Do we really help to establish the truth of our conclusion by introducing a major premise, in our instance proposition 1, which is not known to be true?

This objection is based upon a sound perception. What most interests men in what is popularly called induction is really the process of generalization, the passage from a statement true of some observed instances to a statement true of all possible instances of a certain class. But the question of logic, we must insist, is one of the weight of evidence for such a generalization. We are concerned here not with the undoubted human need for generalization, but with the question what evidence is conclusive, that is, will prove the universal proposition to be true. Obviously many of our generalizations are not true. And the fact that a number of red-headed people have bad tempers is certainly not sufficient evidence for the proposition that all have.

The syllogistic form calls our attention to the real condition which distinguishes valid from invalid generalizations, and that is the homogeneity of the class of which members have been examined. In the actual state of human knowledge such homogeneity cannot be established except with more or less probability. The human need for generalization is so great that we are often impatient with those who point out the logical inadequacy of our ordinary evidence for our generalizations. If we are not to venture beyond what we already know, how shall we ever learn from experience? This is perfectly sound. Nevertheless, mankind also suffers from hasty generalization, of which race prejudice is a notable instance. In any case, scientific procedure requires that even those generalizations which cannot be conclusively proved should have the highest attainable degree of probability.

How can we assure this? That obviously depends upon our knowledge of the given field in which the generalization occurs. Logic can only supply us with a negative precept. We must eliminate the fallacy of selection, that is, the mistake of supposing that that which characterizes observed instances of a class (such as red-headed men) is necessarily true of all possible members of that class. For the red-haired men we have observed may in fact possess peculiar characters, such as being tired, overworked, poor, and so on, which they do not share with the other members of the red-headed group; and their irritability may be due to these peculiar characters. We shall discuss later the rules to help us overcome the fallacy of selection. Here it is sufficient to indicate that our putting the inductive argument in syllogistic form serves to call attention to the real conditions under which valid generalizations can be obtained.

Consequently, whether we know the *truth* of propositions of the form of proposition 1 above or not, the conclusion of the argument *logically depends* upon such propositions. Inductive inferences in so far as they are *demonstrative* must conform to the canons of all valid inference. We must also note that we do not even always know *which* are the premises required as conclusive evidence for the conclusion. But this, once more, does not alter the fact that the conclusion does depend logically upon such unknown premises. In this respect, however, there is no difference in the histories of the mathematical sciences, which are thoroughly deductive, and the natural sciences, which are regarded as inductive. For example, it is a mistake to suppose that the science of plane geometry developed in time by starting with axioms and then demonstrating the theorems. We know, on the contrary, that some of the theorems were known to Thales, who lived in the sixth century B.C. The great contribution of Euclid did not consist in adding new theorems to those already known, but in systematizing the subject by discovering the propositions upon which it depends (the axioms). Similarly, the *systematic basis* for the discoveries of Galileo concerning falling bodies was laid *after* his work was formulated by him. *The order of nature, and the order of logical dependence, are not the same as the order of our discoveries.*

But to return to the reader's objections. In general, not all the premises required logically in an inductive argument are known to be true. For we do not know that the examined instances in which a general proposition is verified are representative or fair samples of the entire class to which they belong. The specific problem of induction is to determine to what extent the samples are fair. Consequently, while induction and deduction are not opposed as forms of *inference,* nevertheless deduction is not concerned with the truth or falsity of its premises, while the characteristic nature of induction is to be concerned with just that. Induction may therefore be viewed as the method by means of which the *material truth* of the premises is established. The proper contrast is not between deductive and inductive inference, but between inferences that are necessary and inferences that are probable. For the evidence for universal propositions which deal with matters of fact can never be more than probable.

§ 2. THE RÔLE OF FAIR SAMPLES IN INDUCTION

Sciences at the early stage of their temporal development must of necessity seek to establish generalizations which are isolated from one another. Biology and the social sciences are still at the stage where the generalizations do not mutually support one another in virtue of their being part of a coherent logical system. We shall first inquire, therefore, how such relatively isolated generalizations as *All red-haired people have bad tempers, All storks are white,* and *All feeble-mindedness is hereditary* may be established.

It is clear that instances verifying such generalizations must be produced. But the mere repetition of verifying instances cannot serve as adequate evidence for the truth of such propositions. Such propositions pretend to state something about all possible instances, of which the actual verifying instances are only a small part. How then may the probability of universal propositions be increased?

We must examine the rôle of repetition of instances in establishing the probability of general propositions. In a well-known passage, Mill remarks that often a very large number of verifying instances is insufficient to establish firmly a generalization (for example, that all crows are black), while a few such instances are sufficient to win our assent to others (for example, that a certain type of mushroom is poisonous). "Why is a single instance, in some cases, sufficient for a complete induction, while in others, myriads of concurring instances, without a single exception known or presumed, go such a very little way towards establishing a universal proposition? Whoever can answer this question knows more of the philosophy of logic than the wisest of the ancients, and has solved the problem of induction." [2]

The answer is not as difficult to find as Mill pretends, although it may not make the reader wiser in logic than the wisest of the ancients. Before discussing it, however, we may also quote a suggestive observation and question of Hume: "Nothing so like as eggs; yet no one, on account of this appearing similarity, expects the same taste and relish in all of them. It is only after a long course of uniform experiments in any kind, that we attain a firm

[2] *A System of Logic,* Vol. I, p. 363.

reliance and security with regard to a particular event. Now where is that process of reasoning which, from one instance, draws a conclusion, so different from that which it infers from a hundred instances that are nowise different from that single one? This question I propose as much for the sake of information, as with an intention of raising difficulties. I cannot find, I cannot imagine any such reasoning. But I keep my mind still open to instruction, if anyone will vouchsafe to bestow it on me." [3]

As a preparation for comprehending what we shall suggest is the answer to Mill, let us recall an earlier discussion. We have pointed out that mankind has invented names for certain classes of objects, but not for others. Thus we have a name for things that are gold, but not for things that are blue. Why is this so? Because "gold" represents a constant conjunction of distinguishable properties, but "blue thing" does not. For example, we may define "gold" by its atomic number and atomic weight; and we find experimentally that these properties are connected with a determinate color, melting-point, boiling-point, solubility in certain acids, specific heat, malleability, and so on. Such is not the case for "blue things." The blue sky, books bound in blue leather, blue veins, blue suits, have not in common a determinate set of properties which are not also shared by things other than blue. Consequently, when we discover that some object answers to the definition of gold, we are pretty certain that it will possess certain other well-known properties. But when we simply know that some object is blue, we cannot tell what other properties it may have.

Now every attempted verification of a universal proposition requires that we should be able to identify some actual case as truly an instance of the universal. But we can do this only if the actual case is one of a *class* of objects with whose constantly conjoined properties we are familiar. Thus when we reason: All diamonds are combustible, this object is a diamond, therefore this object is combustible: we must identify "this object" as a diamond. We are not in the position to do this unless we are familiar with the more or less invariable properties of diamonds. If we are familiar with them, we may infer that since "this object" has, say, a certain luster and hardness, it also has other properties which usually go with them and which are characteristic of diamonds. In such a process of identification we are said to be reasoning *analogically*. Consequently, universal propositions may be safely applied to an actual

[3] *An Enquiry concerning Human Understanding*, Sec. IV, Pt. II.

subject matter only in so far as we are thoroughly familiar with the *type of object* of which the actual case is a *sample.*

We may therefore reply to Mill as follows: While we can never be altogether certain that an examined verifying instance is a fair sample of all possible instances, in some cases the probability that this is true is very high. This is the case when the subject matter of the inquiry is homogeneous in certain relevant ways. But in such cases it is unnecessary to repeat a large number of times the experiment which confirms a generalization. For if a verifying instance is representative of all possible instances, one such instance is as good as another. Two instances which do not differ in their representative nature simply count as one instance.

This, however, is only a partial answer to the problem. A more adequate answer may be given to Mill's query if we bear in mind that a great part of the evidence for a generalization comes from the analogy which its instances bear to instances of other generalizations already well established. Indeed, the entire matter appears in an altogether different aspect when the generalization with which we are concerned is an element in a coherently organized *system* of propositions with far-reaching ramifications. In such a case, the evidence for the generalization comes not only from its own verifying instances, but also from the instances which verify the far-reaching, and often remote, consequences of the system.

Thus when a new chemical compound is discovered, its density is determined by perhaps a single measurement. No chemist doubts that all later measurements will yield approximately the same value. The high probability of this proposition is not due, evidently, to the repetition of measurements on density of this particular compound. It is based upon the assumption that this sample of the compound is homogeneous with all other samples as far as their physical properties are concerned, and upon the general proposition that density is a constant for all homogeneous substances. This latter proposition, however, is part of a comprehensive theory of matter, a theory which is supported by the repeated verification of its logical consequences.

It is convenient, therefore, to distinguish universal propositions that in our knowledge are relatively isolated from one another, from those which mutually support one another in virtue of their being parts of a logically coherent system. The probability of propositions of the first kind depends almost entirely upon repetition of instances, and only slightly upon considerations of analogy. Thus observation of a few bad-tempered redheads may suggest the generalization *All red-headed people are bad-tempered.* This is a hypothesis, which may lead us to examine other red-headed individuals. Such further instances will either confirm the generalization or, as frequently happens, compel us to modify it. As our knowledge of the subject matter increases, we discover ways of obtaining samples so that all possible variations of the subject matter may be exhibited in the samples. Repetition of instances is valuable, therefore, only in so far as the subject is not known to be homogeneous.

If, however, the instances are known to be analogous in certain important respects with other phenomena that are better explored, a generalization which is known to hold for the latter may be adopted for the former with practically no changes. Thus electricity in motion presents certain striking analogies to the behavior of such incompressible liquids as water. The entire theory of hydrodynamics may therefore be extended to electrical phenomena as well. In such cases, the development of the science becomes completely deductive, and seems to require no experimental confirmation of its theorems. But this absence of experimentation is only apparent, since it is experimentation which suggested the fruitful analogies which the theory exploits, and since it is further experiment which acts as a test upon the supposed analogies.

Our ability to conduct fair sampling is thus greatly increased if we can show that the generalization under inquiry is connected with others. For in that case the verifying instances for a universal proposition accumulate more rapidly because the generalizations which they render probable support one another. That is why the deductive elaboration of hypotheses is such an essential part of the method of science. We must examine this point in some detail.

Mechanics is no longer an experimental science. It derives its theorems from the first principles of motion by rigorous reasoning. It is, however, one of the best founded of the *natural* sciences. Why is this so? The answer is not that the first principles of motion are self-evidently true. The answer is that we can conduct a sampling process on a vast scale, when a theory is so stated that its measurably precise consequences have application in very different fields. Newton's theory of matter in motion has consequences verifiable with great precision in the moon's motion around the earth, in the behavior of bodies near the earth's surface, in the motion of the planets, in the behavior of double stars, in the rise and fall of tides, in the phenomena of capillarity, in the behavior of dynamical machines, and so on. The principles of mechanics are highly probable on the evidence consisting of verifications of *random* samples from the set of all consequences which they imply.

Now the probability of the principles of mechanics is a sort of reservoir upon which the many special theorems of the system may draw. Thus all the verifying instances of the theory of the pendulum also verify Newton's laws. And since if Newton's laws are true, the theorems on the moon's motion are also true, experiments on the pendulum help to confirm the lunar theory.

That is the reason why the *repeated* confirmation of Newton's laws in only one domain, say planetary motion, is not as good evidence for their truth as are *fewer* confirmations in *different* domains. For such fewer instances, drawn from diverse domains, serve as more random, and therefore more representative, samples from the possible consequences of the theory than do the instances taken from one domain alone. If in each of the several domains methods of fair sampling are already established, one sample drawn from such a domain is as good as another. Indeed, at an advanced stage in the development of a theory its probability is not influenced noticeably by additional verifying instances. Its probability is affected by the superior or inferior systematizing power of a contrary theory. Thus the theory of relativity is probable not only on the evidence of its own few specific verifying instances, but also on the evidence that it affects a unification of gravitational and electrical theories, so that the instances which verify these theories singly, accumulate to verify the theory of relativity. The theory of relativity thus explains more things than does Newton's theory. The theory is never proved in the sense that all its logically possible alternatives are disproved. But while it may not be *demonstrably* true, it may none the less be verified in random samplings of its consequences. And for the purposes of science, verification (which is practically possible) is of the utmost importance.

The discussion of probable inference is therefore inevitably linked with the discussion of the nature of hypotheses. Hypotheses are valuable for science in proportion as they permit an organized deduction of consequences which are applicable to domains qualitatively different. A hypothesis that is capable of direct refutation or verification is useful enough as a guide to the study of the specific problem which occasioned the inquiry. But it cannot help to organize any wide field of studies. If the reader has mislaid his keys, he may entertain the hypothesis that they are in a suit he wore last week. This hypothesis may be verified or refuted directly by examining the suit in question. But it is not fruitful for any further research. Newton's laws, on the other hand, are hypotheses which cannot be tested directly. But they are eminently valuable in unifying and directing inquiries in diverse fields. In spite of the distance which separates the laws from the observed facts, they are very probable because of the high relative frequency with which samples from their possible consequences are confirmed in great detail by observable phenomena. And the different special theorems of the Newtonian system, applicable to different domains, mutually support one another. Like each separate leg of a tripod when they are spread very widely, no theorem can stand firm by itself; as parts of the system, they not only support the system, they also help each other to stand firmly.

We may state the matter summarily as follows. In the case of observation on chemical compounds we are dealing with a field in which some one hypothesis has been well established, for example: Whatever turns litmus red is acid. This hypothesis agrees with so many facts of observation that we naturally hesitate to form another hypothesis as to why the new compound turns litmus paper red. For the new hypothesis, while explaining the case before us, might not explain some of the numerous cases satisfactorily explained by the established hypothesis. The generalization, therefore, that the new compound will always show an acid character is one that so harmonizes with the body of existing knowledge, and alternative possibilities seem so very precarious, that we naturally think of it as the only legitimate possibility. An inductive argument, while it does not, in the strictest sense, demonstrate a universal proposition, may prove it to be the best evidenced of all suggested hypotheses.

§ 3. THE MECHANISM OF SAMPLING

We have indicated the central rôle of fair samples in the establishment of generalizations. We must not neglect to indicate briefly the technique of obtaining fair samples.

Let us begin with an artificial illustration. Suppose an urn contains 6 balls, 2 white and the rest black. We are to draw out 3 balls at a time, note their colors, replace them, and continue the process, mixing the contents of the urn well after each drawing. Let us name the white balls w_1, w_2, and the black balls b_1, b_2, b_3, b_4. We are required to state the composition of the urn on the basis of the examination of the samples, so that if we draw the sample (w_1,

w_2, b_1), we are to say: Two-thirds of the balls in the urn are white; and so on.

Now all the following twenty samples are theoretically possible:

($w_1w_2b_1$), ($w_1w_2b_2$), ($w_1w_2b_3$), ($w_1w_2b_4$), 4 samples in which ⅔ of the balls are white.

($w_1b_1b_2$), ($w_1b_1b_3$), ($w_1b_1b_4$), ($w_1b_2b_3$), ($w_1b_2b_4$), ($w_1b_3b_4$), ($w_2b_1b_2$), ($w_2b_1b_3$), ($w_2b_1b_4$), ($w_2b_2b_3$), ($w_2b_2b_4$), ($w_2b_3b_4$), 12 samples in which ⅓ of the balls are white.

($b_1b_2b_3$), ($b_1b_2b_4$), ($b_2b_3b_4$), ($b_1b_3b_4$), 4 samples in which none of the balls are white.

If each of the samples is as likely to be drawn as any other, and if the drawings are independent, when we make a very large number of drawings ⅕ of the time we will infer that: Two-thirds of the balls in the urn are white; ⅕ of the time we will infer that: All the balls in the urn are black; and ⅗ of the time we will infer that: One-third of the balls in the urn are white. In other words, we would draw correct inferences ⅗ of the time, that is, we would infer the true proportion of white and black balls in the urn more often than not. Therefore if we did not know the constitution of the urn, but decided to infer it by sampling, this illustration shows that, under the specified conditions, we would hit upon the truth more frequently than not by such a procedure.

The efficacy of this method depends upon two things.

1. The urn must have some determinate nature. Thus balls with distinguishable colors are contained in it, and the ratio of white to black balls is determinate.

2. The samples must be selected at random, so that in the long run each sample will be drawn with the same relative frequency. *If these conditions are fulfilled,* the method is bound to discover for us the approximate constitution of the urn. *For the method is inherently self-corrective.* Were we to select a sample so that we judge the contents of the urn to be all black balls, the method will correct this judgment. It can be demonstrated that if we *continue* drawing samples in very large numbers, the number of them which represent the true constitution of the urn will be overwhelmingly larger than that of those which do not.

This trivial example may serve as a model for the process of sampling in general when conducted on a larger scale in studying nature. The process is of course much more complicated and many special conditions (discussed in advanced treatises on statistics) must be recognized. Sampling consists in judging the nature of a collection to be the same as the nature of the sample. Every experiment we make concerning the connection of characters, is a sample from the inexhaustible source of possible experiments which is nature.

We must notice, however, that large-scale samplings yield only *approximate* conclusions. Thus if a large collection of objects M (say human individuals) have, in a ratio r, a character q (say insanity), then if we were to take a comparatively small number of objects P from it, a ratio r' of these may have the character q. r' is not equal to r in general. But as we draw the samples P, the number having a proportion of character q which resemble the constitution of the collection, will be larger than that of those which do not. We may therefore argue:

A certain proportion (r' per cent) of the samples P have the character q.
The P's are a fair sample of a large collection M.
Hence, probably and approximately, the same proportion (r' per cent) of the collection M have the character q.

The conclusion is probable on the premises, because the manner in which the P's are chosen will *in the long run* lead to their being *representative* samples, even though any *one* sample be altogether atypical.

§ 4. REASONING FROM ANALOGY

Let us examine the nature of what is sometimes called *analogical reasoning.* When we argue that because the planets Mercury and Venus *resemble* the planets Earth, Mars, Jupiter, and Saturn in revolving around the sun in an elliptic orbit, in being nearly spherical, and in shining by reflected light, they also resemble those four planets in rotating around an axis, we are said to be arguing from analogy. Some definitions adopted from J. M. Keynes will be useful in discussing the inference.

When a set of things, say crows, have a common character, say being black, these things will be said to be *analogous* with respect to this character, and this character is an *analogy* for these things. All the known common properties of a set of objects will be called the *known positive analogy,* and all the common properties, whether known or not, will be denoted as the *total positive analogy.* A set of things which have not in common a certain property will be said to be *negatively analogous* with respect to that property. Thus a husband and wife are negatively analogous with respect to the property "being male." The *total negative analogy* and the *known negative analogy* are defined similarly to the corresponding positive analogies. Now when a universal proposition is suggested to us in examining a set of things, what we are considering is the invariable connection of a *part* of the positive analogy with another *part* of the positive analogy. Thus when we examine a number of crows, and are led to consider the proposition *All crows are black*, the anatomical properties which are taken as defining "crows" are part of the positive analogy, and "being black" is another part of the analogy. Let us call the first property the *signifying analogy*, and the second property the *signified analogy*. What we wish to show when we are trying to establish a universal proposition is that a signifying analogy is constantly conjoined with a signified analogy. The proposition *All men are mortal* may therefore be rendered: *All things which have the signifying analogy "being human" also have the signified analogy "being mortal."*

It is quite clear that the signifying and the signified analogies taken together never in fact exhaust the total positive analogy, nor even the known analogy. Being human and being mortal do not exhaust all the properties which human beings have in common. Thus the properties of being located on the surface of the earth and of being less than ten feet tall are not part of the definition of humanity; however, they are part of the known positive analogy. We see therefore that a universal proposition does not "cover" the total positive analogy. The part of the positive analogy not included in either the signifying or the signified analogy is then regarded as irrelevant. Now since not all properties found in common in a finite set of objects are invariably conjoined, it follows that some selections of signifying analogies are unsatisfactory, since they are not constantly accompanied by the signified analogies. It is the purpose of science to discover in a set of observed things those parts of the total positive analogies which are invariably related.

We may now return to the example of analogical reasoning. Mercury and Venus possess an analogy with the other four planets, namely, revolving around the sun in elliptic orbits, being nearly spherical, etc.; this is the signifying analogy. These other planets possess also the common feature of rotating on an axis; this is the signified analogy. We infer that Mercury and Venus possess the remaining, signified, analogy (rotating around an axis). The validity of the inference clearly depends upon the proposition *All planets rotate around an axis*. This proposition, however, is not known to be true. But we have the following evidence for it: *The Earth, Mars, Jupiter, and Saturn are believed to be a random sample from the class of planets; rotation around an axis characterizes these four planets; therefore, rotation around an axis characterizes all planets.* This conclusion is probable on the evidence but not certain, for reasons we have already discussed. We may conclude with the probability, therefore, that since Mercury and Venus possess all the signifying analogies which all planets do, they also rotate around an axis. Analogical reasoning is therefore seen to be a case of probable inference which depends on fair sampling.

CHAPTER XV

MEASUREMENT

§ 1. THE PURPOSE OF MEASUREMENT

Many of the common affairs we conduct daily depend on our being able to distinguish only qualities or characters which are fairly sharply demarcated from one another. This day is "cold," therefore we put on our coats; that day is "warm," therefore we leave them off. This pillow is "hard"; we exchange it for one that is "soft." Some foods are "sweet," others are "sour"; we choose them in accordance with our preferences.

However, it is frequently necessary, even in daily life, to make judgments upon qualities which are not so sharply demarcated from one another. "Take Professor A's course, instead of Professor B's," we may be told; "Professor A is *easier* than Professor B." "Travel by the elevated trains; they are *less crowded* than the subways." "Buy coffee of brand X; it is *fresher* than brand Y." Such injunctions are significant for us because, in spite of the absence of clearly marked distinctions, we nevertheless readily apprehend the difference between "being easy" and "being difficult," between "a crowded train" and "a train not crowded," between "being fresh" and "being stale." In the sciences, too, propositions affirming *qualitative* differences are the first fruits of inquiry. That the planets move among the fixed stars, that iron expands when heated, or that children resemble their blood relatives—these are examples of such qualitative propositions.

Both in daily life and in the sciences, however, it is often essential to replace propositions simply affirming or denying qualitative differences by propositions indicating in a more precise way the *degree* of such differences. It is essential to do so in the interest of *accuracy* of statement, as well as in the interest of discovering *com-*

prehensive principles in terms of which the subject matter can be conceived as systematically related. Thus we may believe that there is more unemployment this year than last, or that the winters during our childhood were more severe than those during the past few years. But it may be important to know *how much* more unemployment there is, or *how much* less severe the winters have become; for if we can state the differences in terms of degrees of differences, we not only guard ourselves against the errors of hasty, untutored impressions, but also lay the foundation for an adequately grounded control of the indicated changes. Similarly, in the sciences we wish to know *how far* the planets are away from us, *how rapidly* they are moving, *how much* iron expands under known conditions of heating, and *how great* is the degree of resemblance between different members of a blood kinship. Such information gives us great practical control over the subject studied; it also makes possible a formulation for it of principles that are capable of *unambiguous* confirmation or refutation.

Theoretical and practical considerations lead us, therefore, to replace *qualitative* distinctions by *quantitative* ones. Quantitative distinctions are employed by many people who would be unable to offer an adequate analysis of what such distinctions mean, or to explain how they may be justified. The mother who says to her friend, "My Johnny is a head taller than your Frankie," very likely has never reflected on the difficulties of analyzing the meaning of her judgment. If pressed to offer the grounds for her assertion, she may stand the boys back to back and note by "looking" that Johnny does top Frankie by a head. But the same mother will be totally at a loss to interpret what she means when she says, "Johnny is twice as good as Frankie in arithmetic."

The employment of numbers to indicate qualitative differences requires a careful examination if it is not to lead us into error and absurdity. If our daily life and the sciences dealt with matters no more complex than that of comparing the heights of children, complicated methods of registering differences would never be used. Measurement, calculation, and the often difficult deduction of consequences from premises, would not require the elaborate techniques which they in fact do require. In every quarter, however, we find it necessary to employ a more intricate machinery of stating, gathering, and estimating evidence than that which is supplied by an untrained look or touch. Very few investigations can be carried through without the introduction of quantitative methods at some point. A study of the method of science must not, therefore, omit the study of the foundations of *applied mathematics.*

§ 2. THE NATURE OF COUNTING

What, then, are the ways of introducing precision into the judgments we make? In many inquiries, *counting* the individuals who possess a certain character is the only possible method of avoiding vague ideas. Are there more children under ten years of age in New York City than in London? Were there more industrial establishments in the United States employing less than ten people in 1900 than in 1920? "General impressions" on such questions are too vague to be reliable. It would be very unsafe to develop a comprehensive social theory (to assert, for example, that the progressive industrialization of a country is accompanied by the elimination of small-sized industries) if no empirical check upon our speculations were possible other than that supplied by vague impressions. But an unambiguous answer can be given to questions such as those cited by making an *actual count* of the individuals who belong to the respective classes.

Counting is undertaken not for its own sake, but because we suspect significant connections between the groups counted. Therefore we do not make a numerical inventory of all the groups of individuals we can find. Enumeration is undertaken on the basis of hypotheses expressing our sense of relevance. Such hypotheses play a controlling rôle at every stage of inquiry. It is clear, moreover, that the comparison of groups by enumerating their members can be made only if the groups are themselves unambiguously distinct from one another. We therefore employ counting to make precise our ideas, *subsequent* to our having acquired sufficient knowledge about a subject to permit us to distinguish various features in it.

Counting is subject to the limitation that only a discrete group, or a subject matter which may be manipulated so as to take on the form of a discrete group, may be counted. We can count the inhabitants of a city, because each inhabitant is distinct from every other inhabitant. We cannot count the drops in a glass of water unless we find some way of separating the drops from one another, and unless we introduce some convention as to what we shall regard as a drop.

The great importance of counting as a method of clarifying our ideas arises from the fact that the *number* of individuals in a group represents an *invariant property* of that group. For suppose we wish to count the apples in a bag. We take out the apples one by one, and place each apple in correspondence with a distinct member of a series of standard objects like our fingers, the numerals, or the letters of the alphabet. Suppose the first apple in the bag is placed in correspondence with the letter *A*, the second with the letter *B*, the third with *C*, and the remaining apple with *D*. We then say that there are *four* apples. The important property of this number is that had we taken out the apples from the bag in a different order and matched them with the letters, the last apple would nevertheless have still been matched with the letter *D*. The number of a collection obtained by counting is, therefore, a *constant* character of that collection; it does not depend on who does the counting, or on the order in which the objects are counted. Applied arithmetic is in part a collection of rules by means of which this invariant property can be most easily found.

Many of the difficulties that accompany the enumeration of groups arise from the difficulty of interpreting *what* it is that is counted. In many inquiries counting can be performed easily and without ambiguity, because the groups enumerated are readily distinguished. We can count the number of men and the number of women in a community, because the different biological functions of men and women makes it impossible to confuse them. But where the lines of cleavage between groups is not so distinct, the interpretation of the numbers obtained by counting is uncertain. Thus it is not easy to draw the line between skilled workers and the unskilled; and while we may count the number of individuals in each group, the result will be infected with all the ambiguity that attaches to the notion of a "skilled worker."

The gathering and interpretation of information about many important social matters is attended by difficulties of a special kind. Such information is generally obtained from written or oral questionnaires submitted to only a part of the population; and it must never be forgotten that the accuracy of the tabulated form of such information cannot exceed the accuracy with which the questionnaires are answered. Allowance must be made for ignorance, dishonesty, and vanity. No amount of mathematical manipulation of the results of counting can eliminate the incalculable inaccuracies in the replies. Thus the United States census for 1890 called for the color of the respondee: whether he or she was black, mulatto, quadroon, or octoroon. Since most people are ignorant as to the *meaning* of these distinctions, and many more are not in a position to know in what classification they themselves belong, it is safe to say that the answers, even if honestly given, were unreliable. The questions asked in a census must be drawn up with great care: they must not appertain to matters about which most people are not accurately informed. Information obtained by means of questionnaires concerning the number of days the respondent has been employed, or concerning the itemized account of his yearly expenditures, is in most cases worthless. A similar evaluation must be made of the growing practice of submitting questionnaires to insufficiently trained students on problems of sex, economics, or politics.

Personal vanity and dishonesty are factors often as important as ignorance. In one British census, information was requested whether the respondent was an employer or an employee. A surprisingly large number of employers was reported, a larger number than was consistent with information based on independent sources. This discrepancy was explained, plausibly at any rate, as arising from the unwillingness of the respondent to suffer the humiliation of appearing before the census-taker as merely an employee. Most enumerations based on answers to questionnaires regarding religious and social beliefs, or the prevalence of physical or mental disorders, are sure to be unreliable, because the answers given are very likely to be influenced by the fear or sense of shame of the respondents.

If the groups we are investigating are large in numbers, or difficult to examine exhaustively, it may be impossible or financially prohibitive to undertake an enumeration of their members. In such cases we resort to taking samples. The limitations of the procedure of sampling we shall consider later. The distinctive feature of this process consists in concluding that the proportion of characters found in the sample is the same as the proportion in the entire collection. It involves the type of argument we have called reasoning from samples, or statistical inference.

§ 3. THE MEASUREMENT OF INTENSIVE QUALITIES

Comparisons based upon counting, as we have seen, depend on our ability to distinguish clearly between different groups or different characters. Frequently, however, characters cannot be sharply distinguished because they form a continuous series with one another. Thus we may wish to distinguish different knives on the basis of their "sharpness," different woods on the basis of their "hardness," different children on the basis of their "alertness." For some purposes it is sufficient to know that one piece of wood is harder than another, employing such rough criteria of the hardness of a wood as the ease with which we can drive a nail into it. But we often want to know just how hard one piece of wood is as compared with any other kind of wood, and we then require a more certain and uniform criterion than the one suggested. We wish, if possible, to assign numbers to indicate the different *degrees* of hardness; and we often do so. The numbers so assigned are said to *measure* the varying degrees of the quality. What principles must we observe in using numbers to denote such differences in qualities?

We must be on guard against a common error. It is often believed that because we can assign numbers to different degrees of a quality, the different degrees always bear to each other the same ratio as do the numbers we have assigned to them. This is a serious mistake, and arises because it is supposed that measurement requires nothing more than the assigning of numbers. As we shall see, not all qualities can be "measured" in the same sense. Thus when we say that one tank contains 100 quarts of water and another 50 quarts, it is legitimate to say, as we shall soon find, that the first tank contains *twice as much* water as the second. In this case, the ratio of the volumes is the same as the ratio of the numbers. But when we say that the temperature one day is 100° and on another 50°, is it permissible to say that the temperature on the first day was *twice as much* as on the second? Or when we find that one student has an I.Q. of 100 and another an I.Q. of 50, is it correct to say that the first student is *twice as intelligent* as the second? An analysis of the conditions of measurement will show that the last two assertions are strictly without meaning.

We must note that numbers may have at least three distinct uses: (1) as tags, or identification marks; (2) as signs to indicate the *position* of the degree of a quality in a *series* of degrees; and (3) as signs indicating the *quantitative* relations between qualities. On some occasions numbers may fulfill all three functions at once.

(1) The numbers given to prisoners or railroad cars serve only as convenient ways of *naming* these objects. Numbers are more convenient than verbal names, because a "name" can be found for any new individual brought into the group by simply taking the number one greater than the last number that has been so employed. When numbers are used for this purpose, most people recognize that no relation between the objects numbered corre-

sponds to the numerical relation between the numbers assigned. The prisoner numbered 500 is not five times as dangerous or wicked as the one numbered 100. It is not even always true that Convict No. 500 entered the prison later than Convict No. 100, since the same number can be assigned several times without confusion.

(2) A scientifically more important use of numbers is when the *order* of numerical magnitude is the same as the *order* of the position of the character studied in a scale or ladder of qualities. Suppose we wish to distinguish bodies from one another with respect to their being harder or softer. We may then accept the following definition of what it means for one body to be harder than another: Diamond is harder than glass if diamond can scratch glass but glass cannot scratch diamond; and one body will be said to be just as hard as a second body if neither can scratch nor be scratched by the other. We may then arrange bodies in a scale of hardness if we can show experimentally that relations like the following hold between every triplet of unequally hard bodies: Diamond is harder than glass, glass is harder than pine wood, diamond is harder than pine wood. The relational property of "being harder than" is then shown to be asymmetrical (if B_1 is harder than B_2, B_2 is not harder than B_1); and transitive (if B_1 is harder than B_2, and B_2 is harder than B_3, then B_1 is harder than B_3. We can then arrange bodies in a linear series of hardness and thus get a scale or "ladder" of this quality.

Suppose now we have 100 different unequally hard bodies B_1, B_2, . . . B_{100} arranged so that B_1 is the hardest and B_{100} is the softest body, in conformity with the above conditions. We may wish to assign numbers to them to indicate their relative hardness in such a way that the order of numerical magnitude is the same as the order of relative degrees of hardness. (This can be done, since the relation of magnitude of numbers is asymmetrical and transitive.) But what number shall we assign to body B_1? We may decide to assign to it the number 0, or 1, or 25, or in fact any number we please. Suppose we decide on 1 for B_1, and also on 100 for B_{100}, and agree moreover to designate 2 as the hardness of B_2, 3 as the hardness of B_3, and so on.

These choices, however, were in no way forced upon us. We may have decided on 1 for B_1, 5 for B_2, 10 for B_3, and so on. *In terms of the procedure* we have followed in arranging the bodies in a scale of hardness, no meaning can be attached, therefore, to the statement that B_{50} is *twice as soft* as B_{25}. This statement has no meaning because the only relations we have defined, in arranging the bodies in the scale, are the relations of transitivity and asymmetry with respect to being capable of scratching. The statement falsely suggests that because one body is "higher up" the scale than another, it "contains more" of something called "hardness." And it falsely suggests, because one body is supposed to contain more of this something, that it contains a unit amount of it *a certain number of times.* Both of these suggestions must be ruthlessly eliminated. They arise from the mistaken idea that hardness is something which can be *added.* But there is nothing in the process of constructing the scale which can justify this. Hardness and softness, like temperature, shape, density, intelligence, courtesy, are *non-additive* qualities. Such qualities are frequently called *intensive.* They can be "measured" only in the sense that the different degrees of the quality may be arranged in a *series.* Concerning them, questions of *how much* or *how many times* are meaningless.

§ 4. THE MEASUREMENT OF EXTENSIVE QUALITIES

We turn to the third use of numbers. They can sometimes be employed to measure quantitative relations in the strict sense, so that answers to the questions, "How much?" and, "How many?" can be given in terms of them. Suppose we consider a set of bodies and that we wish to measure their weights. In order to do this, we must be able, in the first place, to construct a scale or ladder of weights in a manner similar to establishing a scale of hardness. We will agree, for example, that one body, R, is heavier than another body, S, if when R and S are placed in the opposite pans of a beam balance, the pan containing R sinks. We must then establish experimentally that the relation of "heavier than" is transitive and asymmetrical. We will also agree that body R is *equal in weight to* (or is as heavy as) R' if R is not heavier than R' and R' is not heavier than R; this means that neither pan of the balance will sink when R and R' are placed in opposite pans.

We are able not only to construct a scale of degrees of weights. We can also find an interpretation *in terms of some operation upon bodies* for such a statement as that one body is three times another in weight. An interpretation is possible because weights can be *added.* The physical process of addition is *the placing of two or more weights together* in the same pan of the balance. Let us now find three bodies, B, B', B'', which are equally heavy, and place them in one pan; place another body, C, in the other pan so that the beam will balance. The body C is then as heavy as the three bodies

B, B', B'' combined, and is *three times* as heavy as any one of them. This procedure can be extended to define a series of standard weights. In terms of this procedure it becomes significant to say that one object is n times as heavy or $1/n$th as heavy as another.

But we have not yet done enough to be sure that numbers assigned by such a process have all their familiar meanings. We have shown that weight is an additive property as contrasted with hardness, which is not. We must also show, again by experiment, that the numbers so assigned to weights are consistent with themselves. We must make sure that we do not allow *different* numbers to be assigned to the same object. Thus suppose the weight of object A is regarded as the unit or 1, and that we can assign weights to other objects by this process so that A_2 will have weight 2, A_4 weight 4, and A_6 weight 6. Can we be sure that A_2 and A_4 placed together in one pan will just balance A_6 placed in the other? It is very important to note that we cannot be certain of this until we perform the experiment. The proposition that $2 + 4 = 6$ can be demonstrated in *pure arithmetic* without experiment. But until we perform the proper experiments we cannot be certain that the *physical operation* of addition of weights does conform to the familiar properties of pure arithmetical addition. The physical operation of addition of weight possesses the usual formal properties of arithmetical addition only in *some* cases, not in *all:* the beam balance must be well constructed, its arms must be of equal length, and so on.

The method of measuring weights can be employed to measure other properties as well. Lengths, time intervals, areas, angles, electric current, electric resistance, can be measured in the same way. These properties are additive: we can find a process such that combining two objects having a property we obtain an object with an increased degree of that property. Properties which are additive are frequently called *extensive*. They can be measured in accordance with the processes indicated in this section. Such measurement we shall call *fundamental.*

§ 5. THE FORMAL CONDITIONS FOR MEASUREMENT

We may now state the conditions for measurement in abstract language. The minimum requirements for employing numbers in order to "measure" (in the loosest sense of the word) qualitative differences, are stated in the first two conditions.

1. Given a set of n bodies, B_1, B_2, . . . B_n, we must be able to arrange them in a series with respect to a certain quality so that

between any two bodies one and only one of the following relations holds: (a) $B_i > B_j$; (b) $B_i < B_j$; (c) $B_i = B_j$. The sign $>$ and its converse $<$ symbolize the relation on the basis of which the bodies can be distinguished as differing in the quality studied. The relation $>$ must be asymmetrical.

2. If $B_i > B_j$, and $B_j > B_k$, then $B_i > B_k$. This condition expresses the transitivity of the relation.

These two conditions are sufficient for the measurement of intensive qualities, such as temperature or density. They are necessary but not sufficient for extensive measurement. For the latter we require some physical process of addition, symbolized by $+$, which must be shown by experiment to possess the following formal properties:

3. If $B_e + B_f = B_g$, then $B_f + B_e = B_g$.
4. If $B_i = B_i'$, then $B_i + B_j > B_i'$.
5. If $B_i = B_i'$ and $B_j = B_j'$, then $B_i + B_j = B_i' + B_j'$.
6. $(B_i + B_j) + B_k = B_i + (B_j + B_k)$.

Measurement in the strict sense is possible only if all these conditions are satisfied. When only the first two conditions are satisfied, it is nonsense to make statements which imply that all six have been shown to hold. When we assert that one man has an I.Q. of 150 and another one of 75, all that we can mean is that in a *specific* scale of performance (requiring certain specialized abilities) one man stands "higher" than the other. It is nonsense to say that the first man has twice the intelligence or the training the other has, because no operation for adding intelligence or training has been discovered which conforms to the last four conditions necessary to make such a statement meaningful.

§ 6. NUMERICAL LAWS AND DERIVED MEASUREMENT

When we have once established a standard series of measures for any quality of bodies, we measure any further instance of that quality by comparing it with some member of the standard series. A standard series of lengths, for example, is embodied in a platinum meter kept at Paris under certain physical conditions. More or less exact duplications of it are distributed throughout the world. If anyone wishes to know how long a piece of cloth is, he will juxtapose in known ways the cloth and a meter measure or a yardstick. Direct judgments of comparison are therefore required to evaluate the length of the cloth. Similar processes are used for other measurable qualities.

But measurements of qualities are rarely performed for their own sake. They are made in order that precise relations between different properties of bodies may be established. Measurements in laboratories are carried on for the sole purpose of discovering the numerical laws which connect physical properties.

Let us examine one such numerical law. Most people are familiar with the property of liquids and solids which is called their "density." They know in general that it is the density which determines their buoyancies in water. It is not always known, however, what the relation of density is to the other properties of a body. Suppose we wished to measure the densities of the following five liquids: gasoline, alcohol, water, hydrochloric acid, mercury. We will agree to call one liquid, say mercury, more dense than water if we can find some solid body which will float on mercury but sink in water. By experiment we can then show that density so defined is an asymmetric, transitive relation, and that the liquids can therefore be arranged in a series of increasing density. The order of densities will in fact be the order in which we have written down the names of the liquids. We discover, however, that density is not an additive property of a liquid, and that we can measure it only as an intensive quality. We can then assign the numbers 1, 2, 3, 4, 5 to designate the positions of the liquids in the density scale. These numbers, as we have already pointed out, are arbitrary.

The reader may know, however, that altogether different numbers are usually assigned for the densities, numbers which are not arbitrarily chosen. The reason for this is that many intensive qualities can be measured in another way than by simply arranging them serially. And density is one of them.

This other way is fairly well known. It depends on the existence of a numerical law between other properties of the liquids, with which their density property is invariably related. For when we weigh different volumes of a liquid, say water, we discover experimentally that the *ratio* of the numbers measuring the weights and volumes of this liquid is the *same,* no matter how large or small the volume we measure. We thus establish a *numerical law* between the properties of weight and volume of a liquid. This law is that $W = cV$, where W is the measure of the weight, V that of the corresponding volume, and c is a constant for *all* samples of the same liquid but is a different constant for other liquids. By a proper choice of the units of weight and volume, we find that c has the value .75 for gasoline, .79 for alcohol, 1 for water, 1.27 for hydrochloric acid, and 13.6 for mercury. We also make the important discovery that the *order* of these ratios is the same as the order of the density of the liquids when this is determined in the way we did above. This ratio, which is constant for all samples of a homogeneous liquid, can therefore be taken as measuring its density. But we must be on our guard not to say that the density of mercury is 13.6 "times" that of water. For density, no matter how measured, is a nonadditive property. It can be measured precisely, and numbers assigned without arbitrariness to different degrees of it, only in virtue of a connection between weight and volume. This connection can be expressed as a *numerical law* between properties which can be measured by a *fundamental* process. Density can be measured only by a *derivative* method.

Numerical laws play a very important rôle in scientific inquiries. The discovery of numerical laws between qualities which can be measured in the strict sense, that is, by a fundamental process, enables us to measure carefully many intensive properties, such as temperature, density, buoyancy, elasticity, or efficiency of machines. Only by the aid of numerical laws can we measure the temperatures of the distant stars, or the blood pressure in the arteries of living things. But it is important to note that unless some properties were measurable by a fundamental process, numerical laws would be impossible, and the derivative measurement of intensive qualities could not be performed. (However, properties which are measurable by a fundamental process may also be measurable derivatively.) This explains, in part, some of the difficulties in the way of the social sciences. Precise estimates of intensive properties cannot be made because fundamental measurements in social matters are difficult, and because few numerical laws can be found which connect such intensive properties with extensive ones.

Numerical laws represent certain invariable relations between physical properties. Science aims not only at establishing such laws singly, but also at finding how different numerical laws are themselves connected with one another.

Suppose, for example, we let two circular cylinders roll down on two different inclined planes. The cylinders differ in the radii of their right sections, and the planes are inclined to the horizontal at different angles. If we wish to find the law connecting the distance traveled by each cylinder with the time, we may discover that for the first cylinder the law is $d = .20t^2$ and for the second it is $d = .35t^2$. These laws have the same "form." But the numerical constants in them are different, and *seem* to be unrelated to one another.

The science of physics tries to discover some other numerical law which will explain the variation of these numerical constants, as we employ different cylinders and different inclined planes. And physics is remarkably successful. It shows that the numerical law for the behavior of the rolling cylinder can be expressed in the form $d = ft^2$, where f itself is connected with the gravitational constant, the inclination of the plane, the coefficient of friction, the radius of the section of the cylinder, and the distribution of matter in the cylinder. Thus the sciences seek more and more general invariant laws which will account for many special features in a complicated phenomenon. But such search can meet with success only when the different properties of bodies have been distinguished by processes of measurement.

CHAPTER XVI

STATISTICAL METHODS

§ 1. THE NEED FOR STATISTICAL METHODS

In the preceding chapter we have been discussing counting and measuring as ways of making precise our ideas about things. But both processes yield large collections of numerical data, and very soon we may become embarrassed by our riches. We then require some method of handling the multitude of numerical results so that we may perceive and express clearly the significant relations between the properties studied. The method of concomitant variation, when applied to large collections of instances, obviously requires the use of statistical methods.

We may, for example, be interested in the tallness of males in the United States, because we suspect that height is influenced by environment. We may, therefore, measure the heights of several million males. But several million figures could not possibly be compared by us with an equally large collection of data from studies of the environment unless we found some way of compressing each set. We are all psychologically limited, and can attend to only a relatively small number of things at a time.

The physicist may be in a similar difficulty as the result of repeatedly measuring the wave length of a certain line in the solar spectrum. He may use different methods and try each method several times. But in general he will not get exactly the same value each time he measures, and he must therefore find some way of summarizing his results if he is to compare the wave lengths of different spectrum lines.

Although in many fields measurements can be made with some degree of uniformity, the number of independently varying factors may be large. It may therefore be extremely difficult to establish constant relations between them. Sometimes, however, if very large collections of such data are made, certain very general tendencies may be detected. For example, the weather is proverbially uncertain. It depends upon a large group of factors which cannot be isolated one at a time. Nevertheless, although the weather cannot be predicted with accuracy, the comparison of large collections of meteorological data does enable us to find a few helpful correlations. It is important, therefore, to examine the technique employed in compressing and comparing the data obtained from enumeration and measurement. The methods used to evaluate *group phenomena* by an analysis of data supplied by enumeration and measurement comprise the science of *statistics.*

The first step toward simplifying numerical data consists in classifying the information under suitable heads. The nature of the classification depends upon the purpose of the inquiry. Very often *frequency tables* are helpful in giving us an oversight of the material. Thus, we may measure the heights of schoolchildren and find that they vary from 2 feet 6 to 5 feet 6. For most purposes it is not important to know the exact height of each child to a fraction of an inch; we may then find how many children have heights between 2 feet 6 and 2 feet 7, between 2 feet 7 and 2 feet 8, and so on. No general directions can be given as to how large the intervals should be taken in constructing such frequency tables.

The distribution of the frequencies among the different intervals must often be expressed in a much more summary way. For this, two types of statistical numbers are employed. One type is designated as *statistical averages.* In general, these indicate what may be called the position of the distribution, the value around which the different items center. The second type is designated as *dispersion* or *deviation* numbers. They indicate the extent of variation of the items with respect to one of the averages. For two sets of items may have the same central tendency although the amount of deviation of the two sets is very different. Thus in the two sets of numbers 3, 4, 5, 6, 7, and 1, 3, 5, 7, 11, the amount of the dispersion around a common center of distribution is different. Other types of statistical numbers may also be used to characterize a distribution, for example, the symmetry of the distribution around the center; but we shall not be concerned with them.

§ 2. STATISTICAL AVERAGES

How shall we choose the number to represent the central tendency of a group of quantities? What conditions do we wish to impose upon averages, and what significance shall we attach to them? There are several kinds of averages, each with distinct advantages and limitations. No average is good for every purpose, and each average is good for some purpose. In general, however, averages are used for the following reasons: (1) They are required to give a synoptic representation of a group; (2) they are employed as means of *comparing* different groups; (3) they are used to characterize a whole group on the basis of samplings made from it. Consequently, there are some obvious properties we should like them to have.

1. Averages should be defined so explicitly that their numerical value does not depend upon the caprice of the individual who calculates them.

2. Averages should be a function of all the items in the group; otherwise they do not serve as representatives for the whole distribution.

3. Averages should be of a fairly simple mathematical nature, so that they may be calculated with ease.

4. Averages should be capable of algebraic manipulation. If, for example, we know the average of each of two series of heights, we should like to be able to compute from the two averages the average of the larger series obtained by combining the two series.

5. Averages should be relatively stable. If we make several samplings of a group, the averages of the different samples will be different. We frequently desire an average in which such differences will be as small as possible.

The Arithmetic Mean

The most familiar average is the *arithmetic mean*. This is obtained by adding together the set of quantities and dividing the sum by the number of terms. If the number of hours a student sleeps on the successive days of a week is 7, 6, 6, 5, 8, 7, 9, the arithmetic mean is $^{48}\!/_7$ or $6\,^6\!/_7$ hours. The reader will notice that the mean does not correspond to the number of hours the student sleeps on any one night. This clearly indicates that averages represent *group* characteristics, and do not supply information about any *individual* in the group.

The arithmetic mean satisfies the first, second, and third condition for averages we have stated above, and we shall see that the fourth is satisfied as well. But the reader must be cautioned at this point against the appearance of accuracy which arithmetical manipulations may introduce. We can express the mean of the hours the student slept in decimal form, and state the result as 6.85914 hours, or 6 hours, 51 minutes, 25.7 seconds. The *arithmetic* is accurate enough. But the result is misleading if it suggests that the observation of time spent in sleeping is as exact as the mean seems to indicate. The student may have reported the time he spends sleeping roughly in hours. He may have counted an actual time of 6 hours, 15 minutes, as simply 6 hours. Consequently, we must recognize that the precision which is a result of numerical computation is fictitious unless the observations have been made with the same degree of precision.

Is the mean a satisfactory basis for comparing two groups? If the mean income in one community is $1,500 and in another $1,100, is it correct to infer that the members of the first community are better off than those in the second? An example will show that such an inference may be false if the arithmetic mean is not supplemented by other information. Suppose that in one college class its members have the following sums with them: 8 students have 50 cents each, 4 have 75 cents each, 2 have $1.50 each, 1 has $11, and 1 has $27. The average wealth of the class is $3. And suppose further that in a second class, 9 students have a dollar each, 4 have $1.50 each, 1 has $2, and 1 has $3. The average wealth of this class is $1.66⅔. Now although the average of the first class is *greater* than the average of the second, 12 students in the first class (three-fourths of the entire class) have *less* money than *any* of the students in the second class. And if we examine the construction of the arithmetic mean we can understand why it is often an unreliable basis for making comparisons. For its value is very much influenced by extreme variations, and in a case like that above the presence of a relatively few very wealthy individuals in a group may "pull up" the mean altogether out of proportion to their number. This is only another way of saying that two groups may have the same mean although the range of variation within the groups may be very different. The mean supplies no information about the homogeneity of a group. That is why measures of dispersion are also required in statistical work.

In spite of this drawback, the arithmetic mean is an important average because of the ease with which it can be calculated and because of its mathematical properties. It is capable of algebraic manipulation. Thus suppose a student receives the following grades in his subjects one year: 80, 75, 95, 60, 70, so that the average is 76; and 80, 70, 60, 75, 65, a second year, so that the average is 70. What is the average for all his grades for the two years? We can add the ten grades and divide by 10. But we can also add the two averages and divide by 2, so that 73 is the average for the two years. This algebraic property of the mean is a great convenience.

The arithmetic mean is also connected with the mathematical theory of probability. Suppose a chemist makes several hundred measurements on the weight of oxygen. No two measurements yield the same result. What then is the "true value" of the weight of oxygen? If we make certain assumptions about the way the results of the measurements can vary, for example, that all the measurements are made with equal care, the *most probable value* of the weight of oxygen is given by the arithmetic mean.

The Weighted Mean

In many cases the simple arithmetic mean will not do. Thus an instructor may divide the work of a semester into two parts. He may ask a certain student to recite five times during the first half-semester, and give him the grades 10, 9, 8, 10, 8; during the second half-semester, he may call on him only twice, giving him the grades 0, 4. Suppose, now, the instructor were to calculate his *final* grade by finding the mean for the first half-semester, which is 9, the mean for the second half-semester, which is 2, and then obtain the simple arithmetic mean of these. The student's final grade would be 5.5. Would that be fair? If he can assume that the work during each half-semester was equally important and difficult, the student would be right in believing that it is not. He would have a good case in urging that the averages for each half-semester should be *weighted* corresponding to the number of times he was called on to recite. The *true final grade* would then be computed as follows: $\frac{5 \times 9 + 2 \times 2}{5 + 2} = 7$, and the student would receive a passing grade. The numbers 5 and 2 used to multiply the averages are called the *weights.*

It is evident, however, that in this illustration the weighting was not necessary, since the student could have calculated the final mark as a simple arithmetic mean by using all the grades on which the partial averages were based. In such examples weighting is only an arithmetical convenience. A more characteristic use for the weighted mean is found in estimating the change in cost of living during a period of years. We shall consider a slightly absurd illustration. Suppose that for the following five items, taking the price per unit in 1910 as par or 100, the prices in 1920 were: wheat 120, beef 110, iron 105, jewelry 50, hair tonic 40. The arithmetic mean of these items is 85 in 1920. We cannot conclude that the cost of living has gone down, since the articles listed are not generally regarded as "equally important." We may, accordingly, assign different weights to them to indicate our sense of their *relative importance.* Suppose we agree that the following numbers represent the importance of these articles in the order stated: 10, 9, 7, 2, 1. The weighted mean is obtained by calculating the value of $\frac{10 \times 120 + 9 \times 110 + 7 \times 105 + 2 \times 50 + 1 \times 40}{10 + 9 + 7 + 2 + 1}$; it is 105.7, indicating that the relative cost of living has gone up. The determination of the weights in such examples is a very complicated matter, and in the nature of the case a considerable arbitrary element enters into it. For "relative importance" is a nonadditive character; even if we succeed in arranging a set of objects in the *order* of their relative importance (a difficult undertaking in itself), the assigning of numbers to the different items cannot be made without the influence of conventional and subjective factors. However, except when the system of weights is chosen in an unusual way only slight changes in the value of the weighted mean take place when different systems of weights are employed.

The Mode

The *mode* is the item in a group which occurs most frequently. The mode is therefore often regarded as the "typical" representative of the group. Popular references to the "average man" must generally be regarded as denoting the "modal man." The mode of the sums possessed by the members of the first college class mentioned on page 305 is 50 cents.

What are the special merits of the mode? Like all averages, it represents the distribution of characters within a group. But it may represent the nature of the group more successfully than the mean does, since it indicates the largest subgroup in a collection and thus indicates which sort of character will be found most frequently. If the supply-sergeant in a regiment puts in an order for uniforms, he will be guided by the modal measure of the heights and girths of the men who are to wear them. The value of the mode is not influenced by extreme fluctuations in the group and may therefore serve as a fair basis for comparing different groups. If the nature of a collection is determined by samples judiciously drawn from it, the mode may often be used with better results than the mean, since it is a more stable average.

Most of the conditions, however, which we have found useful to demand of averages (page 304) are not satisfied by the mode. In the first place, while the mode is defined unambiguously as the item with maximum frequency, the position of the maximum can sometimes be radically shifted by a different classification of the items in the group. Thus suppose that in an examination of 47 candidates the grades were distributed with the following frequencies:

Grades lying between intervals....	0-20	20-40	40-60	60-80	80-100
Number receiving such grades	4	7	11	15	10

The modal grade lies between 60 and 80, that is, greater than 60 and equal to or less than 80. The intervals may, however, have been chosen differently. Suppose the classification had been as follows:

Grade lying between intervals....	10-30	30-50	50-70	70-90	90-100
Number receiving such grades	8	8	13	14	4

Now the modal grade lies between 70 and 90, that is, greater than 70 and equal to or less than 90. If the passing grade had been determined by the lower boundary of the modal interval, more of the candidates would have been failed by the second method of getting the mode than by the first.

Very frequently no single well-defined type exists in a group, either because the frequency with which *all* of the items occur is very much the same, or because several distinct frequency maxima can be found. If, for example, we are studying wage statistics, we may find two or more wage scales which occur with relatively high frequencies. In such cases we cannot speak of *the* mode. The existence of several "peaks" in the distribution indicates the absence of homogeneity in the group examined. It may be, to continue the illustration, that there are various kinds of grades of labor, for each of which there is a distinct modal wage; but when these different grades are lumped together, the distribution of wages will show several maxima.

Moreover, the mode may in fact not be typical even if it does correspond to the maximum frequency in a group. Thus suppose that in a community the incomes of its members varied considerably. It may happen that a dozen people receive an income of $1,500, while the rest of the community, numbering several hundreds, have incomes no two of which are alike. While $1,500 would then be the modal income, it would by no means be typical.

We must also note that the mode is not a function of *all* the items in a group, so that if several items are eliminated the mode need not be affected. While this is often an advantage, there are occasions when it is desirable that the value of the average should depend on the values of *all* the items. Furthermore, no simple arithmetical process can be found for calculating the mode, so that in practice the determination of the mode is very often difficult and inaccurate. Finally, the mode of a combined group cannot be calculated from the modes of its component groups. This is a serious drawback in theoretical investigations. The chief merit of the mode is its relative stability in the face of repeated samplings. But this is an unimportant advantage when the group is known to be homogeneous, since other averages can be employed in those cases.

The Median

The *median* is the middle term in a series of items when these are arranged in order of magnitude. It follows that a collection with an odd number of members always has a median. The median of the numbers 3, 4, 4, 5, 5, 5, 7 is 5. When the number of members is even, the median is usually defined as the mean of the two middle terms. The median of the group 40, 50, 50, 60, 70, 90 is 55. The median is therefore the term in a series arranged in size order such that there are as many items above it as there are below it in the series.

Unlike the mean, the median is only slightly influenced by the presence of extreme fluctuations in a group. It is therefore a relatively stable average and can be employed to compare ordered groups with respect to the position of their middle term. And, unlike the mode, the median can be determined with accuracy and ease. The chief use of the median, however, lies in those fields where theoretical or systematic considerations have the least force. It has no algebraic properties which would permit the calculation of the median of a group from the medians of its component groups. It has found favor in psychological and social measurements because while it is not often possible to make *fundamental* measurements in these fields, it is quite frequently possible to establish a serial order or a scale of characters. For the median is obtained by the *position* of a term in a series, not by the additive properties of the terms. Thus the arithmetical mean of the I.Q.'s of a group of children represents nothing about the group, and is just nonsense if it is interpreted as denoting the intelligence of the group. But the median may be employed as the basis for comparison in such cases; it is significant if children can be placed in a series of increasing abilities. If, therefore, 95 is the median value of the I.Q.'s in one class and 105 is the median in another, we can say under usual circumstances that there are more children in the second class who can meet a certain specific standard than there are in the first.

The median is sometimes believed to be the value such that the values in the group greater and smaller than it occur with equal frequency. This is not always the case, especially where the characters studied do not form a *continuous* series. Thus when 337 buttercups were examined for the number of their petals, it was found that 312 had 5 petals, 17 had 6 petals, 4 had 7 petals, 2 had 8 petals, and 2 had 9 petals. The median value of the petals is 5. But clearly it is not true that there are as many buttercups in the group having more than 5 petals as there are with less than 5.

§ 3. MEASURES OF DISPERSION

We have seen that groups may differ from one another not only with respect to their central tendencies, but also in the extent of the "spreading" of their members.

The Range

One simple way of indicating the amount of dispersion in a group is to state the *range of variation*. This is the numerical difference between the smallest and the largest item in the group. If incomes in the United States vary from $500 a year to $10,000,000, the range is $9,999,500. But this is not a satisfactory method, since in the first place the extreme values of the variation may be unknown, and in the second place the addition or elimination of a few incomes at either extreme may seriously affect the range. Moreover, the range does not tell us how the different incomes are *distributed* within the range. The two groups of numbers 1, 5, 5, 6, 6, 7, 7, 7, 10 and 1, 2, 2, 2, 2, 10 have the same range, although the form of distribution within the range is different in each case.

The Mean or Average Deviation

We can find more sensitive methods to indicate the extent of variations. Suppose the heights of a group of men, measured in inches, are 61, 63, 64, 65, 65, 66, 67, 68, 69, 72. The mean height is 66 inches. We now calculate the *deviation* or *error* of each measure from the mean by subtracting the mean from each measure. (We may take any average as the base from which to calculate the deviations. For the sake of simplicity, we shall restrict ourselves to the mean.) The deviations are: -5, -3, -2, -1, -1, 0, 1, 2, 3, 6. We may be tempted to take the arithmetic mean of these numbers. But this is useless, since the sum of the deviations from the mean of a group is always zero. We can, however, neglect the signs of the deviations, and then calculate their arithmetic mean. It is called the *average deviation,* or *average error,* or *mean deviation.* The average deviation in this case is $^{24}/_{10}$ or 2.4.

The average deviation assigns the same importance to large deviations as it does to small ones. In general, the smaller the average deviation the more concentrated the items will be around the mean. All the considerations which have been mentioned in the discussion of the arithmetic mean are relevant here also.

We must observe, however, that a "large" average deviation does not necessarily indicate a "large" fluctuation in the values of the group. To be "large" is relative to a standard. If we repeatedly measured the height of a mountain, the mean of the measures might be 5,000 feet and the average deviation 10 feet; compared with the mean, the average deviation is a small number. If, however, we made measurements on the length of a city block, an average deviation of 10 feet would be considerable. For this reason, the average deviation is sometimes divided by the average from which the deviations are taken, and is then called the *coefficient of dispersion.* In the foregoing example of measuring the heights of men, this coefficient is 2.4/66, or .036 +.

The Standard Deviation

For many purposes, especially where considerations based on the theory of probability dominate, the standard deviation is taken as a measure of dispersion. It is obtained by dividing the sum of the squares of the deviations from the mean by the number of items, and then extracting the square root of this quotient. In the example of the measuring of heights, we get $\frac{25+9+4+1+1+0+1+4+9+36}{10}$, or 9, as the mean of the sum of the squares of the deviations. The standard deviation is $\sqrt{9}$, or 3. If $x_1, x_2, \ldots x_n$ are the deviations from the arithmetic mean of n items, σ_x the standard deviation, then $\sigma_x^2 = \frac{\Sigma x^2}{n}$.

The standard deviation is so constructed that it emphasizes the extreme values of the deviations. For by squaring each deviation, the larger errors are given greater weight in the sum than the smaller ones. Nothing of much value can be said concerning the uses of the standard deviation unless the assumptions made about the group of values for which it is calculated are known. In general, however, it is a measure of dispersion which is influenced less by fluctuations in sampling than are other measures. If the group is approximately symmetrically distributed about the mean, and if a distance equal to the standard deviation is marked off on each side of the mean, about two-thirds of the items in the group will lie inside these limits. For our often-mentioned example, these limits are indicated by 66 ± 3. And indeed about two-thirds of the heights do lie between 63 and 69.

The Quartile Deviation

Another simple measure of deviation is obtained by arranging the items in order of magnitude and finding those three items which divide the series into four equal parts. These are called the *first quartile,* the *second quartile* (or median), and the *third quartile.* If Q_1 is the first quartile and Q_3 the third, the quartile deviation is defined as $\frac{(Q_3 - Q_1)}{2}$. It is clear that half the items of the group must lie between the first and third quartiles, so that the extent of the dispersion is roughly measured by the quartile deviation. In our example Q_1 is 64, Q_3 is 68, and the quartile deviation is $\frac{(68 - 64)}{2}$ or 2. If the term halfway between the first and third quartiles is found, one-half of the items will lie between this term and the quartiles. For this reason, the quartile deviation is sometimes called the *probable error.* If we employ the notation 65.5 ± 2 (where 65.5 is the term halfway between the first and third quartiles, and 2 the quartile deviation), there will be as many measures *within* the limits indicated (63.5 and 67.5) as *outside* these limits. It is assumed, in other words, that when we pick at random it is "just as likely" we will pick a measure lying between these limits as that we will pick a measure outside the limits. But this choice of name is unfortunate and confusing, since "probable error" is used to designate other ideas in the literature of this subject.

§ 4. MEASURES OF CORRELATION

The object of all inquiry is the discovery of significant relations within the subject matter studied. The object of statistical studies is to facilitate the discovery and expression of relationships between different groups of characters. We collect vital statistics for the purpose of *comparing* such things as birth rate, death rate, pauperism, in one year with the same things in other years. We collect data on the number of industrial accidents and the hours of employment in the several factories in order to ascertain the relation, if any, between these two sets of phenomena—in order to discover whether they are causally connected or whether they are partially or completely independent of each other.

We have already discussed averages and measures of dispersion which make a more or less precise comparison between groups possible. For many purposes these statistical numbers are all that we require. Thus we can compare incomes in a community during different years by means of one of the averages and a measure of the scattering. Sometimes ratios such as percentages are a useful aid. Has the population of Germany from 1900 to 1910 been growing more rapidly than that of France during that period? The *percentage* increase of population during these years will generally serve as a measure of the growth. Is there a correlation between the hooked or aquiline nose and Jewish descent? The discovery that in fair samples of Jews, only 14 per cent have the "characteristic Jewish nose" is an unambiguous reply.

Occasions arise, however, when none of the statistical numbers so far discussed can be used satisfactorily. Suppose we examined several hundred tree leaves for their corresponding length and width. Is there any connection between the length and the width of leaves? We may entertain the belief, on the basis of general impressions, that the longer the leaf, the wider it is. But when the number of items examined is large, we cannot depend on hasty impressions, since we can neither keep in mind all the items nor detect the significant relations between them. We may then try to place the leaves in order of increasing length, in order to ascertain whether this is also the order of increasing width. If the two orders are the same, we would doubtless conclude that there is some connection between the length and the width of a leaf. If the two orders coincide only in part and not completely, we might still suspect some relation. We would require, however, some *numerical measure* of the correlation between the length and the width of the leaves. Two variables are said to be *correlated,* if in series of corresponding instances of the variables an increase or decrease in one of them is accompanied by an increase or decrease in the other, whether in the same direction or in opposite directions. When the changes in the variables are in the same direction (both increase or both decrease) the correlation is *positive;* when the changes are in opposite directions (one increases, the other decreases) the correlation is *negative.*

Several measures of correlation have been devised. We shall consider only the one known as *Pearson's coefficient.* We must, however, omit the derivation of this coefficient because of the technical nature of the argument required, and simply indicate its definition and use. Let $h_1, h_2, \ldots h_n$ be the values of one variable h, and $w_1, w_2, \ldots w_n$ the corresponding values of another variable w. This means that when h has the value h_1, w has the value w_1, and so on. Further, let $x_1, x_2, \ldots x_n$ represent the deviations of the first variable from the arithmetic mean of its n instances; and $y_1, y_2, \ldots y_n$ the corresponding deviations of the second variable. The symbols σ_x and σ_y will represent, as usual, the *standard deviations* of the two series. Pearson's coefficient is then defined as $r = \frac{\Sigma(xy)}{n\sigma_x\sigma_y}$, where $\Sigma(xy)$ stands for the sum of all the products of the corresponding deviations, so that the formula may be read: the mean of the products of the deviations divided by the product of the two standard deviations.

We shall calculate this coefficient to determine the measure of correlation between the age of husband and wife in a group of twenty couples. The following table supplies the necessary information:

Age of husband	*Age of wife*	*Deviation of age of husband from mean*	*Deviation of age of wife from mean*			
h	*w*	*x*	*y*	x^2	y^2	*xy*
22	18	− 8	− 8	64	64	+ 64
24	20	− 6	− 6	36	36	+ 36
26	20	− 4	− 6	16	36	+ 24
26	24	− 4	− 2	16	4	+ 8
27	22	− 3	− 4	9	16	+ 12
27	24	− 3	− 2	9	4	+ 6
28	27	− 2	+ 1	4	1	− 2
28	24	− 2	− 2	4	4	+ 4
29	21	− 1	− 5	1	25	+ 5
30	25	0	− 1	0	1	0
30	29	0	+ 3	0	9	0
30	32	0	+ 6	0	36	0
31	27	+ 1	+ 1	1	1	+ 1
32	27	+ 2	+ 1	4	1	+ 2
33	30	+ 3	+ 4	9	16	+ 12
34	27	+ 4	+ 1	16	1	+ 4
35	30	+ 5	+ 4	25	16	+ 20
35	31	+ 5	+ 5	25	25	+ 25
36	30	+ 6	+ 4	36	16	+ 24
37	32	+ 7	+ 6	49	36	+ 42
$\frac{\Sigma h}{n} = 30$	$\frac{\Sigma w}{n} = 26$			$\frac{\Sigma x^2}{n} = \frac{324}{20}$ or $\sigma_x = 4.02$	$\frac{\Sigma y^2}{n} = \frac{348}{20}$ or $\sigma_y = 4.17$	$\frac{\Sigma (xy)}{n} = + \frac{287}{20}$

$$\text{Consequently, } r = \frac{\Sigma(xy)}{n\sigma_x\sigma_y} = \frac{287}{20 \times 4.02 \times 4.17} = \frac{287}{335.27} = +\ .856.$$

Pearson's coefficient is so constructed that its numerical value is positive when the correlation is positive, and negative when the correlation is negative. Moreover, its value always lies between + 1 and − 1, + 1 indicating a perfect positive correlation and − 1 a perfect negative correlation. A coefficient of 0 indicates no correlation; in that case, from a knowledge of the way changes occur in one variable we can infer nothing about the corresponding changes in the second variable.

§ 5. DANGERS AND FALLACIES IN THE USE OF STATISTICS

In spite of the great value of statistical numbers, they can be misused and misinterpreted when the assumptions which are required for their use are forgotten. We must therefore caution the reader against the following elementary but frequent blunders:

1. Statistical numbers supply, in a very summary way, information about the characteristics of a *group* of items. They do not supply any information about any *one* item in the group. From the knowledge that male births are approximately one-half of all births, we can infer nothing about the sex of the next child to be born.

2. Statistical averages cannot, without further study, be interpreted as representing *strictly invariable* relations within a group. Buckle in his *History of Civilization in England* argued from the statistics of murders, suicides, marriages, and letters in the Dead Letter Office to the conclusion that "murder is committed with as much regularity, and bears as uniform a relation to certain known circumstances, as do the movements of the tides, and the rotation of the seasons. . . . Suicide is merely the product of the general condition of society, and . . . the individual felon only carries into effect what is a necessary consequence of preceding circumstances. In a given state of society, a certain number of persons must put an end to their own life. This is the general law; and the special question as to who shall commit the crime depends, of course, upon special laws; which, however, in their total action, must obey the large social law to which they are all subordinate. And the power of the larger law is so irresistible, that neither the love of life nor the fear of another world can avail anything towards even checking its operation. . . . Even the number of marriages annually contracted, is determined, not by the temper and wishes of individuals, but by large general facts, over which individuals can exercise no authority. . . . We are now able to prove that even the aberrations of memory are marked by this general character of necessary and invariable order." [1] Buckle's conclusions are not supported by his evidence. The number of suicides per year may remain constant over a period of years; but it does not follow that a certain number of suicides *must* be committed yearly. For we do not know, in the first place, the precise factors (if there are any) which make for suicide; and we do not know, in the second place, that these precise factors will continue to operate every year.

[1] Vol. I, Chap. I.

3. Coefficients of correlation are subject to similar criticism. Pearson's coefficient, for example, is defined so generally that *any* two groups may be examined for their degree of correlation, even if we know on other grounds that the two groups are in fact independent of each other. Thus, in the notation we have employed, x may denote the deviations of the ages of men listed in a dictionary of biography, y the deviations of the number of pages in the books listed in the Congressional Library, the first name listed in the dictionary being made to correspond with the first book in the catalogue, and so on. The coefficient may have a high numerical value. It would not, however, indicate any significant connection between the two groups. With patience, indeed, many correlations may be calculated. The expenditure for the British Navy has been shown to be highly correlated with a growing consumption of bananas, and the spread of cancer in England with the increased importation of apples. But most correlations of this nature are known to be, or are suspected of being, altogether fortuitous and without causal significance.

High coefficients of correlation are not sufficient evidence for invariable connections, since it is frequently difficult to interpret the coefficient because its value may be consistent with more than one hypothesis. Suppose we find that there are an increasing number of arrests during a period of years. Can we infer that this is due to an increase in the number of crimes? May it not be due to an increased severity in the enforcement of the law? Variations in the number of people classified as paupers (because they receive community aid) may be correlated either with the changes in the administration of such aid, or with variations in the age distribution of paupers, or with changes in wages, price level, or employment. Which correlation shall take precedence? Is it safe to affirm a causal connection between climate and the character of a civilization simply on the ground that civilizations of a certain type are found in regions with a certain climate? May not the nature of the civilization be also correlated with the relative freedom of access foreign traders have to that region?

4. It is very easy to commit an error in believing there is a significant connection between two types of events on the basis of the observation that they are frequently associated. Thus, suppose we found that 90 out of 100 red-headed people have quick tempers. May we infer validly that red hair and quick tempers are connected in some special way? Certainly not, unless we also have some information about the relative number of people who are not redheads and who are quick-tempered. For 90 out of 100 people having

other than red hair may also be quick-tempered. Therefore if we wish to discover the connections of an attribute *A* with an attribute *B*, we must not only discover the proportion of *A*'s which are *B*'s, but also the proposition of *a*'s (the absence of *A*) that are *B*. From the knowledge that 29.6 per thousand deaf-mutes are imbecile we cannot validly infer that imbecility and deaf-muteness are dependent attributes unless we know that the proportion of imbeciles in the whole population is less than 29.6, say only 1.5 per thousand. A full moon and fine weather are found frequently together; but if we notice that the absence of a full moon and fine weather are associated as frequently, we will not be inclined to read any significant meaning into the joint occurrence of the first pair of events.

5. High correlations are sometimes obtained simply by mixing two sets of data in which no correlation is found. If, for example, the ages of husbands and wives are uncorrelated in each of two communities, it can be demonstrated rigorously that when the records are mingled, *some* correlation will be found in the new collection, unless indeed the mean of the ages of husbands and wives is the same in both communities. This correlation is the consequence of the purely *mathematical* properties of the two groups, and cannot be taken as evidence for an invariable connection.

This raises the difficulties which come from *sampling*. All correlations can be calculated for groups with a finite number of items only. But almost always we wish to employ the value of the coefficient to indicate the degree of correlation between groups which are more inclusive than those originally examined. But it clearly does not follow that because in one community the coefficient of correlation between the ages of husband and wife is .856, the same relation holds throughout a larger community, or in a community with different social customs. The coefficient, like all statistical numbers, is subject to fluctuations of sampling. And in some cases a relatively high correlation may be altogether casual. Thus if a pair of dice are thrown 100 times and the coefficient of correlation be calculated for the number of points uppermost on each die, *r* may have a value considerably larger than 0, although if the two dice are independent it should be exactly zero. Whether such a greater value of *r* is to be interpreted as indicating some dependence between the dice cannot be determined from *r* alone.

6. In arguing from samples, many fallacies are committed because the samples may have been selected, consciously or not, so that they are not fair or representative of the entire collection. This source of fallacy is especially great where only a few samples have been

taken or where our knowledge of the subject matter and its relevant factors is slight.

There is always a danger from an unwitting selection of material in comparing different groups. A recruiting sergeant will convince most people with the following argument. The death rate in the United States Navy during the Spanish-American War was 9 per 1,000, while the death rate in New York City for the same period was 16 per 1,000; it is therefore safer to be a sailor in the navy during a war than a civilian in New York City. But an examination of the evidence for this conclusion soon shows that the two death rates have not the significance which they appear to have. For the New York City death rate includes the mortality of infants, old people, people in hospitals and asylums; and it is well known that the death rate for the very old and the very young, as well as for the sick, is relatively high. On the other hand, the navy is composed of men between the ages of 18 and 35, each of whom had been judged fit in a rigorous physical examination. It follows that the two death rates do not warrant the conclusion that the navy is a safer place than New York City. Adequate evidence for such a conclusion would require the comparison of two groups which are homogeneous with respect to age, sex, and health.

Many illustrations of this type of fallacy can be cited. At a certain university it was discovered that undergraduates belonging to one racial group ranked higher in college grades and in intelligence than students of other racial groups. Can we conclude that this racial group has greater aptitude for scholarship than other races? Such an inference would be very precarious, especially when it is suspected that racial barriers exist. The differences in achievement can very easily be explained by the more stringent entrance requirement imposed upon members of this race. Consequently, the students at this university who are of this race constitute a highly select group, and would be expected normally to do superior work. Similarly, the discovery that married men in Italy require less medical attention than do the unmarried does not prove that marriage has hygienic values. The difference may arise from the unwillingness of chronically ailing men to marry, and the better health of the married group may then be explained on the grounds of selection of material.

7. A common error in comparing groups is the use of absolute numbers instead of percentages. In a well-known book on socialism, the author tried to show that Marx's prediction of the progressive elimination of small commercial establishments and the growth of large ones was contrary to the facts. He used the numbers of the

following table, without calculating the percentages, to compare the number of commercial establishments in Germany at two periods.

Commercial Establishments	*1882*		*1895*	
	Number	*Percentage*	*Number*	*Percentage*
With no employees	429,825	61	454,540	47.6
With 1 to 5 employees	246,413	35	450,913	47.
With 6 to 50 employees	26,531	3.8	49,271	5.2
With over 50 employees	463	.06	960	.1

The author concluded that, contrary to Marx, the number of small establishments was on the increase. If, however, we calculate the percentages for each type of establishment, the table tells a different story. The numbers at the right in each column state the percentages of that type of establishment for the total number of establishments; these numbers were not given by the author of the book. Clearly the table does not prove what the author supposes it does. For while in 1882 the small shops were 61 per cent of the total, in 1895 they were only 47.6 per cent of the total. Moreover, while the small shops increased in number by 6 per cent, medium-sized shops increased by 83 per cent, large shops by 90 per cent and the very large ones by 107 per cent.

8. Obvious difficulties arise when we make comparisons on the basis of units or classifications which do not retain the same value or meaning for the different groups compared. It is simply useless to compare the number of convictions for larceny in different countries unless we know that the basis for classifying crimes and the principles of court procedure are the same in the countries compared. It does not necessarily follow that men have greater freedom in the twentieth century than they had in the eighteenth simply because the number of democracies is larger; for in spite of the different way governments are classified, in substance they may differ very little.

In comparing incomes over a period of years, allowance must be made for the changes in the "real value" of the monetary unit. From the fact that in 1853, .263 per cent of the population of Germany was taxed for incomes over 3,000 marks, while in 1902, 1.301 per cent of the population was so taxed, it cannot be inferred that the economic lot of Germans has improved. For in terms of what he could buy with that sum a person with an income of 3,000 marks in 1902 may not be better off than a person with a 2,000 mark income in 1853: important changes in the cost of living have taken place meanwhile.

A similar difficulty arises in comparing the wellbeing of a country on the basis of an increase in the number of stocks held. Thus, the number of stockholders at two periods in the United States in several railroads is given by the following table:

Railroad	*1904*	*1908*
Pennsylvania	42,100	59,600
New York Central	11,700	22,000
Union Pacific	14,200	15,000
Southern Pacific	4,400	15,000
Erie	4,300	10,000
Chesapeake and Ohio	1,500	2,600

What does this prove about the *total* number of people who hold stock? Unless we can be sure that a man does not hold stock in several companies, we cannot conclude that a greater number of people held railroad shares in 1908 than in 1904; and we know, in fact, that some financiers hold stock in as many as a hundred different corporations. Moreover, even if there were an actual increase in the number of different shareholders, the concentration of stocks, and therefore of wealth and power, may be greater in the second period than in the first.

9. A very insidious source of error comes from neglecting to differentiate changes in the subject matter with changes in the methods of collecting statistical data. Is cancer on the increase, or is it merely that we have more accurate reports concerning the disease? Is the death rate for heart disease increasing, or is the change noted simply due to the fact that deaths heretofore reported as caused by a different ailment, for example, acute indigestion, have received a new classification? The infant mortality rate is defined as the ratio of $\frac{\text{deaths under one year of age}}{\text{births}} \times 1{,}000$. But since the smaller the denominator, the larger the ratio, it is possible to lower infant mortality without saving a single infant life by improving the birth registration.[2] Can we infer that, because the United States census report for 1900 states a larger percentage of smaller commercial establishments in the country than did the report of 1890, the relative number of such establishments is increasing? No, indeed! The census report for 1900 itself cautions its readers against such

[2] I. S. Falk, *The Principles of Vital Statistics*, 1923, p. 74.

an inference, for it states explicitly that the enumeration of such establishments was more thorough in 1900 than in 1890.

10. Frequently different pictures of a social situation are obtained by employing different units to make comparisons. It is often difficult to know how certain features are to be "measured." Is there a growing concentration of industry, the progressive disappearance of small establishments, and the emergence of a few large corporations? But how shall we measure such a concentration—in terms of the number of employees per establishment, or in terms of the size of output per establishment? With the growth of complicated machinery, the latter may be a better index of concentration than the former. It has been shown, for example, that while between 1904 and 1909 there was an increase in the United States in the *number* of establishments producing goods with an annual value of $20,000 or less, in 1904 these establishments claimed only 6.3 per cent of the *total value* of all the products and in 1909 only 5.5 per cent of the total value.

11. We have already pointed out that the appearance of precision introduced by the mathematical methods of statistics is misleading if we suppose that the accuracy of the data is thereby increased. We do not increase the sensitivity of a measuring instrument by carrying out to a larger number of decimals the numerical value of the mean of several measurements. Nevertheless, statistical methods may show that the original tabulation of data is inaccurate because the data are not consistent with one another. If, for example, we study 1,000 individuals, and find that 550 of them are males and 500 are females, we know that some error *must* have been made. Sometimes, however, the inconsistency of the data is not so obvious, and more elaborate methods of testing for consistency must be used. Thus, we may tabulate the result of a study on a group of 1,000 students at a university as follows:

Freshmen	525	Married freshmen	147
Male	312	Married males	86
Married	470	Married male freshmen	25
Male freshmen	42		

Although there is nothing *apparently* wrong with these figures, it can be shown that they are not consistent. For these figures imply that the number of unmarried female students not in their freshman year should be -57, which is absurd. The discussion of the tests of consistency, however, is too technical for an elementary treatment.

CHAPTER XVII

PROBABLE INFERENCE IN HISTORY AND ALLIED INQUIRIES

§ 1. DOES HISTORY EMPLOY SCIENTIFIC METHOD?

The present, it is often asserted, can be understood only in terms of the past, and the study of the past is thus declared to be the gateway to all rational conduct. This dictum is frequently accompanied by another. History is often contrasted with the natural sciences. The latter aim to discover an abstract law or theory with no specific temporal reference, while the former is the study of individual things and of events having an ineradicable temporal locus. From this it is inferred that history cannot employ the logical methods so successful in the natural sciences, but must develop methods of inquiry which are unique. It is a valid inference from these dicta that if they are true the general logical methods we have been discussing cannot further our understanding of the present.

We must therefore examine whether scientific method is thus limited. Is it true that the study of the past makes no use of the characteristic features of scientific method? Is it true that history has no need of hypotheses, of their deductive elaboration, and of the verification or refutation of some of their logical consequences?

Originally, *historia* meant any learning or knowing achieved through an inquiry, and in the phrase "natural history" it is still used in that sense. In general, however, "history" is limited to the study of past events. We can thus speak of the history of the stars, of the earth, of living organisms, or of the various human arts, sciences, or social institutions. But while we shall understand by "historical knowledge" the knowledge of any sequential temporal subject matter, we shall confine our illustrations to the history of man.

We must now examine the dictum that knowledge of the past is prerequisite to a knowledge of the present. What in fact is the relative temporal order of our knowledge about the present and about the past? If the reader will reflect, it will strike him that the past is not the *original datum,* it is not *given immediately* to the historian. Every study of the past must begin with the study of things existing *contemporaneously* with such a study. The study of the past therefore begins with the present. This is no willful paradox. The events of the past described and interpreted by the historian cannot be observed by him directly; those events are no longer here to be observed, experimented upon, or physically dissected. The past can be reached therefore in only two ways: either (1) through the personal memory of the historian, and in that case it is his *present* memory with which he begins and whose accuracy he must evaluate; or (2) through the interpretation of objects as remains of past events. Such remains of the past which exist in the present may be written testimony (like chronicles, biographies, memoirs, public documents), oral traditions (like songs and tales), or remains of former buildings, implements, monuments, and fragments of long-deceased living beings.

It follows that the statement that the present can be understood only in terms of the past is not the entire truth. The nature of some things in the present must be known before the study of the past can get under way. But we may claim even more. History as the knowledge of the past, and indeed all knowledge, can be achieved only through inference. The premises of inferences concerning the past are obtained from a careful scrutiny of present material interpreted as deposits of the past. The evidential value of this material is determined on the basis of principles of interpretation or hypotheses which must themselves be tested by current events. Therefore the recognition that knowledge of the past must begin with things in the present makes clear not only that knowledge of the past has no exclusive priority over knowledge of the present, but also that the usual canons of scientific method are required in history as well as in the natural sciences.

It is the historian's task to examine the existing remains and testimony for their evidential value concerning the past. The logician examines the types of inferences employed by the historian. We shall study the historian's logical method in some detail, but we may anticipate the sequel in order to characterize it in a general way: The evidence the historian gathers for the past is never complete and conclusive, although it may have the character of yielding "proof beyond a reasonable doubt." In other words, the evidence is of such a nature that the historian's conclusions are *probable* with respect to it.

The theme we have been considering in previous chapters is therefore continued in this. But in applying some of the ideas of the theory of probability to questions of history great caution must be used. Ridiculous results have been obtained by an uncritical use of the mathematical theory of probability. For example, at the end of the seventeenth century John Craig calculated that if the evidence for Christianity had been oral only it would have lost its value by 800 A.D., but since it is also written, it will last until 3150 A.D., when the second coming of Christ will add fresh evidence. A similar remark was made by Mohammedan writers on the argument that the Koran had no evidence derived from miracles. "They say that, as evidence of Christian miracles is daily becoming weaker, a time must at last arrive when it will fail of affording assurance that they were miracles at all: whence would arise the necessity of another prophet and other miracles."[1] We shall see that the truth-frequency theory of probability, which we have previously stated, will find intelligible and interesting applications in the subject matter of the present chapter.

Because the technique of historiography is often not explicitly formulated, the rôle which hypotheses play in history is easily overlooked. The function of hypotheses is, however, not less important because they are implicit or employed tacitly. At what points in the historian's task are hypotheses required? The historian must, in the first place, *select* some period of the past for study, and within such a period select still further those events he regards as significant. Thus he may decide to study the American colonies from 1700 to 1765, but neglect to consider what John Adams had for breakfast on January 1, 1763. Now it is clearly an *assumption,* however justified, that the developments *within* a certain geographical area *during* an interval of time may be studied with little reference to any events outside the limits indicated. And it is also an *assumption* that within such a limited field some events may be neglected as unimportant. Each of these assumptions depends upon theories of social causation and human behavior, theories which color the historian's main results.

Theories determine what the historian selects in another way. For the material which serves as his original data may be frag-

[1] Augustus De Morgan, *A Budget of Paradoxes,* 2d ed., 1915, 2 vols., Vol. I, p. 131.

mentary, so that he can obtain no complete account concerning the subject he is dealing with; this is the case with the study of ancient peoples like the Egyptians. Or the material may be so abundant that the historian is swamped by it; this is the case for the events of the World War. In either case, hypotheses are required which enable the historian either to fill out the scant material with suggestions of connections where none are explicit, or to select what is significant from an overabundant supply.

The ubiquity of hypotheses in historical research will be seen more clearly if we regard the historian's task as a systematic attempt to answer the following questions. (These questions are not, in general, answered in the following serial order, nor can an answer to one be found without aids from answers to the others.)

1. Are the data of research *admissible* as evidence; are the sources *genuine?* This question may also be put as an inquiry into the origin and subsequent career of the data, and deals with such matters as the authorship of documents, and the character and competence of the authors.

2. What is the *meaning* of the assertions contained in the sources; what do the remains *signify?* The inquiry must examine the language, the purpose, and the social setting in which the sources originated.

3. Are the assertions elicited from the data *true;* can we rely upon the sources for information concerning the past? The answers to this question are closely dependent upon the answers to the first, since the authenticity of a document, for example, is often revealed by the information it supplies, and conversely.

4. What are the *explanations* for the past events; what are the systematic connections between the different assertions established as true, in terms of which we achieve an *understanding* of the past?

§ 2. THE AUTHENTICITY OF HISTORICAL DATA

Let us examine some typical illustrations for each of these inquiries, and analyze the type of inference employed.

Since in almost every case the originals of ancient documents have perished, it is important to know whether the copies we possess (which are themselves in turn copies of copies) reproduce authentically their originals. If only a single copy exists, it is very difficult to determine whether any portions of it have been modified—either by accident or by intention—by the copyists. Certain general assumptions are required. Thus if the substance of a manuscript is coherent and intelligible, it is assumed that an obscure passage is due to an error of a copyist. The argument may then be stated as follows: Inconsistencies and obscurities in most documents that are consistent and intelligible in the main, are usually errors. This document, intelligible in the main, has such and such obscure passages. Therefore, *probably* these passages are errors. The ground for our assuming the first premise is the experience with contemporary copyists which verifies it. The particular form of emendation which is then made depends upon further assumptions based on the specific character of the document. Anachronisms in a source, determined to be such on the basis of other information concerning the period when the document is purported to have been written, also offer ground for identifying errors and forgeries. A document purporting to have been typewritten in 1895 can be shown to be false if the paper or type is of later origin.

If more than one copy exists, each must be compared with the other for differences. It is generally assumed that copies containing the same errors (for example, misspellings, anachronisms, inconsistencies) in the same places have been copied either from one another or from a common source. This assumption, once more, is based upon our general experience, verified (in the immediate past and the present) in the behavior of typists, schoolboys, and so on. Such copies cannot be regarded as independent, any more than several hundred books printed by the same press and containing the same misprints would be regarded as so much independent testimony that the author's manuscript contained those errors. Thus if several books were not composed independently of one another, they cannot serve to corroborate one another as to the miraculous events which they report. In comparing *independent* copies, and in trying to "restore" the original from them, use is made of the knowledge we may have from *other* sources of the reputed author, and of assumptions concerning the consistency, style, and reliability of his thought.

In the evaluation of the genuineness of *remains,* such as implements, similar considerations apply. Thus it is a common assumption that things found on the surface of the earth are modern, as are objects found in caves of certain geological structure. These assumptions are based on our knowledge of physical processes, such as the action of glaciers, winds, tides, and so on.

The analysis of a document for its meaning and the discovery of inconsistencies often lays bare the fact that it was written by more than one man. Thus the first five books of the Old Testament have been traditionally regarded as written by Moses, even though no such claim is to be found in their text. But a careful study of the text reveals many inconsistencies. For example, there are two accounts of Creation in the first two chapters of Genesis, two accounts of the Great Flood, and so on. The attempts by orthodox believers to explain away the contradictions have been legion; all sorts of metaphorical constructions have been placed upon words for which no justification could be given other than that they saved the consistency of the traditional account. But an alternative method was open, one which was finally adopted by critical students. Thus in parts of the Old Testament the writer refers to God as Yahweh, in other parts as Elohim. A separation of the texts on the basis of the names attributed to God showed that each account was relatively complete and consistent, while at the same time there were marked differences between them in style, idiom, and general outlook. The assumption that the Pentateuch had a single author was gradually dropped. The Bible was carefully studied on the supposition that different portions of it were written by different men at different times, to be finally synthesized and edited many centuries after the reputed era of Moses. This alternative assumption concerning the authorship of the Bible not only has the advantage of not reading forced meanings into the text, but is also in greater conformity with our knowledge of general history and of the formation of the ancient epics of the Semites.

The determination of the authorship of a document is essential to decide whether the author was in a position to describe what he did or to give an honest and competent account of the matters he wrote about. The ways of fixing authorship are many, and we can illustrate only one. When the *Journal d'Adrien Duquesnoy,* consisting of letters purporting to be from Versailles and Paris during 1789-90, were first published in 1894, there was no clear evidence as to who wrote them. However, some of the known letters of Duquesnoy refer to "bulletins" which may be identified as the *Journal,* although neither of the two extant manuscripts of the *Journal* is in his handwriting. But an examination of the contents showed that all of the *Journal* must have been written by the same person, because the internal cross-references and the use of similar expressions in various parts of it made the theory of a plurality of authors highly improbable. Moreover, the contents showed that it was written by a member of the Third Estate representing Barrois, that he was on familiar terms with the deputies from Nancy, that he was a member of the Committee on Food Supplies, and that he addressed these letters to the people of Lorraine. Could the member of the assembly who satisfied all these conditions be found? Now the Committee on Food Supplies was composed of one representative from each of the administrative divisions of France, and the membership of the committee was printed in the records of the Assembly. An examination of these records showed that Duquesnoy had represented Lorraine. Furthermore, it was discovered that Duquesnoy did in fact represent Barrois in Lorraine, that he had formerly lived at Briey in Barrois, which he had represented in the Third Estate, and that prior to 1789 he had lived in Nancy. The evidence for the authorship of the *Journal* had therefore accumulated, so that "beyond a reasonable doubt" it could be fixed on Duquesnoy.[2] The form of the argument may be stated as follows: If an individual is the author of a document, he will meet certain conditions as to residence, associations, social rank, and so on. Duquesnoy meets a number of the conditions for the authorship of the *Journal*. He was, therefore, the author. The conclusion is probable on the evidence, in the sense specified by the truth-frequency theory of probability.

§ 3. ESTABLISHING THE MEANING OF HISTORICAL DATA

Before any information about the past can be obtained from the testimony of records, an exact determination must be made concerning *what* it is the testimony asserts. Much special equipment is required for this task, but the methods employed are readily comprehensible. We shall illustrate them by a classical example.

1. At the beginning of the nineteenth century, all knowledge of the Egyptian hieroglyphic signs had been lost. The soldiers of Napoleon had accidentally discovered the Rosetta stone, a slab of black basalt, upon which some events in the second century B.C. were detailed in both Egyptian and Greek. But the stone kept its secret until in 1822 Champollion was successful in deciphering the Egyptian with the aid of the accompanying Greek translation. We might examine his procedure, but it will be obviously simpler if we can follow the steps of his argument in another context. Fortunately, the manner in which the cryptogram in Poe's "The Gold Bug" is deciphered bears some striking resemblances to the method used by Champollion.

2 See F. M. Fling, *The Writing of History*, 1920, pp. 62-64.

Legrand in Poe's tale discovered a parchment upon which were traced a series of characters underneath a death's head, and apparently "signed" by a figure of a kid. He entertained the hypothesis that the code was written by Captain Kidd the famous pirate, and that the figure of the kid at the end of the code was intended as his signature. But, Legrand next argued, the message must then be in English, for in no other language could the pun on the name Kidd be appreciated. On the assumption that the code was English, Legrand counted the frequency with which each of the characters occurred, and found that "8" occurred 33 times, the semicolon 26 times, and so on. Now statistical studies on English have shown that the letter "e" occurs most frequently, while the most usual word is "the." He therefore assumed that "8" represented the letter "e," and he also discovered that a combination of three characters in the code, the last of which was "8," occurred more frequently than any other. He therefore ventured to identify the combination ";48" as the word "the," so that the semicolon represented "t," "4" stood for "h," and "8" for "e."

Having identified a single word, Legrand was able to determine the commencement and termination of other words. This he did by substituting successively the letters of the alphabet for the still unknown characters, in order to yield words in English, and so that together these made sense. For example, upon substituting the letters for the known characters, and dots for the unknown ones, he found the expression "thetreethr . . . hthe." This he was able to break up into the words "the tree thr . . . h the," so that the word "through" was clearly suggested, thus giving him the equivalents of three new characters. In this manner Legrand found ten of the most frequent letters of the language, and the remaining ones were easily suggested by the context, on the assumption that all the words were English.

Let us examine part of the argument. The letter "e," for example, was identified as follows: In any English composition of moderate length, the letter "e" occurs more frequently than any other. In this message, presumably English, the character "8" occurs more frequently than any other. Therefore, *probably,* the character "8" stands for "e." The reader should restate the argument and convince himself that it is formally invalid. Next examine the way "r" was identified. This was done by examining the sequence "t.eeth." A survey of familiar English words convinced Legrand that this could never form a word in that language, and he concluded that the final "th" must be the commencement of a new word; but "t.ee" suggested "tree." The argument is: No English word can be obtained by substituting a letter for the dot in "t.eeth"; but this document is in English; therefore "t.eeth" does not form a single word. Moreover, if "t.ee" is a word, the substitution of some letter for the dot will yield an English word, but the substitution of "r" for the dot does give an English word; therefore *probably* the character "(", which has been replaced by a dot, is the letter "r."

Thus on the supposition that the code is in English and that the sequence of words must be intelligible, Legrand solved it by judicious guessing and gradual apprehension of the structure of the cryptogram. The hypotheses which he employed were verified in so far as they explained the peculiarities of the cipher (the relative frequency of the occurrence of different characters, and so on) and in so far as they led to an intelligible set of directions. The hypotheses received further corroboration when the treasure indicated by Legrand's solutions was actually found. But it is logically possible that other solutions besides the one given can be discovered, and that the finding of the treasure was itself a remarkable coincidence. However, the nature of our experiences with coincidences is such that it is improbable (that is, it would happen with a very small relative frequency) that a treasure can be discovered on the basis of a false set of directions.

The reader may note that a great deal of our familiarity with our own and foreign languages is acquired in a way somewhat similar to this. No one consults a dictionary for the meaning of every strange word. Instead, we require to identify only a certain number of key words or phrases. When a sufficient number of identifications have been made, so that we acquire a sense of the *structure* of the language, the unidentified elements are recognized on the basis of an extension of the hypothesis we have made concerning the language.

2. When the *general* sense of the language of the document is known, the task of the critical historian is not over. Great care must be exercised if we do not wish to run the risk of *reading into* a book many of our own prejudices, rather than *reading in it* the actual instruction it contains. The interpretation of the great philosophers, for example, has been too often an exposition of the interpreter's own beliefs. This danger is particularly serious with religious documents. What the Old Testament asserts can be decided only if we know the intimate properties of the Hebrew language, the type of audience for which it was written, the peculiarities of Oriental psychology and literary style, and so on. (In some cases,

as in ancient Persian, it may be impossible to acquire a completely satisfactory knowledge of the idiom of the language, because it has fallen into disuse.) No dictionary, no grammar, no rhetoric, has been handed down to us by the ancient users of this tongue, and many of our interpretations are extremely conjectural. For example, the ancient Hebrew was written without vowels or punctuation, and the vowels and punctuation in use at present have been supplied by later students of the Old Testament. The preconceptions of the latter concerning scriptural interpretation have entered into this work.

The author's intent may sometimes be ascertained if the contents of a document are analyzed under various heads, so that the usage of words concerning a specified subject can be compared. The fundamental maxim to be observed was stated by Spinoza: "We must take especial care . . . not to confound the meaning of a passage with its truth, we must examine it solely by means of the significance of the words. . . ." [8] How the meaning of an obscure biblical passage may be elicited by a comparison of passages is illustrated by him in a noteworthy manner. Indeed, all of the *Tractatus*, and especially Chapter VII, may still serve as a model for textual criticism.

". . . The words of Moses, 'God is a fire' and 'God is jealous,' are perfectly clear so long as we regard merely the signification of the words, and I therefore reckon them among the clear passages, though in relation to reason and truth they are most obscure: still, although the literal meaning is repugnant to the natural light of reason, nevertheless, if it cannot be clearly overruled on grounds and principles derived from its Scriptural 'history,' it, that is, the literal meaning, must be the one retained: and contrariwise if these passages literally interpreted are found to clash with principles derived from Scripture, though such literal interpretation were in absolute harmony with reason, they must be interpreted in a different manner, *i.e.*, metaphorically.

"If we would know whether Moses believed God to be a fire or not, we must on no account decide the question on grounds of the reasonableness or the reverse of such an opinion, but must judge solely by the other opinions of Moses which are on record.

"In the present instance, as Moses says in several other passages that God has no likeness to any visible thing, whether in heaven or

[8] *Tractatus Theologico-Politicus,* in *Works,* tr. by R. H. M. Elwes, Vol. I, p. 101.

on earth, or in the water, either all such passages must be taken metaphorically, or else the one before us must be so explained. However, as we should depart as little as possible from the literal sense, we must first ask whether this text, God is a fire, admits of any but the literal meaning—that is, whether the word fire ever means anything besides ordinary natural fire. If no such meaning can be found, the text must be taken literally, however repugnant to reason it may be: and all the other passages, though in complete accordance with reason, must be brought into harmony with it. If the verbal expressions would not admit of being thus harmonized, we should have to set them down as irreconcilable, and suspend our judgment concerning them. However, as we find the name fire applied to anger and jealousy (see Job xxxi. 12) we can thus easily reconcile the words of Moses, and legitimately conclude that the two propositions God is a fire, and God is jealous, are in meaning identical.

"Further, as Moses clearly teaches that God is jealous, and nowhere states that God is without passions or emotions, we must evidently infer that Moses held this doctrine himself, or at any rate, that he wished to teach it, nor must we refrain because such a belief seems contrary to reason: for as we have shown, we cannot wrest the meaning of texts to suit the dictates of our reason, or our preconceived opinions." [4]

It is clear, however, that Spinoza's conclusions rest upon assumptions he does not make explicit. Thus he assumes that all the portions of the Bible to which he refers have been written by the same person. He also assumes that the writer was consistent in his use of language, and that his thought was consistent as well. The argument may be stated as follows: If a name or phrase has one meaning in one part of a document, it has the same meaning in all other parts where it occurs in relevant contexts; now the name "fire" is applied to anger and jealousy in several passages of the Bible; therefore, in all *remaining* relevant contexts, the name "fire" is applied to anger and jealousy. The argument, the reader will note, is only probable, for the truth of the consequent of the major premise does not invariably follow upon the truth of the antecedent. The evidence for the major premise comes from examination of books written by contemporaries whose intent and meaning may be directly established by questions addressed to them.

The import of a document as a whole may help determine the

[4] *Ibid.*, pp. 101-02.

meaning of some particular passage. What did Christ mean when he said: "Blessed are they that mourn; for they shall be comforted"? This passage does not indicate what mourners he had in mind. However, Christ teaches elsewhere that the highest good is righteousness, and that we should care for nothing else save the kingdom of God (Matt. 6: 33). It is safe to assume, therefore, that by mourners he meant those who mourn for the kingdom of God and the righteousness neglected by man. Lack of these things would be the only cause for mourning by those who despise the gifts of fortune.

The difficulties of the historian in establishing the meaning of a text are in some measure due to the fact that his witnesses cannot be questioned and made to answer. The determination of the meaning of historical data must therefore be more roundabout than the determination of the meaning of a living witness. But the methods do not differ in essential character. We shall devote §7 to the problem of clarifying meanings and weighing evidence in the courtroom.

§ 4. DETERMINING THE EVIDENTIAL VALUE OF HISTORICAL TESTIMONY

Not the least difficult task of the historian is the examination of the testimony for its evidential value concerning the past. In general, the criterion employed for determining the *truth* of the assertions in the testimony is agreement between independent, well-informed witnesses. A large part of the work of the critical historian therefore lies in fixing the identity and competence of the reputed authors; for the character of the authors, and the occasions at which and for which they wrote, are relevant factors in deciding whether any weight can be given to their writings. But other criteria must be employed as well. Are the events described such that they do not contradict well-established principles of the natural sciences? Are the events in accordance with what is known of the psychology of human behavior? Have the events verifiable consequences (for example, the discovery of remains) in the present and the future? Constant appeal is made to canons of evaluation for which the evidence comes from our contemporary experiences with things and with persons. The corroboration of an event by even a respectable number of independent witnesses will generally be held as of no account if the event is known to be contrary to the verifiable body of knowledge called science. As Hume remarked in his *Essay on*

Miracles,[5] the testimony of men is accepted because of "our observation of the veracity of human testimony, and the usual conformity of facts to the reports of witnesses." However, even here care must be exercised. An event reported to have occurred—such as the supposed miraculous healing of the sick—may indeed have occurred. But if we cannot accept the explanations of the reporters, it becomes difficult to know what did happen, since the explanation is generally inextricably bound up with the details of the account.

The difficulties are increased when it is remembered that although a writer may be in general reliable, the time and place at which he made his record may have been unfavorable to accuracy. Even a highly competent person may write about events so much after the date of their occurrence that little credibility can be attached to the record. Moreover, a document may be reliable as a whole but worthless as evidence with respect to certain items; or perhaps the testimony may be valuable in some parts even though incredible as a whole.

One of the most powerful instruments in the hands of the historian is the principle that an honest and competent witness will be consistent with himself and with other witnesses of like caliber. Discrepancies must be noted, and the temptation to make easy compromises must be resisted. "A says two and two make four; B says they make five. We are not to conclude that two and two make four and a half; we must examine and see which is right." [6] How this principle is applied is shown clearly in Strauss's discussion of the Davidical descent of Christ according to the genealogical tables of Matthew and Luke. We shall reproduce his investigation in part.

As is well known, two accounts of Christ's ancestry are given in the Gospels. Matthew declares that from Abraham to Christ there are three groups of fourteen generations each, and he lists this ancestry. But if we actually count the successive generations, we find fourteen names from Abraham to David, fourteen from Solomon to Jechonias, while from Jechonias to Jesus (counting the latter as one) there are only thirteen. How is this discrepancy to be explained? There is little doubt that the deficiency originated with the author himself and is not an error due to some copyists. And no matter how we count the generations, whether by excluding or including the first or last member of each group, some inconsistency remains. Moreover, if we compare the genealogy in Matthew with

[5] Pt. I.

[6] C. V. Langlois and Charles Seignobos, *Introduction to the Study of History*, tr. by G. G. Benny, 1906, p. 198.

the genealogies in the Old Testament, we find that many names recorded in the latter have been omitted by Matthew in order that the number fourteen in each group may not be exceeded. The genealogy in Matthew is therefore suspect. And although we cannot thus far say whether the author of Matthew was merely careless or whether he intentionally changed the older biblical account, the reliability of the author is seriously impaired, at any rate in this one respect. It is not implausible that the author wished to retain three groups of fourteen generations because of the common belief among the Jews that divine manifestations recurred at periodic intervals. "Thus, as fourteen generations had intervened between Abraham, the founder of the holy people, and David the king after God's own heart, so fourteen generations must intervene between the reëstablishment of the kingdom and the coming of the son of David, the Messiah." [7]

If we compare the genealogies of Matthew with those of Luke, our distrust will justifiably increase. Thus while the former has twenty-six generations between David and Christ, Luke has forty-one. But what is more important is that in some parts of the genealogies the two writers cite altogether different ancestors for Christ. With the exception of two names, all the names from David to Joseph the foster-father of Christ are different in the two accounts. Thus in Matthew the father of Joseph is Jacob, in Luke he is Heli; the son of David through whom Joseph is descended is Solomon in Matthew, he is Nathan in Luke; and so on. Many efforts have been made to reconcile these contradictions. Since it is extremely illuminating to exhibit a concrete application of logical methods and the conflict of hypotheses, we shall quote Strauss's examination of one prominent hypothesis attempting to explain away the discrepancies.

"[The hypothesis] is formed upon the presupposition of Augustine, that Joseph was an adopted son, and that one evangelist gave the name of his real, the other that of his adopted father; and the opinion of the old chronologist Julius Africanus, that a Levirate marriage had taken place between the parents of Joseph . . . by one of [the fathers] he was descended from David through Solomon, by the other through Nathan. The farther question: to which father do the respective genealogies belong? is open to two species of criticism, the one founded upon literal expressions, the other upon the spirit and character of each gospel: and which lead to

[7] D. F. Strauss, *The Life of Jesus Critically Examined*, tr. by George Eliot, 2d ed., 1892, p. 111.

opposite conclusions. Augustine as well as Africanus, has observed that Matthew makes use of an expression in describing the relationship between Joseph and his so-called father, which more definitely points out the natural filial relationship than that of Luke: for the former says *Jacob begat Joseph:* whilst the expression of the latter, *Joseph the son of Heli,* appears equally applicable to a son by adoption, or by virtue of a Levirate marriage. But since the very object of a Levirate marriage was to maintain the name and race of a deceased childless brother, it was the Jewish custom to inscribe the first-born son of such a marriage, not on the family register of his natural father, as Matthew has done here, but on that of his legal father, as Luke has done on the above supposition. Now that a person so entirely imbued with Jewish opinions as the author of the first Gospel, should have made a mistake of this kind, cannot be held probable. Accordingly, Schleiermacher and others conceive themselves bound by the spirit of the two Gospels to admit that Matthew, in spite of his *begat,* must have given the lineage of the legal father, according to Jewish custom: whilst Luke, who perhaps was not born a Jew, and was less familiar with Jewish habits, might have fallen upon the genealogy of the younger brothers of Joseph, who were not, like the first born, inscribed amongst the family of the deceased legal father, but with that of their natural father, and might have taken this for the genealogical table of the first-born Joseph, whilst it really belonged to him only by natural descent, to which Jewish genealogists paid no regard. But, besides the fact . . . that the genealogy of Luke can with difficulty be proved to be the work of the author of that Gospel:—in which case the little acquaintance of Luke with Jewish customs ceases to afford any clue to the meaning of this genealogy;—it is also to be objected, that the genealogist of the first Gospel could not have written his *begat* thus without any addition, if he was thinking of a mere legal paternity. Wherefore these two views of the genealogical relationship are equally difficult.

"However, this hypothesis, which we have hitherto considered only in general, requires a more detailed examination in order to judge of its admissibility. In considering the proposition of a Levirate marriage, the argument is essentially the same if, with Augustine and Africanus, we ascribe the naming of the natural father to Matthew, or with Schleiermacher, to Luke. As an example we shall adopt the former statement: the rather because Eusebius, according to Africanus, has left us a minute account of it. According to this representation, then, the mother of Joseph was first married to that person whom Luke calls the father of Joseph, namely Heli. But since Heli died without children, by virtue of the Levirate law, his brother, called by Matthew Jacob the father of Joseph, married the widow, and by her begot Joseph, who was legally regarded as the son of the deceased Heli, and so described by Luke, whilst naturally he was the son of his brother Jacob, and thus described by Matthew.

"But, merely thus far, the hypothesis is by no means adequate. For if the two fathers of Joseph were real brothers, sons of the same father, they had one and the same lineage, and the two genealogies would have differed only in the father of Joseph, all the preceding portion being in agreement. In order to explain how the discordancy extends so far back as to David, we must have recourse to the second proposition of Africanus, that the fathers of Joseph were only half-brothers, having the same mother, but not the same father. We must also suppose that this mother of the two fathers of Joseph, had twice married; once with the Matthan of Matthew, who was descended from David through Solomon and the line of kings, and to whom she bore Jacob; and also, either before or after, with the Matthat of Luke, the offspring of which marriage was Heli: which Heli, having married and died childless, his half-brother Jacob married his widow, and begot for the deceased his legal child Joseph.

"This hypothesis of so complicated a marriage in two successive generations, to which we are forced by the discrepancy of the two genealogies, must be acknowledged to be in no way impossible, but still highly improbable: and the difficulty is doubled by the untoward agreement already noticed, which occurs midway in the discordant series, in the two members Salathiel and Zorobabel. For to explain how Neri in Luke, and Jechonias in Matthew, are both called the father of Salathiel, who was the father of Zorobabel;—not only must the supposition of the Levirate marriage be repeated, but also that the two brothers who successively married the same wife, were brothers only on the mother's side. The difficulty is not diminished by the remark, that any nearest blood-relation, not only a brother, might succeed in a Levirate marriage,—that is to say, though not obligatory, it was at least open to his choice (Ruth iii. 12. f. iv. 4 f). For since even in the case of two cousins, the concurrence of the two branches must take place much earlier than here for Jacob and Eli, and for Jechonias and Neri, we are still obliged to have recourse to the hypothesis of half-brothers: the only amelioration in this hypothesis over the other being, that these two very peculiar marriages do not take place in immediately consecutive generations. Now that this extraordinary double incident should not only have been twice repeated, but that the genealogists should twice have made the same selection in their statements respecting the natural and the legal father, and without any explanation,—is so improbable, that even the hypothesis of an adoption, which is burdened with only one-half of these difficulties, has still more than it can bear. For in the case of adoption, since no fraternal or other relationship is required, between the natural and adopting fathers, the recurrence to a twice-repeated half-brotherhood is dispensed with; leaving only the necessity for twice supposing a relationship by adoption, and twice the peculiar circumstance, that the one genealogist from want of acquaintance with Jewish customs was ignorant of the fact, and the other, although he took account of it, was silent respecting it." [8]

As a consequence of such analyses, Strauss concludes that the genealogies are contradictory to one another and to the statements in the Old Testament, as well as to our well-based knowledge of social behavior and natural events. Neither table has any advantage over the other. Neither can be regarded as historical, for it is highly improbable that the genealogy of a family as obscure as that of Joseph should have been preserved during a period of exile and after. Strauss concludes, in answer to the question what historical result may be gleaned from the genealogies: ". . . Jesus, either in his own person or through his disciples, acting upon minds strongly imbued with Jewish notions and expectations, left among his followers so firm a conviction of his Messiahship, that they did not hesitate to attribute to him the prophetical characteristic of Davidical descent, and more than one pen was put into action, in order, by means of a genealogy which should authenticate that descent, to justify his recognition as the Messiah." [9]

The reader will note that Strauss's analysis was carried out on the assumption that Matthew and Luke were *independent* chroniclers of certain events, and that *some* of their testimony—for example, that Christ was a historical person, that he did have a foster-father, and so on—was *veridical*. These assumptions do not invalidate his analysis, although they gave rise to a long and important controversy in which an even more radical stand than Strauss's was taken toward the historicity of the Gospels. In any case, however, this

[8] D. F. Strauss, *op. cit.*, pp. 113-15.
[9] *Ibid.*, p. 118.

example of historical analysis makes clear the rôle of scientific method in history. Hypotheses are used (in this case to explain contradictions) which are deductively elaborated (in this case by the critic, to discover some consequences not compatible with the subject matter) and put to the test (in this case some are refuted, others confirmed). The conclusion which Strauss obtains is admittedly probable. Thus at one place his argument may be stated as follows: The genealogies of most obscure families are not preserved; but the family of Nathan is obscure; therefore, *probably,* the genealogy of his family has not been preserved. We constantly appeal, as in the present case, to the confirmation by contemporary experience of the major premises in such inferences.

In some cases the truth of a proposition concerning the past may not only be corroborated by independent witnesses, but also be verified on the basis of some calculable consequence which a past event may have in the present. Did Captain Kidd bury any treasure on the mainland of North America, as a certain document testifies? If he did, and if we dig in the places indicated, the discovery of a treasure would verify the proposition that he did. Was Tutankhamen's mother-in-law beautiful? If she was, the discovery of her mummy may verify that fact. Did the reader mail the letter he wrote two days ago? If he did, the receipt of a reply would confirm his doing so, just as the discovery of the letter in his pocket would refute it.

Principles of social behavior and the natural sciences may also help to confirm propositions concerning the past. Thus the city of Salamis is reputed to have been founded by Phoenicians. We may offer the following argument to support this: The name of a city is generally in the language of the founders; the city of Salamis bears a Phoenician name; therefore *probably* Salamis was founded by Phoenicians. So also tradition claims that Thales stopped a war by predicting an eclipse. Is this true? We may not be able to decide definitely, but we can corroborate the fact of the eclipse. Modern astronomy enables us to calculate backward as well as forward, and we find that in May, 585 B.C., an eclipse did occur which was observable in Greece, and therefore by Thales.

§ 5. SYSTEMATIC THEORIES OR EXPLANATIONS IN HISTORY

The confirmation of isolated propositions is no more sufficient in history than it is in the natural sciences. The propositions dealing with the past must be so connected that they form a coherent whole. Systematic theories in terms of which the past is "explained," or made intelligible, are essential.

Moreover, the reader is laboring under a very false impression indeed, if he thinks that ability of extraordinary nature is not required in the historian's attempt to make the past understandable. Sensitivity to significant meanings and connections is as imperative here, and requires the same type of genius, as in the case of the natural scientist who develops a fertile hypothesis where others see only insignificant events.

Let us illustrate this from the work of the legal historian Maitland. He discovered that the title of the kings and queens of England had ended (for two hundred and fifty years until abolished during the reign of Victoria's grandfather in 1800) with the words "Defender of the Faith and so forth," or the last phrase in Latin, *"et caetera."* The phrase, as it stands, is meaningless. Has it always been so?

Maitland found that the first sovereign who bears an *"et caetera"* in her title is Queen Elizabeth. Why was this included in her title? "Now let us for a moment place ourselves in the first day of her reign. Shall we not be eager to know what this new queen will call herself, for will not her style be a presage of her policy? No doubt she is by the Grace of God of England, France, and Ireland Queen. No doubt she is Defender of the Faith, though we cannot be sure what faith she will defend. But is that all? Is she or is she not Supreme Head upon earth of the Church of England and Ireland?" [10]

Now one of the statutes of her father Henry VIII declared that the headship of the church belonged to the Crown by the very word of God. But one of the statutes of her sister Mary declared that Henry's ecclesiastical supremacy was null and void, and that the title of such supremacy could not be conferred by Parliament but rested with the Pope. What was Elizabeth to do? Should she treat her sister's statutes as void, or should she declare herself against the statutes of her father? "Then a happy thought occurs. Let her highness etcetrate herself. This will leave her hands free, and then afterwards she can explain the etcetration as occasion shall require. Suppose that sooner or later she must submit to the pope, she can still say that she has done no wrong. . . . There are always, so it might be said, some odds and ends that might conveniently be packed up in 'and so forth.' . . . And then, on the

10 F. W. Maitland, "Elizabethan Gleanings," in *Collected Papers,* 1911, 3 vols., Vol. III, p. 157.

other hand, if her grace finds it advisable, as perhaps it will be, to declare that the Marian statutes are null, she cannot be reproached with having been as bad as her sister, for we shall say that no reasonable man, considering all that has happened, can have doubted that the '&c.' signified that portion of King Henry's title and King Edward's title which, for the sake of brevity, was not written in full. . . . Therefore, let her be 'defender of the faith, and so forth.' He who knows what faith is 'the' faith will be able to make a good guess touching the import of 'and so forth.'"[11]

This is the hypothesis to account for the *"et caetera."* Then, in a masterly fashion, Maitland considers the evidence which confirms it and the evidence which apparently contradicts it. He shows that his explanation of *"et caetera"* is in good agreement with all the known facts, and that the consequences which one should expect if the theory is true are in fact verified.

Maitland's problem was to find an explanation for the harmless-looking phrase "and so forth." What made him examine the phrase? No rule can be given to direct our attention to the significant factors in a situation. However, Maitland's argument may be stated as follows: If a person uncertain of his future behavior does not wish to bind himself, he will be as vague as he can be in his commitments about the future; now Queen Elizabeth employed a flexible phrase in her title; therefore *probably* she was uncertain, at her accession, what her position concerning the Church would be. In virtue of our familiarity with the psychology of persons in power, Maitland's discussion of the purpose and motives which led to the introduction of the *"et caetera"* into Elizabeth's title are "beyond a reasonable doubt." As a consequence, a number of isolated facts have been woven into a coherent pattern.

It must be admitted, none the less, that theories as comprehensive as those of the natural sciences, and those on the basis of which the development in time of social institutions could be predicted, have not been, and perhaps cannot be, achieved in human history. This is largely so because the subject matter of human history is more complicated, that is, it involves more factors, and therefore theories in human history cannot be formulated as precisely as in the natural sciences. They cannot be explored very easily in a deductive fashion, and consequently they are not capable of definite verification and refutation.

Moreover, the complexity of the subject matter of human history

[11] F. W. Maitland, *op. cit.*, pp. 157-60.

is so great that comprehensive theories which are believed to be opposed to one another are in fact merely supplementary. Thus the theory that the development of society is the work of great men does not necessarily contradict the theory of economic determinism. The two theories may simply indicate different factors in a complicated domain, and call our attention to the fact that the same events may be viewed under different but compatible aspects or connections. And it may well be the case that one theory of history will explain some events which another theory cannot explain as well, but at the same time be unable to make understandable what another theory makes clear. Thus far, at any rate, pluralism as regards theories in history has done less violence to facts than all the attempts to lay the past upon the Procrustean bed of a single theory.

It may not be amiss to inquire why histories need to be rewritten. Some answers are obvious. The discovery of new documents, new remains, the discovery of mistaken inferences in previous historians, the discovery of unnoticed connections between old data, and the application of hitherto unused or unknown principles of natural science—any of these may enable us to reconstruct the past in a manner which is new and illuminating.

But there are other reasons. We have already pointed out that a historian must necessarily select his material. It is impossible to write a history that will exhaust every phase of a subject. One historian may study the biographical details of the principal actors, another the political alignments, another the religious practices, another the economic and social factors, and so on. Each of these aspects provides a new vantage ground from which to survey again a field already explored. But each of these aspects is necessarily partial, and in selecting his material the historian is guided by what happens to interest him, by his social outlook, and by his general philosophical beliefs. The events studied, their grouping and the significance attributed to them, thus reflect the interests of the student. The conclusions obtained, as we have amply seen, are problematic. This point has been very clearly stated by Santayana. He says of "a natural science dealing with the past": "The facts it terminates upon cannot be recovered, so that they may verify in sense the hypothesis that had inferred them. The hypothesis can be tested only by current events; it is then turned back upon the past, to give assurance of facts which themselves are hypothetical and remain hanging, as it were, to the loose end of the hypothesis itself. . . . Inferred past facts are more deceptive than facts prophesied, because while the risk of error in the inference is the same, there is no possibility of discovering that error; and the historian, while really as speculative as the prophet, can never be found out."[12]

One further reason should be mentioned. Events have consequences in their own future, and the significance of an event depends upon the consequences to which these lead. The present therefore contains the past in the sense that there are consequences or traces of the past in the present. Since, however, the present is ever disappearing into the past, past events are seen in the light of new connections into which their consequences enter. Hence the significance of the past, the relations which past events bear to other events, is continually changing. Therefore if the aim of the historian is not only to chronicle the past but also to understand it, his task is never over as long as events have consequences.

§ 6. THE COMPARATIVE METHOD

The popularity of the theory of biological evolution has brought to the forefront a method frequently employed to support the thesis of organic development according to fixed stages. This method has also been applied in social anthropology in order to show that societies and their institutions, like biological species, continually evolve and pass through a series of stages or forms. With the truth of these theories we are not concerned, and nothing in what follows is intended to impugn them. But the method, known as the *comparative method,* deserves to be examined.

A frequently used argument for organic evolution is based upon comparative anatomy. For it has been found that plants and animals may be arranged in groups according to certain resemblances, and the groups arranged in a series of increasing or decreasing complexity of anatomical structures. For example, the legs of horse, sheep, dog, monkey, and man may be arranged in such a series. It is often concluded that the *species have arisen in time in the order presented in the series.* It is this inference which must be examined.

[12] *The Life of Reason: Reason in Science,* 1906, p. 50. The last statement is not strictly true, for archeological research may, and sometimes does, turn up remains of the past which prove the historian to have been wrong. Still, our means of verifying statements about the past are much more limited than are those about the future.

Let us see how the method works in social anthropology. A widely accepted theory at the beginning of this century was that mankind has not always had its present institutions of marriage. On the contrary, it was believed that the institution of marriage has passed through several stages, the latest one of which is the most "advanced" or civilized. The stages assumed are as follows: (1) *promiscuity,* in which there is no definite structure of sexual behavior, and sex intercourse is unregulated; (2) *group marriage,* in which a group of women, whether related biologically or not, are regarded as the wives of a group of men, whether these are related or not; (3) *clan,* in which the tribe is divided into hereditary social units, the children counting their descent from the mother's clan; (4) *gens,* a similar arrangement, in which, however, the children belong to the clan of the father; (5) *individual family,* which is regarded as the basic form of social organization, and the most developed stage of which is believed to be a monogamous arrangement. This theory has been supported by the following type of argument: Different primitive tribes are studied, and are found to have some one or other of the above forms of the institution of marriage. In some tribes, perhaps, the *temporal* development of the marriage institution is also studied, and found to have passed through the stages mentioned and in the order given. (There is no evidence, however, for the existence of a tribe without some regulation of sexual relations.) It is concluded that the marriage institution in all tribes has passed or is passing through a similar series of stages, and that the form of institution a tribe has *at present* indicates the stage of its development.

The argument may be represented diagramatically. Let the numbers 1, 2, 3, 4, 5 represent the five types of marriage institutions, and let *A, B, C, D, E* represent five tribes.

	A	*B*	*C*	*D*	*E*
1	×	—	—	—	—
2	—	×	—	—	—
3	—	—	×	—	—
4	—	—	—	×	—
5	—	—	—	—	×

The crosses in the appropriate places show that the tribe has at present the indicated form of institution, the dashes that it has not. It is inferred that these stages represent *temporal sequences,* so that stage 1 must precede in time stage 2, and that tribe *C,* which at present is in stage 3, must have passed through stages 1 and 2, and so on.

What shall we think of this? Suppose the reader, Yankee fashion, is whittling in the back yard one afternoon, and that he later collects the shavings and arranges them in order of size. Does it follow that the order of the shavings according to *size* is the *temporal* order in which the reader cut them? It clearly does not, and to suppose that it does is to confuse a logical order with an order in time. But this confusion is just the one committed by Spencer, Morgan, and other uncritical users of the comparative method. For the institutional forms discussed above belong to different historical series, each one having a determinate position in the history of the tribe in which it is found. Unless, therefore, we know beforehand the *truth* of the evolutionary theory according to which the stages must succeed each other in a given order, we cannot employ the fact that different societal forms can be arranged *logically* in the order required by the evolutionary theory as evidence *for* the theory. Unless we assume, in other words, that the sequence of stages in each of the different tribes is identical, we have no relevant evidence at all. But since this assumption is equivalent to the theory of social evolution, the argument is circular.[13]

In this connection, another fallacy easily committed may be mentioned. In comparing the institutions of different peoples, untrained observers often mistake superficial resemblances for significant similarities. In this way social phenomena that are hardly comparable may be arranged according to some desired pattern. Thus it is often claimed that all peoples develop ideas and ceremonials concerning life and death, and that in this respect mankind is alike. But the forms which such ideas take may make them significantly different. One group may believe that the human soul continues to exist in the physical form its owner had at death; another that the soul will be reborn in the same family; another that souls enter the body of animals; a fourth that souls continue human pursuits after death, and wait to be led back to the world. Other examples of faulty comparisons are still more striking. Is it possible to compare

[13] In point of fact, we know some tribes that have followed a different order, and changed the reckoning of kinship through the father to the reckoning of kinship through the mother.

different social groups on the basis of their valuation of human life? But in one group it is permissible to kill a father before he is decrepit, so that he may live a vigorous life in the world to come; in another, a father may kill a child as a sacrifice for his people; in a third, to kill a personal enemy is honorable; in a fourth, personal feuds are not tolerated although to kill members of foreign groups is extolled on occasions. The motives involved in each of these acts are so different that direct comparison of them is of little significance. Ideas and acts which when stated formally seem to be alike may represent fundamental psychological and cultural differences, and cannot be used as a basis for comparison.

§ 7. THE WEIGHING OF EVIDENCE IN COURT

We have already examined the historian's procedure in evaluating the testimony of documents and remains. An essentially similar procedure is followed in the courtroom when the testimony of witnesses is weighed and judged. For the fact to be proved in a court is of the past, while the testimony or the evidential facts are of the present.

The law distinguishes between two degrees of proof: one in which a proposition is established simply with a probability of over $1/2$, and which is called *preponderance of evidence;* the other requiring a degree of probability differing from certainty by so little, that anyone who acts upon that difference would be regarded as unreasonable—this degree of probability is called *proof beyond reasonable doubt.* The first degree of probability is sufficient in civil cases, while proof in criminal law requires the second.

The evidence presented in court is usually classified as (1) *testimonial evidence,* or the assertion of a human being as to the existence of the facts at issue, and (2) *circumstantial evidence,* or the production or citation of any other fact by inference from which the facts at issue are to be decided. Both of these kinds of evidence may vary in the degree of directness or remoteness with which they bear on the point at issue, and sometimes some evidence is rejected as too remote—which means that its probability is too slight to be used. However, most of the technical rules of evidence employed in our courts limit the admissibility of evidence which is logically probative. This is done for practical reasons, such as economy of time or the promotion of confidence between certain parties like priest and penitent, doctor and patient, man and wife.

The veracity of a witness is judged partly on the basis of the confirmation of his testimony by independent evidence, and partly on the basis of his character. If the testimony of the witness is contrary to his own interests, it is probable that he is telling the truth; if the witness stands to gain by having his testimony accepted, his veracity is suspected. The evaluation of circumstantial evidence often requires specialized knowledge of some science. Thus if the fatal bullet is offered as evidence that it came from the defendant's gun, expert knowledge of ballistics is required to settle the issue. As a consequence the evidence is frequently highly complicated, and we can indicate no more than the bare form of the arguments employed.

Suppose that a piece of cloth from a coat is found in the hand of a murdered man, and that a coat with a piece torn off and belonging to the defendant is found in his own room. The argument may then be stated as follows:

1. If the scrap came off the coat of the accused, it will fit the tear of the latter.
   This scrap of cloth does fit this tear.
   ∴ The scrap came off the coat of the accused.

2. The owners of coats are most often their wearers.
   The accused was the owner of this coat.
   ∴ The accused wore this coat at the time of the assault.

3. If the wearer of the coat was the assailant, the victim of the struggle would tear at the assailant's clothing.
   A piece of the defendant's coat is torn off.
   ∴ The defendant is the assailant.

Each of these inferences, the reader will note, is only probable on the evidence. The first argument proceeds from the affirmation of the consequent. Obviously, it is abstractly possible that some other coat made from the same cloth might be torn in exactly the same way. But this is so infrequent that our inference has a very high probability.

The second argument is also only probable, for the major premise asserts what is only generally but not universally true. But it is highly probable because the frequency with which one wears one's own coat is very high.

The reader should analyze in similar fashion the third argument.

The objective of the defense is, of course, to minimize the evidence of the prosecution by offering alternative explanations of the apparently damning evidence, or by offering evidence which contradicts the assertions of the prosecution. Thus the defense may point out that this scrap may not have come from this coat; or that the coat was not worn at the time of the murder; or that the defendant is not its owner; or that even if he is the owner he was not the wearer at the crucial time; or that even if he was the wearer he was not the assailant, but simply a bystander coming to the victim's assistance.

It is usually the case that no single bit of evidence or no single line of argument is sufficient to "prove beyond a reasonable doubt," but that the combined evidence makes the conclusion highly probable. Thus the Lowell Committee appointed by Governor Fuller of Massachusetts to advise him on the granting of a new trial to Sacco and Vanzetti declared: "As with the Bertillon measurements or with finger prints, no one measure or line has by itself much significance, yet together they may produce a perfect identification, so a number of circumstances—no one of them conclusive—may together make a proof clear beyond a reasonable doubt." [14] And they enumerated the circumstances which, when combined, convinced them of the guilt of the defendants. We can indicate the form of such complex evidence in a schematic way. Let T_a represent the testimony of a witness to the fact *a*, C_a represent circumstantial evidence for the fact *a*, and so on, and let *P* represent the proposition to be proved. The arrows go from the evidence to that which it is evidence for. Thus in column 1, C_{a_1} is supported by the circumstantial evidence C_{a_2} in column 2, which in turn is testified to by a witness T_a in column 3, and so on. For example, the finding of bullets of a certain type upon the accused is a circumstance which may be evidence for his guilt, and which must be evaluated by means

3	2	1	
$T_a \rightarrow$	$C_{a^2} \rightarrow$ $T_b \rightarrow$	$C_{a^1} \searrow$ $C_b \rightarrow$ $T_c \nearrow$ $T_d \nearrow$	*P*

of the principles of circumstantial evidence. But the fact of this finding must be supported by some person's testimony, and the

[14] Fraenkel, *The Sacco-Vanzetti Case*, 1931, p. 175.

rules of testimonial evidence must be used to weigh his assertion. Age, sex, race, emotion, experience, physical capacity for vision and hearing, are some of the factors which may enter into the evaluation of testimony.

It is often forgotten, however, that similarly, while no single item in the argument of the defense may make reasonable the defendant's innocence, the cumulative effect of the defendant's arguments may raise a reasonable doubt concerning his guilt. How fatal the neglect of this principle may be was pointed out by John Dewey in his analysis of the Report of the Lowell Committee.[15] He showed that the recommendation of the committee against a new trial was a consequence of its systematic use of this principle to prove the defendant's guilt, but a systematic neglect of it when the innocence of the accused men was considered.

The general nature of the argument is always of the form: If X did the deed, then the phenomena m_1, m_2, . . . m_n should be observed; but the phenomena m_1, . . . m_i are observed; therefore X did the deed. The argument is not conclusive, and on three counts:

1. It proceeds by the affirmation of the consequent and fails to prove that the phenomena could not be observed if X did not do the deed.

2. It is easy to contest, because difficult to prove, that if X did the deed the specific phenomenon must *always,* rather than sometimes, follow.

3. It is always a possibility that the actually observed phenomena are not precisely those which would be observed if X did the deed.

The reader might suggest that sometimes the argument might be put in a logically conclusive form, as follows: If X did not do the deed, the phenomena m_1, . . . m_n would be observable; but these phenomena are not observable; therefore, X did the deed. The weakness of this argument lies in the difficulty of negative evidence, of arguing from what has not been observed to the conclusion that such expected things have not in fact taken place. For often later observation reveals these phenomena.

The reader will note that the evaluation of evidence requires an indefinite number of material assumptions, either with respect to the relevance of testimony or with respect to the invariable connection of phenomena. In most cases, perhaps, the material assumptions are supported by nothing better than guesses. Nev-

[15] *Characters and Events,* Vol. II, p. 526.

ertheless, the general form of the argument is clear. A general law or principle is assumed, from which all sorts of consequences can be deduced with the aid of logic and other material assumptions; some, if not all, of the consequences are verified empirically; and it is concluded, with probability, that the principle is applicable to the case at hand.

CHAPTER XVIII

LOGIC AND CRITICAL EVALUATION

§ 1. ARE VALUATIONS BEYOND LOGIC?

Discussions of logic and scientific method are usually confined to propositions about natural or other forms of existence. There are, indeed, a great many writers who believe that scientific method is inherently inapplicable to such judgments of estimation or value, as "This is beautiful," "This is good," or "This ought to be done." Now if we agree that all judgments of the latter type express nothing but feelings, tastes, or individual preferences, such judgments cannot be said to be true or false (except as descriptions of the personal feelings of the one who utters them). Few, however, are willing to maintain this view consistently. Even those who urge the maxim *De gustibus non disputandum est* are not willing to maintain that there is neither truth nor falsity in the judgment which denies any beauty in the works of Shakespeare or Beethoven, or indiscriminately condemns as immoral all the doctrines of such diverse teachers as Confucius, Buddha, the Hebrew Prophets, Socrates, Epicurus, Mahomet, Nietzsche, and Karl Marx. Almost all human discourse would become meaningless if we took the view that every moral or esthetic judgment is no more true or false than any other.

This reflection is of course not in itself a logical proof that there is an element of objective truth in moral and esthetic judgments, but it points to the necessity of a closer examination of the issue. There are, to be sure, great differences of opinion with regard to moral and esthetic issues. But this is also true, though to a lesser degree, about questions of existence in nature and in human affairs. Indeed there are no questions of natural science about which our information is so complete as to eliminate all differences of opinion. But the fact that certain issues cannot as yet be definitely settled does not mean that any opinion is as good as any other. Though we do not know the cause of cancer, we may know enough to say that some opinions on this point certainly have less evidence or rational ground than others. Hence even if moral and esthetic judgments are largely matters of opinion, may not logical tests enable us to clarify our opinions, discover their implications, and find out whether some of these opinions have more evidence in their favor than others? To answer this question is the object of the present chapter.

§ 2. MORAL JUDGMENTS IN HISTORY

From the fact that scientific method in history is directed to the discovery and narration of what actually happened many have argued that the historian must avoid passing moral or esthetic judgment on the personalities or events within his purview. Others, on the other hand, have argued that such an attitude is humanly impossible, that no one ever completely avoids passing such judgment implicitly, if not explicitly, and that moreover a history of human affairs without judgment on the human values which enter into them would be devoid of significance. In the course of this discussion three issues have been raised with regard to the application of scientific method:

1. That scientific method applies only to the discovery and proof of what actually happened, and that since human acts are the necessary results of causal laws, all judgments of approval or disapproval are irrelevant, if not meaningless.

2. That judgments of value are entirely subjective, arbitrary, or vary with the individual historian.

3. That there is no way of bringing adequate evidence for or proving such judgments.

1. It is obviously beyond the competence of logic itself to decide whether the events of human history are or are not subject to necessary laws. It is pertinent, however, to note in passing that empirically ascertainable history does not and cannot prove the existence of universal or necessary laws. No one has in fact formulated any generally accepted laws of history, and it ought to be obvious that our fragmentary and often disputable knowledge of the past is not adequate to prove what must and will be in the future. The assumption that human events are connected by laws or invariant relations may be defended as a requirement or postulate of scientific method. But it is well to note that any laws applic-

able to the subject matter of history can be only the laws of social science, which connect not complete totalities but only certain abstract phases of human life, laws such as that of supply and demand in economic activities. But history, dealing with concrete totalities, must include reference to the unique and unrepeatable. Some other individual may be as ambitious as Caesar, as skillful a general, as generous a foe. But the unique person who crossed the Rubicon at a given time and drove Pompey and the senatorial party out of Italy will not return if the temporal course of events is a reality.

If then the ascertainable laws of human events are of the same type as the laws of all natural events, they are repeatable patterns of abstract relations. If events *A* and *B* are intimately connected in a way formulated by some law, then the historian must regard *B* as caused or determined by *A*. But for that very reason it is legitimate and relevant to ask what would have happened in any given situation if some factor or circumstance present had been different. What, for instance, might have happened if Alexander of Macedon had decided not to invade Bactria and Sogdiana but Italy, as his cousins of Epirus did later? It is no objection to this question to urge that all the factors having been what they were, his final decision was inevitable. One of the factors in the case was undoubtedly the fact that the wealth of India and the unwarlike character of its population made it more tempting than anything offered by Italy. Such questions as to what policies were on the human horizon do not at all contradict the demands of scientific determinism, but must rather be faced in order to understand the significance of what did happen.

When the historian thus places himself in the position of the man in a given historical situation and tries to envisage the causes that attracted or repelled human effort, he cannot ignore the moral feelings and standards prevalent at the given time among the people with whom he is dealing, for such feelings are a genuine part of the history which it is his business to ascertain.

2. The objection that moral judgments are subject to partisan or sectarian bias which distorts our view of history has certainly a good deal of historical evidence in its favor. Many of the actual judgments which historians have passed have been not only provincial and bigoted but, what is perhaps more important for logic, really irrelevant to the subject matter treated. It is irrelevant for the understanding of the conduct of American Indians to judge them by the code of Confucius, and it is not always illuminating

to judge the past, for example the conduct of Philip II of Spain, by our contemporary moral attitudes. But this is an argument against inadequate historical knowledge and insight, not one against the relevance of the moral element in history. Certainly, every human group known to us in history professes explicitly or implicitly certain ideals of conduct, and judges certain things as noble or ignoble, admirable or contemptible. And such ideals or moral standards are a genuine part of history.

3. So far as the scientific historian is concerned, then, norms or moral standards are objective facts to be discovered on the basis of the best available evidence. That the gathering of adequate evidence for such facts is very difficult, cannot be denied. The accidents of birth and early training make us feel that our language and our modes of conduct are *the* natural ones, and it is difficult for us to understand foreign forms of moral estimation. It is hard, for instance, for the modern historian to realize that some of our religious and moral evaluations are or were not felt by Hindus, Chinese, ancient Hebrews, or even by Italians and Spaniards in the fifteenth and sixteenth centuries. On the other hand, the human craving for novelty, for the exotic and the bizarre, often makes us exaggerate the moral differences between peoples of different epochs, climates, and social organizations. The historian therefore has to be on guard against being either insensitive to subtle differences or blind to obvious agreements of human nature. If the eighteenth century rationalistic historians failed in the first respect, more romantic modern historians, zealous for the discovery of minute differences, have certainly sinned in the latter respect.

There are other circumstances which make it difficult to find adequate evidence for the prevalence of moral standards or norms in any group or epoch. There is always the question how far any views or acts are typical of the given period, country, or class. But that, after all, is the kind of difficulty which the historian must constantly face if he wishes to describe the life of a whole people by narrating the deeds of the few who typify them.

We may dispute as to what part the moral indignation of the Romans played in bringing Nero to his death and the Julian line of emperors to extinction, or as to the extent to which King John's lack of moral integrity led to the rebellion of his barons and to the loss of Normandy to England, as well as to the granting of Magna Carta. But surely such questions are not meaningless, and are to be answered by consideration of evidence. Moral prestige and abhorrence are certainly factors in human affairs which the historian cannot rightly ignore. That Alexander the Great was intensely admired, that Mary Stuart was morally condemned by the Protestant leaders, that certain peoples view with horror the eating of certain animal food, and that some American Indians regard it as intensely immoral for a man to be seen anywhere near his mother-in-law—these are all facts of history for which the evidence is of the kind that we accept for other facts. We can thus establish on the basis of the explicit teachings of the early Buddhists, of their ritual practices and other forms of conduct, that though they were most successful as missionaries, they did not believe in persecuting nonbelievers, and that though they emphasized sympathy with human suffering, they did not attach as much value to life as we do. Ideologically, this may be connected with their belief in an enormously long period during which the soul migrates from one body to another, so that our final and eternal salvation does not have to be won in one brief life. But the historian may also seek to connect these moral views with the character and social condition of life in India and in the lands to which Buddhism spread.

We have so far discussed the fact that the historian cannot eliminate the history of morals and moral judgments from the content of the history which he wishes to portray as adequately as possible. But can or should the historian always avoid the application of his own moral standards? That it is difficult to avoid passing judgment is shown by the fact that historians who vigorously maintain that it is their business to narrate and not to pass judgment have emphasized conservative or revolutionary moral values in a naïve and unavowed manner. Some have thus unduly eliminated religious motives, and others have ignored the element of indignation or resentment against injustice which is never completely absent in human affairs. We cannot justly leave all these matters out, and it is extremely difficult to think of them in nonmoral categories. But leaving aside the question of difficulty, we must face the question, *Should* the historian refrain from all moral judgments? Now if we wish to understand the effects of the moral standards of any epoch, we must inevitably take a point of view somewhat wider than that age itself and must therefore pass judgment on the adequacy of its ideas and standards. And this is itself a moral question, which the historian answers on the basis of his own moral assumptions. Thus if we wish to understand the effects of the crusading zeal of the Spaniards of the fifteenth and sixteenth centuries, or the docile and submissive attitudes of various oriental countries, we must enter into these ethical views of life, and this we can hardly do without passing judgment on their value. We may do so implicitly and unavowedly, or we may deliberately reflect on what is involved in these judgments. In the latter case we are concerned with the science of ethics—the theory of moral judgments. Those who deny that there can be a science of ethics (as distinct from a science of what actually exists) are willing to apply the term "art" to the effort to construct a consistent body of moral judgments. Let us then consider whether logic is applicable to such an art.

§ 3. THE LOGIC OF CRITICAL JUDGMENTS ON ART

While all people pass judgment on works of art, there are those who give special attention to such judgments. We call them critics. In their writings we may see the ordinary judgments developed and made more or less coherent and systematic. We may distinguish three tendencies in the criticism of works of art: (1) the impressionistic, (2) the historical or philological, and (3) the esthetic.

1. *Impressionistic Criticism.* Impressionistic criticism frankly confesses itself to be a partial biography of the critic himself, an account of his likes and dislikes, his impressions when he heard the music, saw the painting, or read the poem. This view is often defended by the argument that we cannot know anything except our own impressions. In this, however, its adherents are not entirely consistent, for they assume that we do and therefore *can* know that there are objects which give us these impressions. Indeed these critics generally go further and assume that these objects were created by certain individuals with a history of their own. In any case critics of this school cannot tell us anything about the work criticized unless their account of their impressions is in some way connected with the nature of the artistic work which produces these impressions. They must therefore undertake some analysis of the object judged.

The subjectivistic philosophy often alleged as the ground of this view does not, however, do it justice. A much sounder reason for this type of criticism is its rejection of the "classic" procedure, which judges every work of art by fixed standards and thus completely overlooks or despises what is novel, individual, and distinctive in it. As a certain effect of freshness, freedom, and spontaneity is demanded of fine art, and as the latter must assume new forms if it is to serve life under new conditions, it seems essential that the critic shall cultivate a high sensitiveness, so that new works of art may have their distinctive merits more readily recognized. Unfortunately, however, the critics of this school in their violent reaction against classicism or academicism fail to note that pure novelty, or work that is distinguished by nothing save its being different from everything that was ever created before, is only odd and, in the etymological sense of the word, idiotic. Indeed, if a man is so original that he entirely cuts himself off from human communication, we shut him up in an insane asylum. A great work of art is original in the sense that it is a new revelation, but a revelation of something which is deeply and widely human, and in a manner which finds a response in men and women far beyond the time and place at which it was produced. (We thus distinguish between the great and the minor artist on the ground that the latter is of significance only for his own time and school or sect, while the former is not so limited by time and by peculiar or local tastes.) We can thus understand why as a rule the world's great artists (as well as its great critics) have carefully studied the rules of their art, and the greatest innovators among them have made liberal use of the traditions in which they were trained. The rules of any art are more or less successful summaries of past experience in it. They therefore generally have a certain relevance and applicability even if in the course of time they become hardened and inflexible. And in point of fact the leading critics of the impressionist school (such as Anatole France) are generally thorough traditionalists. It is by the mastery of classic tradition that we gain freedom and power to use the tradition in new ways.

The limitation of impressionistic criticism—of the view that assertions such as those about beauty are merely personal, arbitrary, and beyond proof or disproof—can be seen if we consider the elements common to both the industrial and the fine arts. The work of the shoemaker, the carpenter, or the mason, like that of the painter, the musician, the sculptor, or the architect, is an instance of human skill modifying nature in accordance with some human desire. The judgment on the workmanship of the artist is thus in many respects the same whether he be the writer of a play or the baker of bread. While the standards by which we determine the question whether the work has been well done are different in these two cases, both judgments are logically of the same type and generally involve analysis of both the standard and the achievements. Such analysis, if it is to be consistently carried through, is clearly subject to logical standards.

Another phase of this last point is that art can be, and is in some measure, taught. And when this is possible, there are certain directions, rules, or reasons why things should be done in one way rather than another. To be successful, we must obviously be consistent in the application of such rules or reasons. (Indeed, in Latin the word *ars,* as in *ars poetica,* was used for the body of doctrine to be applied by the artist.) This does not mean that the artist must go through a process of conscious reasoning before he does anything, though moments of deliberation as to whether this or that should be done are surely not absent from any of the higher arts, such as painting, composing music, or writing dramatic poetry. Now if the art critic wishes to understand or to make others understand the work of art, he must state the artist's problem in some respects more explicitly, and even analytically, in order to understand how the artist achieved or failed to achieve his objective. The critic who wishes to understand the actual or historical work before him must ask, What did the artist try to achieve? What means were there before him? And what ways did he follow? These are all in a definite sense questions of history, that is, questions as to what has actually happened, and are clearly subject to rules of evidence. It is this which constitutes the strength of the historical or philological school of criticism.

2. *The Historical or Philological Type of Criticism.* This type of criticism is based on the fundamental assumption that intelligent judgment on any work must be based on a close study of its actual character and the art involved in it. Even the art of music, dealing with such an intangible subject matter, has become the subject of many diverse branches of learning which have received the collective name of musicology. For our purpose, however, it will be simpler to consider the judgment or criticism of poetry. While the charm of poetry is inseparable from its verbal or phonetic medium, it is obvious that we cannot judge it without understanding its meaning. Poetry is more than meaningless euphony. The first question is therefore, What does the poem mean? In answering this question, we can clearly distinguish between scientific procedures and those which are not.

Unscientific Interpretation. Unscientific, though edifying, are the various allegorical methods of interpretation which find hidden moral or spiritual teachings in poems like Homer's Iliad, or the Rubáiyát of Omar Khayyám, or in the stories about the domestic life of the patriarchs of the Old Testament. These methods are unscientific because they are arbitrary and unverifiable. Thus according to Philo the author of the Pentateuch (supposed by him to be Moses) meant to teach the philosophy later taught by Plato, and the various characters in the Book of Genesis were intended to signify abstract qualities or virtues. Thus the two sons of Joseph, Manasseh and Ephraim, really represent recollection and memory. But this does not at all explain the biblical account of the relations between the two tribes called by those names. The land of Canaan is assumed to represent piety; but this is altogether inconsistent with the fact that in biblical usage "Canaanite" was a term of opprobrium, and that the Israelites were commanded by the Lord to exterminate all the Canaanites. Such inconsistencies can of course be explained away by further metaphorical interpretation. But if so it becomes obvious that the whole procedure is purely arbitrary, for the purpose of edification, and is not concerned with the question of evidence or proof that the given interpretation was ever actually intended. All sorts of conflicting spiritual interpretations are thus equally legitimate or illegitimate.

Equally unscientific are the cabalistic attempts to find the hidden meaning of the sacred text by systems of numerology, that is, by assigning numerical values to the different letters of a word, or by regarding ordinary words as composed of letters each of which is the initial letter of a word or phrase. By means of such methods all sorts of conflicting meanings can be read into a text, and no real evidence is produced for any one of these meanings. The interpreter is simply giving prestige to his own views or statements by clothing them with a sacred text that may have nothing to do with them. There are, of course, cases in which writers do indicate that they wish to convey two meanings, one a literal and the other an allegorical one. This seems to be clearly the case with the *Divine Comedy,* not only on internal evidence but according to Dante's expressed testimony in his letter to Can Grande. But even here we get involved in interminable controversies as to whom the three beasts in the first Canto represent. Meanwhile Dante's poetry, in its literal interpretation, continues to charm readers.

Philological Interpretation. The first task of a scientific philology is to determine by a close scrutiny of the text the precise meaning of the actual language used. Most of us read hastily and are not prepared to note every significant feature; hence different readers carry away different vague impressions. An adequate understanding must be able to account for every word and clause as well as for the total plan. This does not deny that the meaning of a word is determined by the context in which it occurs. Indeed in some cases the meaning of a sentence is so clear apart from a given word in it that we can tell that this word is a mistake or a misprint, just as we can recognize a word that is misspelled. Nevertheless we cannot dispense with a close study of the words and phrases which constitute a literary masterpiece. Sometimes we mistake the meaning of a passage because some word in it has what is to us an unfamiliar connotation. We must know the Shakespearean meaning of the word "let" in order to understand truly the line, "I will make a ghost of him who lets me." Similar remarks may be made about Shakespeare's phrase "Man's glassy essence," or Milton's use of the word "dear" as a synonym of "bitter." Such historical differences of meaning have to be established by comparisons of texts. More often it is the figurative meaning of a word or phrase which we must examine carefully if we are to establish the poet's meaning. What, for instance, does Milton mean when in his famous sonnet he calls Shakespeare "Child of Memory"? We can answer the last question only if we know that the Muses, who were regarded as the inspirers of poetry, were supposed to be the daughters of Memory.

Obviously we cannot tell what an author meant unless we know something about the subject matter about which he was writing. We must, in other words, know the historical conditions to which the text is to be applied. Without knowing the habits of the Greeks we cannot know, for instance, what Homer means by "washing the corn." We need to know that Jews from many different lands, and therefore with different currencies, came to the Temple, and had to buy the various birds or animals offered as sacrifices, to understand the reference to the "money-changers" in the Temple.

Philological and historical interpretation may, and sometimes does, lose itself in minor details, and thus loses sight of the main character of the work of art. This leads to the ancient remark that we must pay more attention to the spirit than to the letter. And there are critics who think that the prime requisite for understanding a great work of art is the exercise of our own creative imagination or intuitive understanding. But while no one can fully understand any work of art unless he imaginatively passes through something of the creative experience of the artist, there is no proof of our ever attaining any understanding unless we can thereby explain the details. The spirit of a poem that cannot be supported by the letter of its language might as well be the spirit of some other poem. If logic and scientific weighing of evidence are not the whole of intelligible artistic judgment, they are necessary ingredients of it.

3. *Complete Esthetic Criticism.* Actual historical or philological criticism often undertakes too much in some respects and too little in others. In trying to explain a work by the life of the artist, the

critic is apt to forget—and often does—that not all biography, not all that happened to an artist or writer, is relevant to the understanding of his work. We certainly do not know enough of the causes of artistic ability and peculiar genius to support with any logical cogent evidence the numerous vague generalizations by which psychoanalytical critics try to explain works of men like Leonardo da Vinci or Shakespeare. Nor is history alone adequate to take the place or function of criticism. The analysis of an actual work of art, the effect that it can produce, involves more than history.

Philological criticism, bent on analyzing the music or poetry, is apt to neglect the study of its effect on the hearer or reader, and that effect is essential to full esthetic judgment. The weakness of impressionism is not in being concerned with the effects of the work of art on the beholder, but in failing to realize that an adequate account of such effects involves a study of the object. Even the mere enjoyment of music or dramatic presentation is enhanced if we know more of the subject matter of the art and learn to discriminate the various elements in its structure. When we see more in a painting or a poem, there is more to enjoy. The uninstructed person is satisfied to say that the music or play was "grand" or "rotten." But cultivation enables us to attain greater discrimination. To declare that a given work is beautiful is, then, not merely to assert a simple feeling or an irreducibly simple quality, but to assert that the object has a certain form which produces certain effects. We may not always be able to analyze it in a way satisfactory to all observers. But it is something to be discovered, and assertions about it are subject to logical rules.

To determine what makes an object beautiful, sublime, or possessed of what has been called esthetic form, is the problem for the study or theory of art, of which esthetics is a part, though the latter also studies natural beauty that is not the object of art. The logician is interested only in noting that such a study involves both factual considerations, experimentally determined, and purely logical considerations of consistency.

§ 4. THE LOGIC OF MORAL AND PRACTICAL JUDGMENTS

Moral judgments usually take the form of imperatives. We ought to honor our father and our mother, to be loyal to the interests of our country, to tell the truth, to refrain from murder, and the like. In what sense, if any, do such judgments involve propositions that are true or false, so that logical principles can be applied to them?

The Distinction between What Is and What Ought to Be. As a historical fact, there can be little doubt that these maxims are in general part of our social inheritance. They are prescriptions for conduct that have come down from time immemorial and are hallowed by traditions and by the precepts of religious teachers and sages. For the most part, they express that which we wish to see put into practice, and which in a large measure is actually in effect. People do honor their parents, tell the truth, and refrain from taking life not only through compulsion but by a preference which seems natural. Yet natural inclination, even when reënforced by the various forms of organized social pressure or sanctions, is not sufficient to remove the divergences between moral maxims and actual conduct. Indeed, if in any community, no matter how large, no one for a long time acted counter to any of these rules and no preaching of them were regarded as necessary, we might be inclined to view them as natural laws in the sense of invariant relations or uniformities of actual conduct; but it would still be logically possible to ask, Why should I always respect other people's property? Why should I honor my parents? And the very form of these questions shows a difference between the maxims often called moral laws and the laws of natural science. If we ask whether all men do respect human life, it is relevant to point out that some do not. But if we raise the issue as to whether we *ought to* or *should* refrain from murder, the mere fact that some men do not is not directly relevant. The validity of a moral command is not denied by the fact that some do not follow it. Moral maxims, as imperatives concerning what ought to be, differ from natural laws as uniformities of existence. The ordinary man attending to his daily affairs and a good many philosophers who call themselves positivists are agreed not only that there is this difference, but that there is no science of moral imperatives.

The ordinary man regards moral imperatives as rules to be accepted unquestioningly. One who reasons about them is suspected of being open to immorality. The purely authoritarian view of moral judgments cannot, however, be strictly carried out, for men do find new or difficult situations in which they cease to be certain as to what it is that the moral law requires. Should a doctor tell a lie to a patient who has heart disease? Does the command "Thou shalt not kill" prohibit war? The honor or support due to our parents may conflict with our duties to our country, to our religion, or to humanity. In cases of this sort, men often realize the uncertainties and inadequacies of individual conscience, and consult religious or moral teachers. The latter generally view morality, even when it has a supernatural sanction, not as a body of arbitrary commands or prescriptions, but as a set of rules for which there are good reasons. Moral doctrines thus take the form of systems of propositions logically concatenated.

The thinkers whom we have called positivists do not, however, admit that any logically concatenated body of propositions can properly be called a science. Science, they contend, must be restricted to what exists in the natural world. We can have a science of actual uniformities in human conduct and even of the standards which actually prevail or are operative in human affairs. But the mere deduction of specific rules from some general moral postulates cannot be called a science. At best, they argue, this is an effort to systematize or rationalize our moral judgments, and may be called an art, or even a rational art, but not a science.

It is obvious that if the word "science" is by definition restricted to existential propositions, ethics as a logical system of moral judgments is not a science. The reader may note in passing that if we accept this definition, we must also rule out pure mathematics from that realm. However, disputes about the name "science" are of no logical importance to the extent that they are concerned about the prestige which the word "science" carries with it. What is of importance to us is to see the extent to which logical procedure in ethics is similar to, and to what extent it is dissimilar from, the procedure of the natural sciences.

The Existential Element in Moral Valuation. Many moral imperatives are similar to the rules of art. When a man is building a house we tell him to make sure that the ground is firm and dry, and that his timber is seasoned; our directions are based on the observed uniformity that without the existence of the means recommended the end, namely, a house firmly set and dry, cannot exist. A similar observation may be made as to the fine arts. The theory of harmony or counterpoint, that of the composition of colors, or that of proper proportion in architecture, consists of imperative rules which experience and reflection have shown are necessary means if the artist is to attain his objective. This is also the case in regard to the rules of conduct when judged by the standards of efficiency, economy, prudence, hygiene, courtesy, or propriety in speech, dress, or manners. Certain courses of conduct lead to losses of time or of money, or make us fail to achieve the ordinary ends of our practical endeavors. If we wish to attain our ends we must follow certain rules of efficiency, economy, or prudence. We advise people to avoid drafts or not to overfeed their children, to adopt certain modes of expression, dress, or manner, because we assume that certain ends (namely, good health, or conduct which is regarded as courteous or socially correct) cannot otherwise be attained.

The Bearing of Comprehensive System on Moral Judgment. Customary moral rules consist largely of such practical imperatives. But modern philosophers have tried to draw a sharp distinction between the two on the ground that moral rules are absolutely binding, while the others are conditional. The rules of economy, of prudence, or of hygiene are conditional in the sense that we must follow them *if* we wish to achieve certain ends. But we may not care for these ends, and in that case we do not need to follow the rules. Moral rules, such as that against lying, are thus absolute in the sense that we must follow them under all circumstances without regard to likes or preferences. Now there is no doubt that many moral rules are felt by a good many people to be absolute in this sense. Thus American Indians have felt that the use of machinery for the cultivation of corn was highly immoral, even when they realized that a larger supply of food could be obtained by its use. But reflective people who are in the habit of asking for the reasons for customary moral rules do not always accept their absolute character. Few, for instance, agree with the German philosopher Kant in regarding the rule against lying as absolute in the sense that it must not be violated under any circumstances whatsoever, not even to save human life.

The distinction between moral and practical rules can be put on the ground of comprehensiveness rather than on that of absoluteness. The various practical ends which we pursue, and the various practical rules which they involve, sometimes come into conflict, and a comprehensive and consistent set of judgments as to the proper conduct of life must seek such general axioms or postulates, so that all situations can be properly judged. Most men, for instance, regard health as a good and may even look with amazement at the suggestion that in some cases it might not be so. Yet there are numerous occasions when we do deliberately sacrifice health to achieve other ends. Sometimes such sacrifices, like those made for the sake of temporary pleasures, wealth, honor, or reputation for beauty, may in retrospect be regretted and pronounced foolish. But at other times, as when we sacrifice health for the sake of those we dearly love, or for some cause like country or religion without which we deem life to be not worth much to us, we look back on such sacrifices with approval. This is also true of other ends, such as wealth, reputation, and the like. Moral rules, then, according to this view, enable us to discriminate between ultimately wise and ultimately unwise choices. Thus the fact that people generally desire something does not prove that they should do so, for by following their desires they may bring on themselves an evil even greater than that of struggling in vain to get what they want—the evil of getting it and experiencing regret and disappointment, if not worse calamities. What people *ought to* desire is what they *would* desire *if* they were enlightened and knew both what they really wanted and what natural means would bring it about. Morality is thus wisdom applied to the conduct of life, and yields rules which we would follow if we thought out all the implications of our choices and knew in advance their consequences. Obviously, however, this is not completely attainable so long as our knowledge is imperfect. But it is an ideal which illumines the nature of our choices whenever we reflect on them.

The foregoing view of moral judgments is not universally accepted. The objection is raised that if we regard morality merely as a matter of wisdom, a higher sort of prudence, there is no absolute authority for moral rules when people differ in opinion. We must, however, discriminate between the social necessity of agreeing on some rules which will make it possible for men to live together and the logical possibility of proving that only one set of moral judgments can be rationally proved.

We may agree on the necessity of obeying laws in the political sense, even if we do not agree as to the wisdom of these laws, for the consequences of lawlessness may be worse than the evil of some specific law. But this social necessity of making and obeying laws does not logically prevent the continuance of differences of opinion as to what is right and what is wrong. Ultimately, all systems of ethics start from certain fundamental assumptions, and unless two men agree on the same assumption there is no way of removing differences of opinion. But a community in which free inquiry into moral issues is allowed, is bound to discover more stable bases of moral judgments and conduct, just as free discussion in the field of natural science is bound to eliminate arbitrary opinions. For so long as men live in a common world and have common elements of human nature, their choices and judgments will agree in proportion as they are enlightened and recognize this common nature.

We may conclude by noting that while the distinction between what *exists* and what *should be* cannot be eliminated, the two have been shown to be intimately connected. Unless there were human beings or beings of a similar nature, questions of morality would be devoid of meaning.

The Function of Logical Form in Critical Evaluation. In this connection we should note that the view of esthetic and ethical forms in this chapter is an application of the view of formal logic expounded throughout this book. The rules of logic, as we have seen, are not external rules arbitrarily imposed on us. It is not, in other words, necessary to assume the rules of logic in order to draw proper conclusions, though these rules are a help in isolating those features which make one proposition relevant to another. To suppose that the rules of reasoning are premises without which specific inferences would lose their validity is a logically untenable view. Similarly, the rules of the industrial and the fine arts are not arbitrary authorities, but abstract formulae stating that certain ends result from following certain specific ways. The beauty, sublimity, or other esthetic form is not itself an object or thing, like a marble bust, a painting, or a song. It is rather what characterizes objects of a certain sort in certain relations. So are the rules or principles of ethics, likewise, formulae; these formulae indicate what it is in specific courses of conduct that makes them relevant or conducive to specific human ends which constitute a system that can be the object of our entire devotion. Ethical formulae are not the *sources* of our actual moral preferences or repugnances, but they explain what it is about certain moral judgments which makes them obligatory or not, as the case may be.

§ 5. THE LOGIC OF FICTIONS

While every proposition is either true or false, it would be a mistake to believe that people use language primarily to communicate truth. In its origin as well as in its daily rôle, language is a means for expressing emotion. Other considerations than those which arise from a desire to report the truth accurately dictate the form which our language takes. Moreover, even in the sciences, while it is literalness in statement at which we aim, we rarely begin with an accurate perception of the similarities and differences between things, and hence we express our confused insights by means of language that is highly metaphorical. For the understanding of the element of truth present in popular expressions, and as a safeguard against rejecting any part of science simply because it employs linguistic forms not capable of a literal interpretation, we must examine the rôle played by such expressions in daily life and in the sciences.

1. *The Emotive Use of Language, Ceremonial Expressions, and Euphemisms.* Just as the ritual of social life demands certain forms of dress, certain steps or gestures, regardless of convenience, so it demands certain accredited expressions regardless of their literal truth. Thus the rules of courtesy among the Chinese and some other peoples require the host to say always that his house is "mean," that his guest is "distinguished" and confers an "honor," and so on. Even those who do not take such expressions literally may be offended at departures from the social modes.

The rôle of ceremonial expressions in the outer forms of make-believe is as important in social life generally as in the games of children and of primitive man. The social life of a country like England may be viewed as a game that requires among other things that people should speak of his Majesty's army, navy, treasury (though the debt is "national"), or that the actual leaders of the government should speak of "advising" the king where the latter has no choice but to obey. Similarly it is the fashion to speak of the United States as a democracy where the law is the will of the people made by its representatives, even though few know what laws are being made or have much control over those who make them.

Ceremonial expressions are often attacked as conventional lies when they are intended not to deceive but to express the truth euphemistically. Courtesy or politeness demands the elimination not necessarily of the truth, but of certain unpleasant expressions that are for some reason or other taboo. This is readily explained by the fact that words have emotional effects on their own account. Thus it is permissible to refer to a female (or lady) dog, but bad taste to use the single-syllabled word. It is proper for the stage pirates in Peter Pan to refer to a future meeting "below," but they would shock the audience if they used the more realistic and theologically canonical word.

2. *Metaphorical and Habitual Shorthand Expressions.* To appreciate the intellectual or scientific function of metaphors the reader had better begin with an experiment. Let him pick out a page or two of philosophical prose in any classical treatise or modern discussion. Let him read this extract carefully and mark the number of passages in which the meaning is suggested metaphorically rather than literally. Let him then read the passage a second time and reflect how many of the passages first taken as literal truths

are really metaphors to which we have become accustomed. We mean such expressions as "the root of the problem," "the progress of thought," "the higher life," "falling into error," "mental gymnastics." Indeed, whenever we speak of the mind as doing anything, collecting its data, perceiving the external world, and the like, we are using the metaphor of reification (Latin *res* = thing), just as we use the metaphor of personification whenever we speak of bodies attracting and repelling each other. The third stage of the experiment is to try to rewrite the passage in strictly literal terms without any metaphors at all. The result of such an experiment will confirm the conclusion that to eliminate all metaphors is impossible. This is especially clear when we try to express general considerations of a novel or unfamiliar character. For how can we apprehend new relations except by viewing them under old notions? At any rate, the experiment will make more plausible the view that metaphors are not merely artificial devices for making discourse more vivid and poetical, but are also necessary means for the apprehension and communication of new ideas.

It would be an error, however, to regard every metaphor as an explicitly formulated analogy, in which the words of comparison, "like," "as," and so on, are omitted. This presupposes that the recognition of the literal truth precedes the metaphor, which is thus always a conscious transference of the properties of one thing to another. But history shows that metaphors are generally older than expressed analogies. If intelligence grows from the vague and confused to the more definite by the process of discrimination, we may well expect that the motion common to animate and inanimate beings should impress us even before we have made a clear distinction between these two kinds of being. Thus it is not necessary to suppose that the child who kicks the chair against which it has stumbled personifies the chair by a process of analogy. The reaction is clearly one arising on the undifferentiated level.

Metaphors may thus be viewed as expressing the vague and confused but primal perception of identity, which subsequent processes of discrimination transform into a conscious and expressed analogy between different things, and which further reflection transforms into the clear assertion of an identity or common element (or relation) which the two different things possess. This helps us to explain the proper function of metaphors in science as well as in religion and art, and cautions us against fallacious arguments for or against views expressed in metaphorical language.

The fact that metaphors express the primal perception of a thing with something of its undifferentiated atmosphere gives these metaphors an emotional power which more elaborate and accurate statements do not have. This is perhaps best seen in the profoundly simple metaphors of the New Testament. "Feed my sheep" is more potent than "Teach my doctrine," because it carries with it the atmosphere of suggestion which those genuinely moved to preaching feel before they can formulate it—tender sympathy for the helpless, the distress of the spiritually hungry, shown especially in the tense, open-mouthed faces of an oriental audience. The same is true of the simile "sowing the seed of truth," or St. Paul's metaphors of preaching as "edification," of the righteous life as "girding on the armor of light," "garrisoning or fortifying the heart." Goethe's metaphor "Gray is all theory" is a vivid expression of what it would require considerable reflection to formulate in purely literal terms. In practical affairs the personifying of cities or nations, as the likening of the state to a ship ("Don't rock the boat!") or of changes of attitude to "the swing of the pendulum," contains a potency which literal statements do not have.

These considerations will illumine the nature of fictions as they appear in the field of their greatest development, that is, in the law. Here fictions appear clearly as assertions that contain an element admittedly false, but convenient and even indispensable to bring about certain desired results. Though fictions resemble myths, they can be distinguished from those which are genuinely believed, and from pious frauds which are intended to deceive in aid of a good cause. Thus when a deed or a mortgage is recorded, a really innocent purchaser is said to have had notice, and he is not allowed to prove the contrary. For this really means that the act of recording makes the rights of all purchasers, innocent or not, alike, so that the fact of actual ignorance is irrelevant.

Why, however, does not the law use accurate expressions instead of asserting as a fact that which need not be so? Why assert that a corporation *is* a person, instead of saying that a certain group of rights and duties are analogous in some ways to those of a natural person? Why say that the United States Embassy in China or on a boat at sea is on American soil, when we mean that certain legal relations in or concerning it are to be treated according to the law of the United States? The answer is partly that the practical convenience of brevity outweighs the theoretical gain of greater accuracy. But more important is the fact that at all times the law must grow by assimilating the new to old situations. And in moments of innovation we cling all the more to old linguistic forms. The latter minister to the general feeling of security, especially where the prevailing myth or make-believe is that the judge merely declares the law and cannot change or extend it. That the law can be obeyed even when it grows is often more than the legal profession itself can grasp.

From the point of view of social policy, fictions, like eloquence, are important in giving emotional drive to propositions that we wish to see accepted. They can be used to soften the shock of innovation (as when courts protect a man's vines by calling them trees), or to keep up a pleasant veneration for truths which we have abandoned (as when we give new allegorical or psychological meaning to old theological dogmas that are no longer tenable). But if fictions sometimes facilitate change, they often hinder it by cultivating undue regard for the past. If the social interest in truth were to prevail, we should in our educational and social policies encourage greater regard for literal accuracy, even when it hurts national pride and social sensibilities. But no one has seriously suggested penalizing rhetoric and poetical eloquence in the discussion of social issues. The interest in truth is in fact not so great as that in the preservation of cherished beliefs, even though the latter involve ultimate illusions whose pleasantness is more or less temporary.

3. *Abstractions and Limits.* Various fallacies result from the inadequate realization of the metaphorical character of many propositions and of the symbolical nature of all language. Words are counters or symbols, and it would be a grave error to identify a symbol with what it stands for or represents. Indeed, all thinking proceeds by noting certain distinguishable features in things, symbolizing such selected features by appropriate counters, and then reasoning upon such abstracted features by means of the symbols. In dealing intellectually with some concrete, specific situation, we do not pay attention to all the infinitely complex relations which it has, or to all of its qualities. On the contrary, we neglect almost all of the qualities and relations which a thing has, and note only those features which enable us to view that thing as an instance or example of indefinitely repeatable *patterns* or *types* of situations. Thus our knowledge of things involves abstraction from the infinitely complex and perhaps unique properties which situations have. We regard two objects as being "tables," because they each possess certain properties which may be found elsewhere; it does not follow that because a thing is a table it may not be something else also, or that it may not have other properties which differen-

tiate it from other objects that are tables. Through this process of abstraction we develop the notions of limiting or ideal patterns of structure and behavior. We thus arrive at the concepts of a perfectly straight line, of a frictionless surface, of a pure economic motive, of a rigid body, and so on, each of which represents a phase of some situation or other, but none of which can be identified with the whole concrete nature of anything.

It is a serious and widespread error, however, to suppose that because the objects of all discursive thinking are selected abstract phases of things, and not things in their concrete undifferentiated totality, that therefore science is fallacious and fictional.

For we should realize that the abstract objects of thought such as "numbers," "laws," "perfectly straight lines," and so on are real parts of nature (even though they do not exist as *particular* things but as the *relations* or *transformations* of such particulars). Because numbers or ratios are abstractions it does not follow that there is anything fictional in the assertion that the earth has "one" moon, or that the "rate" of infant mortality has recently decreased. The contrary supposition arises from a false notion of scientific procedure and its results. It arises from forgetting that abstractions are real parts, phases, or elements in things or their relations, even though they are not identical in all respects with the things.

The results of the processes of abstraction and classification have been called neglective fictions because, it is claimed, the class "man" does not exist and only individuals do exist. But it cannot be denied that such statements as "John is a man" can have significance only if the predicate denotes something really common to a number of individuals. Even an artificial classification of governments, such as that of Aristotle, cannot be called fictional merely because actual governments do not conform to it. For existing governments may be mixed forms or combinations of the elements of monarchy, aristocracy, and democracy and their perversions, and our classification helps us to recognize such mixed forms because of the elements they contain. The fact that certain elements always occur in conjunction with others, and never in isolation is no more an argument against their reality than the fact that no one can be a brother or a creditor without being other things is an argument against the possibility of having these abstract characteristics. Science must abstract some elements and neglect others, because not all things that exist together are relevant to each other. Hence there is no fiction in talking about purely economic motives if we remember, as Adam Smith surely did, that in actual life these are associated with other motives.

If we recognize the reality of abstractions, then there is nothing fictional (in the sense of false) about perfectly straight or circular lines, perfectly free bodies, frictionless engines, and other entities which seem imaginary and indeed are known to be incapable of separate existence. For the *relation* of distance between things exists in nature where the things are and is independent of the thickness of cord or chain by which we measure it. While there exist no free bodies (that is, bodies not acted upon by any forces), all existing bodies do move in such a way that we can find the part played by inertia (what would happen if all other forces ceased to act); similarly, while no actual engine is frictionless, we can from certain data compute the part that friction plays in the total work of any engine. It is not true that "artificial" lines of latitude or longitude are fictional merely because such lines are not actually marked on the earth. They do represent certain actual geometric relations on our earth. No map is ever a perfect picture of the country it represents. It must neglect all except a very few traits. But it may be perfectly accurate, truthful, within the required limits.

Incompleteness characterizes all human discourse. Statements such as "Jones is a good father" or "Smith is an efficient electrician" are true or false only if qualified. But these qualifications are either understood in the context in which they occur or they may be unnecessary for the degree of accuracy required. This is likewise true of such statements as "The rent of stores depends upon their volume of business."

Another way of looking at neglective fictions such as perfectly rigid bodies, perfect distribution, and the like is to view them as ideal limits. No *one* thing in nature corresponds to these, but things do differ in degrees of rigidity or homogeneity, and may be arranged into series according to the degree of rigidity or homogeneity they possess. Perfect rigidity would then be the character which all the members of the series possessed in some degree, and on the basis of which they are ordered in the series. It is the principle of order of such a series.

If there is no inherent fiction in abstraction, there is none in scientific "construction" out of such elements, as, for example, a typical vertebrate animal, a typical river valley or factory as an economic unit, an ideal government under limited conditions. Much abuse has been heaped on the "social contract" as a fiction. If it is taken as a historical fact it is a myth. This, however, does not apply to the great thinkers of the seventeenth and the eighteenth centuries. To them it was rather a logical device for analyzing actual complex social processes. If we apply the term "state of nature" to human conduct apart from the influence of laws, we can regard our actual social relations as those of a state of nature modified in certain ways analogous to the way our conduct is modified by contracts. The analogy is helpful only to the limited extent to which it is true.

In its search for the truth science must formulate some anticipation of what it expects to find. Such anticipation is clearly not fictional even if it turns out to be false, provided it has been held as a hypothesis to be tested. In trying to visualize the unknown, the imagination must clothe it with attributes analogous to the known. Thus electricity was first conceived as a fluid, then as lines or tubes of force, and now as a current of mutually repellent "electrons." So the mind was viewed by British psychologists as an associated group of "mental states," and by James as a "stream of consciousness." Each of these, like the various mechanical models of the ether or of various unknown physical processes, suggests verifiable analogies and thus directs research. If these directions turn out false, our analogy has acted like a false hypothesis.

The typical fiction which is often cited is the so-called imaginary number, $\sqrt{-1}$. As in ordinary algebra there are no numbers whose square can be negative, this is triumphantly adduced as a clear example of a useful devise based on a logically impossible entity. Modern mathematics, however, has made it clear that $\sqrt{-1}$ is no more imaginary or self-contradictory than $\sqrt{2}$, which is still called irrational, surd, or absurd. Starting with certain useful conventions as to pair of numbers, $\sqrt{-1}$ becomes a most useful clue to the properties of certain fields of forces. Logically, similar considerations hold with regard to the argument that self-contradiction inheres in the notion of infinite number, or infinitesimal magnitude. Modern mathematics has removed the basis for such arguments. One hears nowadays that the ether is a fiction which involves contradictory qualities. This also is simply not true. The ether is a hypothetical entity the existence of which follows from certain assumptions such as the law of the conservation of energy. Some of its properties are undoubtedly very unusual, and modern electromagnetic theory makes most of the mechanical models or analogies of it useless. But it is not at all self-contradictory—certainly not when it is in any way a useful explanation.

Neutral hypotheses, those of which the subject matter can never be directly proved or disproved, are very numerous in all sciences. Thus the old-fashioned books on economics begin by imagining one or more people landed on a desert island, just as the older theories of law and politics begin with an imaginary social contract, or modern mathematical physicists ask you to imagine a creature in a one- or two-dimensional space. Reasoning from such imaginary constructions is often confusing, because we do not always form a very clear idea of what it is that we are asked to imagine. But there is nothing fallacious in the method of such arguments. Concepts of this sort are like the auxiliary lines in a drawing or the parallels of latitude and longitude which we use in drawing maps. If one were to tell us that to draw a map of North America we should begin with drawing a certain triangle, then draw certain other lines, and so on, it would be absurd to object that North America is not and never was a triangle. The triangle can, in truth, represent the relations between a point in Greenland, one in Alaska, and one at the Isthmus of Panama; and by beginning with these points the relation of others to them could be indicated in the manner directed. The map will never be a complete picture of North America, but it can be perfectly true on the scale indicated. Fictions, like maps and charts, are useful precisely because they do not copy the whole but only the significant relations. These relations are identical in analogous cases; and we perceive and master the flux of phenomena only when we see running through it the threads of identity.

CHAPTER XIX

FALLACIES

§ 1. LOGICAL FALLACIES

It has been customary for books on logic to contain a separate section or chapter on *fallacies,* defined as *errors in reasoning.* These fallacies have generally been classified as: (*A*) purely logical or formal: (*B*) semilogical or verbal; and (*C*) material.

A. Formal Fallacies

These are arguments which fail to conform to the type of valid inference. There is no necessity for a separate treatment of them here, since they have been considered in the body of the book in connection with the various tests or rules which differentiate valid from invalid forms of reasoning. When put in hypothetical form all such fallacies are instances of arguments that proceed either from the affirmation of the consequent or from the denial of the antecedent; or else they assert an implication or logical connection where there is none. An instance of the latter is the syllogism in which the middle term is undistributed. This, as we saw (p. 77), reduces itself to an argument with four terms in which the premises give us no ground for, or proof of, the conclusion. To this kind of fallacy also belongs the next type.

B. Semilogical or Verbal Fallacies

These all seem to conform to valid forms of inference, but on careful examination are seen not to do so—the appearance being due to an ambiguity, that is, to the use of the same word or verbal sign for two different terms. The argument seems to be of the form: *A* is *B,* and *B* is *C,* therefore *A* is *C;* but in fact, it is: *A* is *B,* and *D* is *C,* therefore *A* is *C.*

It is well to note that not every instance of an ambiguity is a fallacy. If we consult any scholarly dictionary of the English language, we can note that there are few words which do not have more than one meaning. While there is only one way of writing "yes," there are many ways of saying it, which have different shades of meaning. A fallacy takes place only when one asserts that certain premises necessitate a given conclusion and when this claim is false because of the absence of real connection, an absence covered up by the use of the same word for two different things.

While all these fallacies result from the ambiguous use of words, certain special forms of them are worth noting because of their relative frequency.

1. The *fallacy of composition* frequently occurs when we reason from the properties of elements or individuals to the properties of the wholes which they constitute. For the same word may have a different significance when applied to a totality than it has when applied to an element. Thus the fact that the soldiers of a given regiment are all "strong" does not justify the conclusion that the regiment which they constitute is "strong." The word "strong" does not mean the same in the two cases. The fact that the soldiers are Irish fails to prove that the regiment is Irish. It may be part of the British or even of the French army.

2. The *fallacy of division* is the converse error of reasoning as if the properties of any whole are always properties of each part. If we say that mankind generally attains its end after trial and error, it does not follow that any individual or group will finally be successful by this method. The fact that the Roman Senate was a wise body does not prove that any individual in it was wise. Nor does the foolishness of assemblies, as judged by their resolutions, prove that the individual members were foolish. Men have different characters when in a group than when alone.

3. The *fallacy of accident* (also called *a dicto simpliciter ad dictum secundum quid)*. It is illustrated by the argument: You eat today what you bought yesterday and you bought raw meat yesterday; therefore you eat raw meat today. The two assertions do imply that the meat which was raw and bought yesterday is eaten today, but not that it is eaten raw. The particular form in which we eat it is not implied in the premises. In other words, the adjective which characterizes the condition of the meat when bought does not apply necessarily to the form in which we eat it. The premises of our argument do not, for instance, preclude the fact that the meat has grown one day older between the two operations. This seemingly trivial example illustrates many serious errors, for ex-

ample, when people argue from the rational nature of man to the rationality of any particular transaction, or from the fact that men are inherently curious to the explanation that kissing originated in curiosity.

This fallacy is widely prevalent amongst rigorous moralists, legalists, educationalists, and other social theorists who try to deduce the answer to specific human issues from some absolute moral, legal, educational, or other social rule. From the proof that lying is bad, that justice should be extended to all alike, and that property should be protected, men like the Stoics, Kant, and Blackstone have deduced such results as that one may not tell a lie to save the life of an innocent human being, that a criminal must never be pardoned, and that the state may not for public purposes take away a man's property against his will even if it pays him the market value. The rules involved in these cases are proved to be generally desirable on abstract considerations. But from this it does not logically follow that in specific cases one consideration of high importance may not be counterbalanced by others. The mutual confidence necessary for human intercourse condemns lying, but the saving of human life may outweigh that consideration.

4. The *converse fallacy of accident* (also called *a dicto secundum quid ad dictum simpliciter)*. This is the fallacy frequently illustrated when we try to refute a universal proposition, such as the law of supply and demand, by the argument that it does not hold in the case of a certain individual or a specific transaction. That which is true of individuals in certain specific situations is not necessarily true of them in general or abstract relations. Many "accidental" truths are irrelevant to certain general or abstract relations.

The avoidance of ambiguity is an extraordinarily difficult task. Scientific procedure seeks to avoid ambiguity by the use of technical terms, and by persistently searching for instances which illustrate the truth of the premises and the falsity of the conclusion. If the latter can be found, the argument is invalid.

C. Material Fallacies

In popular usage any argument which leads to a false conclusion is said to be fallacious or to "contain some fallacy somewhere." Now if we hold fast to the view that logic is not identical with all knowledge, and cannot guarantee the material truth of all conclusions, we cannot admit that logic alone can tell which conclusions are in fact false. And if a conclusion is actually false, the reasoning by which it was deduced (from a false premise) may be perfectly sound. It follows, therefore, that only mistakes in reasoning properly belong to logic. Hence we cannot consistently speak of false assumptions or false observations as logical fallacies. However, we certainly fail to prove the material truth of a proposition when we deduce it from one that is false. And we may speak of material fallacies to denote false claims or illusions of proof. Whether the A that follows event B is caused by it, is a question of fact and not merely of logic. But the assumption that whatever follows an event is therefore caused by it *(post hoc ergo propter hoc)* is false, and all arguments based on it fail to prove their point. It is thus also a fallacy to claim to have proved a proposition at issue if it has been smuggled, in some more or less disguised form, into our premises. (This is called *begging the question, petitio principii.)* To assume a proposition as a premise is not the same as to prove it.

1. A special form of this fallacy is called *arguing in a circle.* It consists in introducing into our premises a proposition that depends on the one at issue. Thus it would be arguing in a circle to try to prove the infallibility of the Koran by the proposition that it was composed by God's prophet (Mahomet), if the truth about Mahomet's being God's prophet depends upon the authority of the Koran. There is a sense in which all science is circular, for all proof rests upon assumptions which are not derived from others but are justified by the set of consequences which are deduced from them. Thus we correct our observations and free them of errors by appeal to principles, and yet these principles are justified only because they are in agreement with the readings which result from experiment. In other words, science cannot rest on principles alone. Nor can it rest on experimental observations regarded as all free and equal. Each is used to check the other. But there is a difference between a circle consisting of a small number of propositions, from which we can escape by denying them all or setting up their contradictories, and the circle of theoretical science and human observation, which is so wide that we cannot set up any alternative to it.

2. The *fallacy of the false question* also called the *fallacy of many questions.* To the extent that a question asks rather than gives information, it is not a proposition and cannot be true or false. We have seen, however, that the meaning of questions depends upon assumptions involved in them. Thus the question: Why do boys resemble their maternal uncles more than their paternal ones?, assumes it as a fact that they do so. Why was Esau wrong to sell his

birthright?, assumes that he was wrong. Taking advantage of this, we often smuggle false propositions into our question and then proceed to prove other propositions by their aid. Such proofs are seen to be illusory and to have no logical force when we realize the false assumption in the question. But lawyers on cross-examination frequently trip witnesses into testifying to and thereby proving (to the jury) a false proposition by making it part of a question; in that case, either an affirmative or a negative answer will imply an admission which the witness would not make if the point at issue were directly raised.

3. The *fallacy of the argumentum ad hominem,* a very ancient but still popular device to deny the logical force of an argument (and thus to seem to prove the opposite), is to abuse the one who advances the argument. Thus the fact that a man is rich or poor, married or single, old or young, is frequently used as an argument to disprove the truth of the proposition he affirms, or to lend force to its contradictory. This has received a great impetus in recent times from popular psychoanalysis. Any argument whatsoever can be refuted in this way by inventing some unfavorable psychogenetic account of how or why the proponent of the argument came to hold that view. Thus attempts have been made to refute some of Spinoza's arguments as to the nature of substance, or as to the relation of individual modes to that substance, on the ground that they were advanced by a man who had separated himself from his people, a man who lived alone, was intellectualist in temper, and so on. Now it is true that certain motives weaken our competence and our readiness to observe certain facts or to state them fairly. Hence the existence of such motives, if such existence can be proved in any given case, is relevant to determine the credibility of a witness *when he testifies as to what he himself has observed.* But the individual motives of a writer are altogether irrelevant in determining the logical force of his argument, that is, whether certain premises are or are not sufficient to demonstrate a certain conclusion. If the premises are sufficient, they are so no matter by whom stated. The personal history of Gauss is entirely irrelevant to the question of the adequacy of his proof that every equation has a root; and the inadequacy of Galileo's theory of the tides is independent of the personal motives which led Galileo to hold it. The evidences for a physical theory are in the physical facts relevant to it, and not in the personal motives which led anyone to take an interest in such questions.

§ 2. SOPHISTICAL REFUTATIONS

The word "sophist" which originally denoted a wise or learned man (like the word "savant") has, through historic accidents, come to mean one who argues to make the worse seem the better cause. To a certain extent in the heat of argument almost all human beings are more concerned to gain their point than to find the truth. But, leaving all questions of motive aside as irrelevant to logic, we may call attention to the fact that in addition to the examples in the previous section, certain other arguments are frequently used as if they were logically cogent, though no one consistently pretends that they are. They are generally used as refutations and may therefore be called *sophistical refutations.*

The most common of these is to pooh-pooh an argument, especially by the use of the word "mere." Thus the value of pure mathematics is often denied by saying that it gives us "mere" consistency, or, according to Huxley, "no experimental verification." The value of theoretical economics is thus denied because it does not enable us to predict the variations of the stock exchange or other markets. But one may as well pooh-pooh honesty because it does not guarantee good crops or does not supply us with water in a desert.

A variant of this is to disparage an argument, or make it seem ridiculous, by exaggerating its claims. This is often done not directly but by innuendo. Thus when in 1910 the Liberal Government of Great Britain threatened to create enough peers to override the veto of the House of Lords, the Conservative press countered by picturing the absurdities of every cab-driver or butler being a lord. Thus also one charged with certain literary or other deficiencies tries to refute the charge by pretending that he is accused of deadly sins or heinous offenses which it would be ridiculous to attribute to him.

Most sophistries consist in using words or raising issues of an emotional character that are logically irrelevant to the question in point. They are thus instances of what has been called the fallacy of irrelevance, more popularly known as confusing the issue. Thus, when some thinkers urge that on certain questions of vital importance we do not and perhaps cannot have sufficient knowledge to enable us to give a definite answer, they are often met with the query, What is the good of ignorance? Can man live on doubts? What is the value of a mystery or enigma? But no one considers that the valuable character of knowledge or the urgent need of it proves that we have it, any more than the urgent need for other things proves that we have them.

One of the commonest forms of sophistry is to make an argument seem ridiculous by confusing it with a part of that which it denies and thus rendering it self-contradictory and ridiculous. An amusing instance of it is given by De Morgan. In a debate in the House of Commons on the value of the decimal system of coinage, one member caused great laughter by showing the absurdities of a poor apple-woman trying to give change out of a shilling in the decimal system. Similar in character to this are many refutations of the relativity theory of motion by showing the absurdities to which it leads when older ideas (inconsistent with it) are tacitly assumed with it. Sydney Smith has collected a number of typical sophistical fallacies and put them in the form of an oration.

The number of such sophistical devices is legion, and it is no part of the task of logic to give an exhaustive account of them. But it is well to note that the rules of logic itself are frequently used in a sophistical way to refute valid arguments. Thus philosophical critics frequently argue as if an opponent's failure to give an explicit definition of his terms invalidates his argument. But it is clearly false to assume that all words can be defined. Similarly, critics frequently argue as if the use of words having more than one meaning vitiates an argument. But the fact is that few words are devoid of several meanings, and the actual meaning in any given passage is best determined by the text, and not by previous definitions or resolutions as to how a word should be used.

§ 3. ABUSES OF SCIENTIFIC METHOD

There are many ways in which the rules of logic are used to give the appearance of rigor to arguments which fail to prove their conclusions. We have indicated in the body of the book the proper scope of these rules; their right use is a matter of training or habituation. It would be impossible to enumerate all the abuses of logical principles occurring in the diverse matters in which men are interested. There are, however, certain outstanding abuses of scientific method which it will be profitable to note.

Fallacies of Reduction

Scientific method is largely concerned with the analysis of objects into their constituent elements. Thus the physicist, the chemist, the geologist, and the biologist each seeks to find the constituents of the objects that he studies; psychology, the social sciences, and philosophy try to do the same. It is understandable, therefore, how the misconception arises that science identifies objects with their *elements*. Science, however, does *not* do so, but analyzes its objects into *elements that are related* to each other in certain ways, so that if the same elements were related in different ways they would constitute *other* objects.

This misconception gives rise to two erroneous views: (1) that science denies the reality of the connecting links or relations, and (2) that science is a falsification of reality or the nature of things. Instances of the former are arguments which depend upon regarding, say, scientific books as nothing but words, animate or inanimate nature as nothing but atoms, lines as nothing but points, and society as nothing but individuals—instead of holding books, nature, lines, and society to be constituted by words, atoms, points, and individuals, respectively, *connected* in certain ways.

Building on this first mistake, many argue that science is therefore a falsification of reality. The motive for this conclusion appears very naïvely in a dialogue between two popular philosophers, Mutt and Jeff. When the former asks the latter whether he has heard that water is composed (by weight) of eight parts of oxygen to one part of hydrogen, the latter replies, "What! Is there no water in it?" Jeff's difficulty arises from a misapplication of the sound logical principle of identity, that water is water and not something else. What we ordinarily mean by water is a fluid with definite, familiar properties, which are not those of oxygen or hydrogen in isolation; and it seems clear that eight pounds of oxygen and one pound of hydrogen is not the same as nine pounds of water. Nevertheless, not only can water be constituted by or be broken into just such elements in just these proportions, but this fact enables us to understand many of the perceptible properties of water, and has enabled us to discover others which we would not have otherwise suspected. Water *is* hydrogen and oxygen *combined* in a certain way, just as a sentence is a group of words *ordered* in a certain way.

Similarly, many philosophers object to the analysis of ideas into their elements, on the ground that our primitive apprehension of ideas does not reveal them to have the logical structure that analysis reveals them to have. Here again a kernel of truth is misapprehended. There can be no doubt that taken as psychological events, our primitive perceptions do not apprehend the elements which logically constitute them. But we must not confuse the deliverances of primitive perception with the full meaning of what is apprehended. Thus the great mathematician Poincaré objected to a certain analysis of the number one, on the ground that the resulting complex of elements could not be recognized as the number one by children learning elementary arithmetic. Obviously, this argument is fallacious, since children cannot be expected to understand the full meaning of the ideas with which they begin to operate. Again, philosophers have objected to the analysis of the causal nexus as a certain relation between a number of states or configurations. For this, they have argued, leaves out the element of "efficiency." But what is this efficiency? If our analysis is sound, it *is* the *complex* of relations which *connect* one state of nature with another in a certain way.

The Fallacy of Simplism or Pseudo-Simplicity

Science aims at the simplest account which will systematize the whole body of available knowledge. This does not mean, however, that of any two hypotheses, the simpler is the true one. Systems with the simpler initial premises may turn out to be more complicated in the end. For example, Einstein's physics, assuming non-Euclidean geometry, turns out to be simpler than the Newtonian physics, which begins with the assumption of Euclidean geometry. In any case, we must guard against identifying the true with the apparently simple. And in fact hasty monism, the uncritical attempt to bring everything under one principle or category, is one of the most frequent perversions of scientific method. This is certainly true of popular forms of materialism, economic and other forms of determinism, subjective idealism, panlogism, and all other monistic doctrines according to which the absolute totality of all things is exhausted by some one category. Thus popular materialism thinks it is scientific in arguing that there is nothing in the world but matter, because everything we can talk about intelligibly contains matter or reference to it. But obviously, erroneous views exist in this world, and the materialist cannot argue that errors are themselves material. Errors do not exert electric or gravitational influence. And if he argues that only matter has real existence, he has only given us an implicit definition of "real existence"; he has not effectively denied that there are other elements in this world besides matter.

The same is true of the popular Berkelian idealism, which denies the reality of matter and insists that all is mind, perception, or idea. The monism which this doctrine pretends to establish is illusory. For the difference between the loaf of bread which exists only as an idea in my mind and the loaf of bread which I actually eat to satisfy my hunger is the same under the Berkelean idealism as under materialism. Both doctrines simply stretch old words to cover what are usually regarded as their opposites. But the difference between day and night remains even if both are said to constitute a day. The Hindu mystic insists that only the Atman (the self) truly exists and all the rest is illusion. But his vehemence in rejecting his critics' view that there is no illusion shows that the reality of illusion as opposed to the Atman is a necessary part of his view. The monism is verbal, not real.

More closely related to logic itself is the false assumption that logic requires a unique and irreversible order between any two concepts or propositions, so that if *A* presupposes *B*, the converse cannot be true. This ignores the possibility that there may be two factors which continually modify each other. Thus ignorance may be a cause of poverty without poverty's thereby ceasing to be a cause of ignorance. Increased production may be a cause of increased consumption, and conversely. Fallacious therefore are the arguments of those who dispute as to which is more fundamental, religion or economics, experience or reason, and the like.

To the last, which may be called the *fallacy of absolute priority* (which assumes that there must be an absolute first term in every series) we may add the *fallacy of exclusive linearity,* that is, the assumption that a number of factors are so related as to form necessarily a linear series. This is seen in the attempts of philosophers like Kant to arrange human faculties and other existences in linear series.

In general, before an object or concept is analyzed, it may frequently present an appearance of great simplicity and of lacking completely in internal structure. But such simplicity is most often the consequence of the fact that our attention has been directed to it either in a perfunctory manner or in a way to obtain the maximum of esthetic enjoyment and the minimum of rational knowledge. For example, the idea of number or that of motion seems unique and unanalyzable to the common man. Analysis, however, may reveal many complexities in the preanalytically simple object or concept. If then the preanalytical object is compared with the postanalytical one, and if the deliverances of "common sense" and esthetic appreciation are regarded as of superior value, there seems to be an air of artificiality about the outcome of analysis. It is often believed, in consequence, that analysis inevitably falsifies and distorts. But such a belief is based upon nothing else than an annoyance with the fact that reflective analysis does not take the preanalytical object at its face value. But there is no good reason to suppose that the "common-sense" view of things (which is generally unreflective and uncritical) is sounder or more profound than the views which are the outcome of arduous intellectual labor.

An allied form of error is the *fallacy of initial predication.* It frequently happens that some familiar characteristic of a thing, or some characteristic known sooner than others, is taken as definitive of its nature. There is, however, no good reason to believe that every trait of an object defines its nature adequately. On the contrary, serious errors arise from such beliefs. Thus the familiar fact about the sun that it rises in the east and sets in the west cannot be taken, on pain of error, as expressing adequately the nature of solar motion. Philosophers have been guilty of this fallacy in concluding that the essence of a thing is that it be known, since the only way in which we can think of things is as objects of knowledge.

A special form of simplism is the *fallacy of false opposition* or *false disjunction,* that is, the erroneous logical assumption that all alternatives are mutually exclusive, so that if A is B it cannot also be true that A is C. Thus it has been argued that there can be no harmony of interests between workmen and their employers, because they have conflicting interests in the relative shares of the industrial product that go to wages and profits. But while this conflict is real, there may also be an identity of interests in a protective tariff against a foreign industry. Conversely, the existence of a harmony or identity of interests does not deny diversity or conflicts in other respects. Thus also the proof of some evil or disorder in the body (physical or political) does not demonstrate the desirability of some proposed remedy. For the particular remedy may be worse than the disease, and there may be other alternatives, so that some other remedy may be preferable. Similarly, the untenability of some theory or the inadequacy of some remedy fails to prove the truth of some other theory or the desirability of the existing state. We must not hastily assume that the known alternatives exhaust the field of possibility.

A most important instance of the fallacy of false disjunction is the way people frequently argue that things cannot be constant if they change, or vice versa. It is obvious on reflection, however, that there is no change without some constancy and no constancy except relative to change. Of course we must discriminate between the phase in which things change and the phase in which they remain the same. But it is obviously a fallacy to deny that an individual owes the same debts if he has changed with respect to age, or that a mountain is the same despite the process of weathering. In general, anything which changes contains some element of identity which makes us see the various states as those of one entity. Yet this obvious observation is ignored in the fallacious argument used by many contemporary philosophers that there can be no constant laws of nature because things are constantly changing. Clearly, the statement that things are constantly changing is itself meant as an unchanging account of changeable nature.

A very widespread form of this fallacy is the confusion of, or the failure to discriminate between, the concrete and the real, and thus to jump to the conclusion that the abstract is unreal. This leads to the view that abstract science is a falsification of reality. Now abstract science does not pretend to describe the whole of reality. It always isolates certain common or invariant features of a group of events. Thus, to take obvious examples, the theories of physics take account of relations between mass, length, time, and the like, and neglect those aspects of a subject matter which may be chemical, biological, psychological, and so on. It follows that while a theory may treat adequately certain traits of a group of things which are common to all members of the group, it does not treat *exhaustively* the properties of any one member of such a group. It is a serious error, therefore, to suppose that a theory, which is an abstraction, is an adequate substitute in every context for that of which it is an abstraction, or that it is a falsification. We invite nothing but confusion if we suppose that any theory can do justice to *all* the properties of a subject matter, or if we imagine that it cannot illumine the nature of any property at all.

It is convenient to distinguish, as a special form of this error, the *fallacy of exclusive particularity.* It is often mistakenly assumed that a term which stands in one relation within one context cannot stand in any other relation within the same or other contexts. Elementary illustrations of this fallacy occur when from the fact that a person is honest or competent on one occasion it is inferred the same person cannot be dishonest or incompetent on other occasions. Nor does it follow that because a penny is round (when viewed from one position) it cannot also be elliptical or rectangular (when viewed from another).

A more complex and dangerous form of the fallacy is committed when it is supposed that because a given theory expresses any important truth about a subject, every other theory must be false. If social institutions and customs are a function of the prevailing means of economic production, it does not follow that these have not geographical, psychological, or political determinants as well.

Another manifestation of the fallacy of simplism or false economy is the confusion between necessary and sufficient conditions. A proposition p states a *sufficient condition* for another proposition q if "p implies q" is true. A proposition p states a *necessary condition* for another proposition q if "*not-p* implies *not-q*" is true (or, what is the same thing, if "q implies p" is true). These distinct relations of propositions are frequently confused. Thus the existence of sexual desires is sometimes said to be the cause of the family as a human institution on the ground that in the absence of sexual desires there could be no marriage. Evidently, however, all that is thus shown is that the existence of sexual desires is a *sine qua non* or *necessary* condition for this institution. But in order to explain adequately the family in terms of sex, it must be shown that man's sexual nature is by itself a *sufficient* condition for the existence of that institution, and this is not true if we can find sex expression without family life.

Many of the fallacies mentioned under other headings may also be analyzed as illustrating the failure to discriminate between necessary and sufficient conditions. Thus the proposition that a body or society is sick is necessary for the proof of the desirability of some remedy or reform, but clearly not sufficient. To prove the desirability we need further knowledge as to how the remedy or reform will work. Thus, also, it is a fallacy to argue, as many courts have, that the mere fact that an act of A threatens irreparable harm to the property of B is sufficient ground for enjoining it. The interest of a just or well-ordered community demands that the judge consider whether the issuance of an injunction may not do more harm than good by depriving those enjoined of their fundamental civil rights, the right of free assembly, free speech, or the like.

Another variety of simplism or failure to make proper discrimination may be discussed under a new heading.

The Genetic Fallacy

1. One form of this fallacy takes a logical for a temporal order. Our previous discussions ought to make it clear now that the facts of history cannot be deduced from logic alone, that factual data are needed to confirm or verify any speculation as to the past. This truth condemns all attempts current in the eighteenth century, and still widely popular, to reconstruct the history of mankind

prior to any reliable records, on the basis of nothing but speculations as to what must have been. The theories as to the origin of language or religion, or the original social contract by which government was instituted, which were based on empirically unsupported assumptions as to what "the first" or "primitive" man *must* have done are all historically untenable. It is clearly a logical error or fallacy to assume that actual history can be so constructed or discovered. Not much different, however, are those speculative *a priori* histories which under the name of social evolution attempt to deduce the stages which all human institutions must go through and therefore actually have gone through. In all of these attempts to trace the history of the family, industry, the state, and the like, the earlier stages are assumed to have been simpler, and the later stages more complex.

Such attempts appeal to us because we can understand the present complex institutions better if we see them built up out of simpler elements. But it is an inexcusable error to identify the temporal order in which events have actually occurred with the logical order in which elements may be put together to constitute existing institutions. Actual recorded history shows growth in simplicity as well as in complexity. Modern English, for instance, is simpler as regards inflection than Old English, and our legal procedure became less complicated when the old forms of action were abolished. *A priori* evolutionists had no doubt that the matriarchal family must precede the patriarchal form, and that the nomad state of society must precede the agricultural form. This, however, cannot prevent an actual Indian tribe from changing from the patriarchal to the matriarchal form. Nor can it prevent the Peruvians from skipping the nomad stage because the western slopes of the Andes could not provide them with sufficient cattle to serve as a basis of social organization. Indeed, the supposed law of development from the simple to the complex is too vague to enable us to deduce any specific historical events from it. That which seems simple in one state of knowledge or ignorance is seen to be more complex after increased knowledge or on closer examination. And many things bewilderingly complex at first become simpler to us after systematic study. Genetic accounts or theories which attract us by their *a priori* plausibility thus cease to do so when we discriminate between the intelligible and the temporal order, when we subject theories of what actually happened to the test of verifiability.

2. The converse error is the supposition that an actual history

of any science, art, or social institution can take the place of a logical analysis of its structure. When anything grows by additions or accretions, a knowledge of the order of such successive additions is a clue to the constitution of the final result. But not all growth is of that form. Science, for instance, as well as art and certain social organizations, is sometimes deliberately changed according to some idea or pattern to which previous existence is not relevant.

To suppose that the history of any science can take the place of a logical analysis of it involves also a confusion between our knowledge and the nature of what is known. Any history of physics, biology, astronomy, or geology is concerned with the growth of human knowledge. But the subject matter of these sciences is something which they themselves assume to have existed before any human knowledge, and indeed before human beings appeared on the earth. But even if we ignore the physical universe and restrict our view to science as a human achievement, it is still an error to identify the temporal order according to which any science has historically grown with the logical order in which its propositions are at any stage concatenated. We have already noted that many of the theorems of geometry were discovered before the systematic connection of the theorems was even suspected. The logical priority of the axioms to the theorems is therefore not identical with temporal priority in our apprehension or knowledge. We have also seen that the premises which are required to validate so-called inductive conclusions are logically prior to the latter, even though in the order of time we may discover them subsequently to the conclusions. The temporal order in which we learn or acquire our knowledge is not, in general, the same as the logical order of the propositions which are constituents of that knowledge.

CHAPTER XX

CONCLUSION

§ 1. WHAT IS SCIENTIFIC METHOD?

In the introductory chapter to Book II we asserted that the method of science is free from the limitations and willfulness of the alternative methods for settling doubt which we there rejected. Scientific method, we declared, is the most assured technique man has yet devised for controlling the flux of things and establishing stable beliefs. What are the fundamental features of this method? We have already examined in some detail different constituent parts of it. Let us in this final chapter bring together the more important threads of our discussions.

Facts and Scientific Method

The method of science does not seek to impose the desires and hopes of men upon the flux of things in a capricious manner. It may indeed be employed to satisfy the desires of men. But its successful use depends upon seeking, in a deliberate manner, and irrespective of what men's desires are, to recognize, as well as to take advantage of, the structure which the flux possesses.

1. Consequently, scientific method aims to discover what the facts truly are, and the use of the method must be guided by the discovered facts. But, as we have repeatedly pointed out, what the facts are cannot be discovered without reflection. Knowledge of the facts cannot be equated to the brute immediacy of our sensations. When our skin comes into contact with objects having high temperatures or with liquid air, the immediate experiences may be similar. We cannot, however, conclude without error that the temperatures of the substances touched are the same. Sensory experience sets the *problem* for knowledge, and just because such experience is immediate and final it must become informed by

reflective analysis before knowledge can be said to take place.

2. Every inquiry arises from some felt problem, so that no inquiry can even get under way unless some selection or sifting of the subject matter has taken place. Such selection requires, we have been urging all along, some hypothesis, preconception, prejudice, which guides the research as well as delimits the subject matter of inquiry. Every inquiry is specific in the sense that it has a definite problem to solve, and such solution terminates the inquiry. It is idle to collect "facts" unless there is a problem upon which they are supposed to bear.

3. The ability to formulate problems whose solution may also help solve other problems is a rare gift, requiring extraordinary genius. The problems which meet us in daily life can be solved, if they can be solved at all, by the application of scientific method. But such problems do not, as a rule, raise far-reaching issues. The most striking applications of scientific method are to be found in the various natural and social sciences.

4. The "facts" for which every inquiry reaches out are propositions for whose truth there is considerable evidence. Consequently what the "facts" are must be determined by inquiry, and cannot be determined antecedently to inquiry. Moreover, what we believe to be the facts clearly depends upon the stage of our inquiry. There is therefore no sharp line dividing facts from guesses or hypotheses. During any inquiry the status of a proposition may change from that of hypothesis to that of fact, or from that of fact to that of hypothesis. Every so-called fact, therefore, *may* be challenged for the evidence upon which it is asserted to be a fact, even though no such challenge is actually made.

Hypotheses and Scientific Method

The method of science would be impossible if the hypotheses which are suggested as solutions could not be elaborated to reveal what they imply. The full meaning of a hypothesis is to be discovered in its implications.

1. Hypotheses are suggested to an inquirer by something in the subject matter under investigation, and by his previous knowledge of other subject matters. No rules can be offered for obtaining fruitful hypotheses, any more than rules can be given for discovering significant problems.

2. Hypotheses are required at every stage of an inquiry. It must not be forgotten that what are called general principles or laws (which may have been confirmed in a previous inquiry) can be ap-

plied to a present, still unterminated inquiry only with some risk. For they may not in fact be applicable. The general laws of any science function as hypotheses, which guide the inquiry in all its phases.

3. Hypotheses can be regarded as suggestions of possible connections between actual facts or imagined ones. The question of the truth of hypotheses need not, therefore, always be raised. The necessary feature of a hypothesis, from this point of view, is that it should be statable in a determinate form, so that its implications can be discovered by logical means.

4. The number of hypotheses which may occur to an inquirer is without limit, and is a function of the character of his imagination. There is a need, therefore, for a technique to choose between the alternative suggestions, and to make sure that the alternatives are in fact, and not only in appearance, *different* theories. Perhaps the most important and best explored part of such a technique is the technique of formal inference. For this reason, the structure of formal logic has been examined at some length. The object of that examination has been to give the reader an adequate sense of what formal validity means, as well as to provide him with a synoptic view of the power and range of formal logic.

5. It is convenient to have on hand—in storage, so to speak—different hypotheses whose consequences have been carefully explored. It is the task of mathematics to provide and explore alternative hypotheses. Mathematics receives hints concerning what hypotheses to study from the natural sciences; and the natural sciences are indebted to mathematics for suggestions concerning the type of order which their subject matter embodies.

6. The deductive elaboration of hypotheses is not the sole task of scientific method. Since there is a plurality of possible hypotheses, it is the task of inquiry to determine which of the possible explanations or solutions of the problem is in best agreement with the facts. Formal considerations are therefore never sufficient to establish the material truth of any theory.

7. No hypothesis which states a general proposition can be demonstrated as absolutely true. We have seen that all inquiry which deals with matters of fact employs probable inference. The task of such investigations is to select that hypothesis which is the most probable on the factual evidence; and it is the task of further inquiry to find other factual evidence which will increase or decrease the probability of such a theory.

Evidence and Scientific Method

Scientific method pursues the road of systematic doubt. It does not doubt *all* things, for this is clearly impossible. But it does question whatever lacks adequate evidence in its support.

1. Science is not satisfied with psychological certitude, for the mere intensity with which a belief is held is no guarantee of its truth. Science demands and looks for logically adequate grounds for the propositions it advances.

2. No single proposition dealing with matters of fact is beyond every significant doubt. No proposition is so well supported by evidence that other evidence may not increase or decrease its probability. However, while no single proposition is indubitable, the body of knowledge which supports it, and of which it is itself a part, is better grounded than any alternative body of knowledge.

3. Science is thus always ready to abandon a theory when the facts so demand. But the facts must really demand it. It is not unusual for a theory to be modified so that it may be retained in substance even though "facts" contradicted an earlier formulation of it. Scientific procedure is therefore a mixture of a willingness to change, and an obstinacy in holding on to, theories apparently incompatible with facts.

4. The verification of theories is only approximate. Verification simply shows that, within the margin of experimental error, the experiment is *compatible* with the verified hypothesis.

System in the Ideal of Science

The ideal of science is to achieve a systematic interconnection of facts. Isolated propositions do not constitute a science. Such propositions serve merely as an opportunity to find the logical connection between them and other propositions.

1. "Common sense" is content with a miscellaneous collection of information. As a consequence, the propositions it asserts are frequently vague, the range of their application is unknown, and their mutual compatibility is generally very questionable. The advantages of discovering a system among facts is therefore obvious. A condition for achieving a system is the introduction of accuracy in the assertions made. The limit within which propositions are true is then clearly defined. Moreover, inconsistencies between propositions asserted become eliminated gradually because propositions which are part of a system must support and correct one another. The extent and accuracy of our information is thus increased. In fact,

scientific method differs from other methods in the accuracy and number of facts it studies.

2. When, as frequently happens, a science abandons one theory for another, it is a mistake to suppose that science has become "bankrupt" and that it is incapable of discovering the structure of the subject matter it studies. Such changes indicate rather that the science is progressively realizing its ideal. For such changes arise from correcting previous observations or reasoning, and such correction means that we are in possession of more reliable facts.

3. The ideal of system requires that the propositions asserted to be true should be connected without the introduction of further propositions for which the evidence is small or nonexistent. In a system the number of unconnected propositions and the number of propositions for which there is no evidence are at a minimum. Consequently, in a system the requirements of simplicity, as expressed in the principle of Occam's razor, are satisfied in a high degree. For that principle declares that entities should not be multiplied beyond necessity. This may be interpreted as a demand that whatever is capable of proof should be proved. But the ideal of system requires just that.

4. The evidence for propositions which are elements in a system accumulates more rapidly than that for isolated propositions. The evidence for a proposition may come from its own verifying instances, or from the verifying instances of *other* propositions which are connected with the first in a system. It is this systematic character of scientific theories which gives such high probabilities to the various individual propositions of a science.

The Self-Corrective Nature of Scientific Method

Science does not desire to obtain conviction for its propositions in *any* manner and at *any* price. Propositions must be supported by logically acceptable evidence, which must be weighed carefully and tested by the well-known canons of necessary and probable inference. It follows that the *method* of science is more stable, and more important to men of science, than any particular result achieved by its means.

1. In virtue of its method, the enterprise of science is a self-corrective process. It appeals to no special revelation or authority whose deliverances are indubitable and final. It claims no infallibility, but relies upon the methods of developing and testing hypotheses for assured conclusions. The canons of inquiry are themselves discovered in the process of reflection, and may themselves

become modified in the course of study. The method makes possible the noting and correction of errors by continued application of itself.

2. General propositions can be established only by the method of repeated sampling. Consequently, the propositions which a science puts forward for study are either confirmed in all possible experiments or modified in accordance with the evidence. It is this self-corrective nature of the method which allows us to challenge any proposition, but which also assures us that the theories which science accepts are more probable than any alternative theories. By not claiming more certainty than the evidence warrants, scientific method succeeds in obtaining more logical certainty than any other method yet devised.

3. In the process of gathering and weighing evidence, there is a continuous appeal from facts to theories or principles, and from principles to facts. For there is nothing intrinsically indubitable, there are no absolutely first principles, in the sense of principles which are self-evident or which must be known prior to everything else.

4. The method of science is thus essentially circular. We obtain evidence for principles by appealing to empirical material, to what is alleged to be "fact"; and we select, analyze, and interpret empirical material on the basis of principles. In virtue of such give and take between facts and principles, everything that is dubitable falls under careful scrutiny at one time or another.

The Abstract Nature of Scientific Theories

No theory asserts *everything* that can possibly be asserted about a subject matter. Every theory selects certain aspects of it and excludes others. Unless it were possible to do this—either because such other aspects are irrelevant or because their influence on those selected is very minute—science as we know it would be impossible.

1. All theories involve abstraction from concrete subject matter. No rule can be given as to which aspects of a subject matter should be abstracted and so studied independently of other aspects. But in virtue of the goal of science—the achievement of a systematic interconnection of phenomena—in general those aspects will be abstracted which make a realization of this goal possible. Certain common elements in the phenomenon studied must be found, so that the endless variety of phenomena may be viewed as a system in which their structure is exhibited.

2. Because of the abstractness of theories, science often seems in

patent contradiction with "common sense." In "common sense" the unique character and the pervasive character of things are not distinguished, so that the attempt by science to disclose the invariant features often gives the appearance of artificiality. Theories are then frequently regarded as "convenient fictions" or as "unreal." However, such criticisms overlook the fact that it is just certain *selected invariant relations* of things in which science is interested, so that many familiar properties of things are necessarily neglected by the sciences. Moreover, they forget that "common sense" itself operates in terms of abstractions, which are familiar and often confused, and which are inadequate to express the complex structure of the flux of things.

Types of Scientific Theories

Scientific explanation consists in subsuming under some rule or law which expresses an invariant character of a group of events, the particular events it is said to explain. Laws themselves may be explained, and in the same manner, by showing that they are consequences of more comprehensive theories. The effect of such progressive explanation of events by laws, laws by wider laws or theories, is to reveal the interconnection of many apparently isolated propositions.

1. It is clear, however, that the process of explanation must come to a halt at some point. Theories which cannot be shown to be special consequences from a wider connection of facts must be left unexplained, and accepted as a part of the brute fact of existence. Material considerations, in the form of contingent matters of fact, must be recognized in at least two places. There is contingency at the level of sense: just *this* and not *that* is given in sense experience. And there is contingency at the level of explanation: a definite system, although not the only possible one from the point of view of formal logic, is found to be exemplified in the flux of things.

2. In a previous chapter we have enumerated several kinds of "laws" which frequently serve as explanations of phenomena. There is, however, another interesting distinction between theories. Some theories appeal to an easily imagined *hidden mechanism* which will explain the observable phenomena; other theories eschew all reference to such hidden mechanisms, and make use of *relations* abstracted from the phenomena actually observable. The former are called *physical* theories; the latter are called *mathematical* or *abstractive* theories.

It is important to be aware of the difference between these two kinds of theories, and to understand that some minds are especially attracted to one kind, while others are comfortable only with the other kind. But it is also essential not to suppose that either kind of theory is more fundamental or more valid than the other. In the history of science there is a constant oscillation between theories of these two types; sometimes both types of theories are used successfully on the same subject matter. Let us, however, make clear the difference between them.

The English physicist Rankine explained the distinction as follows: There are two methods of framing a theory. In a mathematical or abstractive theory, "a class of objects or phenomena is defined . . . by describing . . . that assemblage of properties which is common to all the objects or phenomena composing the class, as perceived by the senses, without introducing anything hypothetical." In a physical theory "a class of objects is defined . . . as being constituted, in a manner not apparent to the senses, by a modification of some other class of objects or phenomena whose laws are already known." [1]

In the second kind of theory, some visualizable model is made the pattern for a mechanism hidden from the senses. Some physicists, like Kelvin, cannot be satisfied with anything less than a mechanical explanation of observable phenomena, no matter how complex such a mechanism may be. Examples of this kind of theory are the atomic theory of chemistry, the kinetic theory of matter as developed in thermodynamics and the behavior of gases, the theory of the gene in studies on heredity, the theory of lines of force in electrostatics, and the recent Bohr model of the atom in spectroscopy.

In the mathematical type of theory, the appeal to hidden mechanisms is eliminated, or at any rate is at a minimum. How this may be done is graphically described by Henri Poincaré: "Suppose we have before us any machine; the initial wheel work and the final wheel work alone are visible, but the transmission, the intermediary machinery by which the movement is communicated from one to the other, is hidden in the interior and escapes our view; we do not know whether the communication is made by gearing or by belts, by connecting-rods or by other contrivances. Do we say that it is impossible for us to understand anything about this machine so long as we are not permitted to take it to pieces? You know well

[1] W. J. M. Rankine, *Miscellaneous Scientific Papers,* 1881, p. 210.

we do not, and that the principle of the conservation of energy suffices to determine for us the most interesting point. We easily ascertain that the final wheel turns ten times less quickly than the initial wheel, since these two wheels are visible; we are able thence to conclude that a couple applied to the one will be balanced by a couple ten times greater applied to the other. For that there is no need to penetrate the mechanism of this equilibrium and to know how the forces compensate each other in the interior of the machine." [2] Examples of such theories are the theory of gravitation, Galileo's laws of falling bodies, the theory of the flow of heat, the theory of organic evolution, and the theory of relativity.

As we suggested, it is useless to quarrel as to which type of theory is the more fundamental and which type should be universally adopted. Both kinds of theories have been successful in coördinating vast domains of phenomena, and fertile in making discoveries of the most important kind. At some periods in the history of a science, there is a tendency to mechanical models and atomicity; at others, to general principles connecting characteristics abstracted from directly observable phenomena; at still others, to a fusion or synthesis of these two points of view. Some scientists, like Kelvin, Faraday, Lodge, Maxwell, show an exclusive preference for "model" theories; other scientists, like Rankine, Ostwald, Duhem, can work best with the abstractive theories; and still others, like Einstein, have the unusual gift of being equally at home with both kinds.

§ 2. THE LIMITS AND THE VALUE OF SCIENTIFIC METHOD

The desire for knowledge for its own sake is more widespread than is generally recognized by anti-intellectualists. It has its roots in the animal curiosity which shows itself in the cosmological questions of children and in the gossip of adults. No ulterior utilitarian motive makes people want to know about the private lives of their neighbors, the great, or the notorious. There is also a certain zest which makes people engage in various intellectual games or exercises in which one is required to find out something. But while the desire to know is wide, it is seldom strong enough to overcome the more powerful organic desires, and few indeed have both the inclination and the ability to face the arduous difficulties of scientific method in more than one special field. The desire to know is not often strong enough to sustain critical inquiry. Men generally are

[2] *Op. cit.*, pp. 290-91.

interested in the results, in the story or romance of science, not in the technical methods whereby these results are obtained and their truth continually is tested and qualified. Our first impulse is to accept the plausible as true and to reject the uncongenial as false. We have not the time, inclination, or energy to investigate everything. Indeed, the call to do so is often felt as irksome and joy-killing. And when we are asked to treat our cherished beliefs as mere hypotheses, we rebel as violently as when those dear to us are insulted. This provides the ground for various movements that are hostile to rational scientific procedure (though their promoters do not often admit that it is science to which they are hostile).

Mystics, intuitionists, authoritarians, voluntarists, and fictionalists are all trying to undermine respect for the rational methods of science. These attacks have always met with wide acclaim and are bound to continue to do so, for they strike a responsive note in human nature. Unfortunately they do not offer any reliable alternative method for obtaining verifiable knowledge. The great French writer Pascal opposed to logic the spirit of subtlety or finesse (*esprit géometrique* and *esprit de finesse*) and urged that the heart has its reasons as well as the mind, reasons that cannot be accurately formulated but which subtle spirits apprehend none the less. Men as diverse as James Russell Lowell and George Santayana are agreed that:

> "The soul is oracular still,"

and

> "It is wisdom to trust the heart . . .
> To trust the soul's invincible surmise."

Now it is true that in the absence of omniscience we must trust our soul's surmise; and great men are those whose surmises or intuitions are deep or penetrating. It is only by acting on our surmise that we can procure the evidence in its favor. But only havoc can result from confusing a surmise with a proposition for which there is already evidence. Are all the reasons of the heart sound? Do all oracles tell the truth? The sad history of human experience is distinctly discouraging to any such claim. Mystic intuition may give men absolute subjective certainty, but can give no proof that contrary intuitions are erroneous. It is obvious that when authorities conflict we must weigh the evidence in their favor logically if we are to make a rational choice. Certainly, when a truth is questioned it is no answer to say, "I am convinced," or, "I prefer to rely on this rather than on another authority." The view that physical science is no guide to proof, but is a mere fiction, fails to explain why it has enabled us to anticipate phenomena of nature and to control them. These attacks on scientific method receive a certain color of plausibility because of some indefensible claims made by uncritical enthusiasts. But it is of the essence of scientific method to limit its own pretension. Recognizing that we do not know everything, it does not claim the ability to solve all of our practical problems. It is an error to suppose, as is often done, that science denies the truth of all unverified propositions. For that which is unverified today may be verified tomorrow. We may get at truth by guessing or in other ways. Scientific method, however, is concerned with verification. Admittedly the wisdom of those engaged in this process has not been popularly ranked as high as that of the sage, the prophet, or the poet. Admittedly, also, we know of no way of supplying creative intelligence to those who lack it. Scientists, like all other human beings, may get into ruts and apply their techniques regardless of varying circumstances. There will always be formal procedures which are fruitless. Definitions and formal distinctions may be a sharpening of tools without the wit to use them properly, and statistical information may conform to the highest technical standards and yet be irrelevant and inconclusive. Nevertheless, scientific method is the only way to increase the general body of tested and verified truth and to eliminate arbitrary opinion. It is well to clarify our ideas by asking for the precise meaning of our words, and to try to check our favorite ideas by applying them to accurately formulated propositions.

In raising the question as to the social need for scientific method, it is well to recognize that the suspension of judgment which is essential to that method is difficult or impossible when we are pressed by the demands of immediate action. When my house is on fire, I must act quickly and promptly—I cannot stop to consider the possible causes, nor even to estimate the exact probabilities involved in the various alternative ways of reacting. For this reason, those who are bent upon some specific course of action often despise those devoted to reflection; and certain ultramodernists seem to argue as if the need for action guaranteed the truth of our decision. But the fact that I must either vote for candidate X or refrain from doing so does not of itself give me adequate knowledge. The frequency of our regrets makes this obvious. Wisely ordered society is therefore provided with means for deliberation and reflection *before* the pressure of action becomes irresistible. In order to assure the most thorough investigation, all possible views must be canvassed, and this means toleration of views that are *prima facie* most repugnant to us.

In general the chief social condition of scientific method is a widespread desire for truth that is strong enough to withstand the powerful forces which make us cling tenaciously to old views or else embrace every novelty because it is a change. Those who are engaged in scientific work need not only leisure for reflection and material for their experiments, but also a community that respects the pursuit of truth and allows freedom for the expression of intellectual doubt as to its most sacred or established institutions. Fear of offending established dogmas has been an obstacle to the growth of astronomy and geology and other physical sciences; and the fear of offending patriotic or respected sentiment is perhaps one of the strongest hindrances to scholarly history and social science. On the other hand, when a community indiscriminately acclaims every new doctrine the love of truth becomes subordinated to the desire for novel formulations.

On the whole it may be said that the safety of science depends on there being men who care more for the justice of their methods than for any results obtained by their use. For this reason it is unfortunate when scientific research in the social field is largely in the hands of those not in a favorable position to oppose established or popular opinion.

We may put it the other way by saying that the physical sciences can be more liberal because we are sure that foolish opinions will be readily eliminated by the shock of facts. In the social field, however, no one can tell what harm may come of foolish ideas before the foolishness is finally, if ever, demonstrated. None of the precautions of scientific method can prevent human life from being an adventure, and no scientific investigator knows whether he will reach his goal. But scientific method does enable large numbers to walk with surer step. By analyzing the possibilities of any step or plan, it becomes possible to anticipate the future and adjust ourselves to it in advance. Scientific method thus minimizes the shock of novelty and the uncertainty of life. It enables us to frame policies of action and of moral judgment fit for a wider outlook than those of immediate physical stimulus or organic response.

Scientific method is the only effective way of strengthening the love of truth. It develops the intellectual courage to face difficulties and to overcome illusions that are pleasant temporarily but destructive ultimately. It settles differences without any external force by appealing to our common rational nature. The way of science, even if it is up a steep mountain, is open to all. Hence, while sectarian and partisan faiths are based on personal choice or temperament and divide men, scientific procedure unites men in something nobly devoid of all pettiness. Because it requires detachment, disinterestedness, it is the finest flower and test of a liberal civilization.

APPENDIX

APPENDIX[1]

EXAMPLES OF DEMONSTRATION

§ 1. WHAT DOES A DEMONSTRATION ESTABLISH?

According to an age-old tale, Hiero, the tyrant of Syracuse, commanded a votive crown of pure gold to be placed in a temple of the immortal gods. But gossip concerning the goldsmith led him to suspect that silver had been mixed in its construction, and he requested Archimedes to determine, without injuring the crown, whether or not this was the case. While taking a bath, Archimedes noticed that his limbs were unusually light when in the water, and that in proportion as his body was immersed in the tub, water ran out of it. A method of resolving the problem forthwith became evident to him, and leaping out of the tub in great joy, he returned home naked, shouting as he ran, "Eureka! Eureka!"

The reader may know that the solution of the problem depends on the proposition: *A solid denser than water will, when immersed, suffer a loss in weight equal to the weight of the displaced water.* But how can we, or how did Archimedes, demonstrate the truth of this key proposition? The incident in the bath cannot be regarded as conclusive evidence for it, even though it may have *suggested* the proposition to Archimedes.

How would the reader go about demonstrating it? If he is empirically minded and very modern, he may perhaps believe that all that is necessary is to make a series of careful measurements on the weight of bodies in and out of the water by suspending them from a spring balance. Archimedes, however, was too wise a scientist, too well acquainted with the requirements of a *demonstration,* to do any such thing. For in the first place the confirmation of the proposition by measurement would never be more than *approximate.* No two measurements will yield exactly the same weight lost

[1] To be read by more advanced students after Chapter I.

by the solid, or show that the weight lost is precisely equal to the weight of the displaced water. And in the second place, no number of measurements can show that the proposition is true *for all possible cases*—for the unexamined instances which occurred in the past, as well as for the instances when a solid will fall into water in some distant future. How can one be sure, on the evidence supplied by such measurements, that if the solid exceeds a definite size, or if the quantity of water is increased sufficiently, the relation asserted by the proposition will still hold? The reader will agree that the method of experimental confirmation cannot guarantee that exceptions may not occur.

How, then, did Archimedes *prove* this proposition? Fortunately, the demonstration he found adequate is included in the extant portions of his treatise *On Floating Bodies*. It has been for centuries a model of what a demonstration should be, and has served as inspiration for the work of such men as Kepler and Galileo. It consists in exhibiting necessary relations between the *nature* or *definition of fluids* and the *nature of the behavior of solids* immersed in them. Let us examine it in detail, and so discover for ourselves the essential features of deductive reasoning.

Archimedes begins his treatise with a postulate, or assumption, which serves to define the nature of fluids. He then demonstrates six propositions by means of this postulate and the theorems of geometry which have already been demonstrated in treatises on that subject. In order to prove the seventh proposition, however, only the postulate and two of the preceding propositions are required. We shall simply state these, and then reproduce the demonstration of the seventh theorem. (We shall omit quotation marks here and abridge somewhat.)

The *postulate* reads: Let it be supposed that a fluid is of such a character that, in all uniform and continuous positions of its parts, the portion that suffers the lesser pressure is driven along by that which suffers the greater pressure. And each part of the liquid suffers pressure from the portion perpendicularly above it if the latter be sinking or suffer pressure from another portion.

Proposition 3. Solids which have the same density as a fluid will, if let down into it, be immersed so that they do not project above the surface but do not sink lower.

Proposition 6. If a solid lighter than a fluid be forcibly immersed in it, the solid will be driven upwards by a force equal to the difference between its weight and the weight of the fluid displaced.

Proposition 7 and its proof are as follows: A solid denser than a fluid will, if placed in it, descend to the bottom of the fluid; and the solid will, when weighed in the fluid, be lighter than its true weight by the weight of the fluid displaced.

Proof. 1. The first part of the proposition is obvious, since the part of the fluid immediately under the solid will be under greater pressure than the parts of the fluid under this part; and therefore these other parts will give way until the solid reaches bottom.

2. Let A be a solid heavier than the same volume of the fluid, and let $(G + H)$ represent its weight, while G represents the weight of the same volume of the fluid.

Take a solid, B, lighter than the same volume of the fluid, and such that the weight of B is G, while the weight of the same volume of the fluid is $(G + H)$. (That is, B is to be so chosen that its volume is equal to that volume of the fluid which is equal in weight to the body A.)

Let A and B be now combined into one solid and immersed. Then, since $(A + B)$ will be of the same weight as the same volume of fluid, both weights being equal to $(G + H) + G$, it follows that $(A + B)$ will remain stationary in the fluid.

Therefore the force which causes A by itself to sink must be equal to the upward force exerted by the fluid on B by itself. This latter is equal to the difference between $(G + H)$ and G. Hence A is depressed by a force equal to H, i.e., its weight in the fluid is H, or the difference between $(G + H)$ and G.

The reader should examine this proof carefully and repeatedly. When he has, let him consider the following questions.

1. In what sense does the "proof" demonstrate the proposition, assuming that it is conclusive?
2. Is the proof conclusive?
3. Upon what factors or aspects of the subject matter does the conclusiveness depend?

These questions must be squarely met if we are to avoid confusion concerning the philosophy of proof.

1. If the proof is cogent, then for all possible solids and for all fluids *conforming to the conditions stated in the postulate,* the relations specified in the proposition *must* hold. No exceptions to the proposition are possible, and no empirical examination of liquids is required in order that we be certain of this. The proposition can be asserted without fear of contradiction by any future experiment—*if the postulate is assumed.* But this qualifying "if" clause is

extremely important. It reminds us that we have *not* proved the proposition to be *materially* true. We have not shown that in any *actual* volume of water a denser solid will sink to the bottom—*unless,* indeed, the actual water is a fluid for which the postulate is true. What we have shown, therefore, is that *if* water is a fluid whose nature is expressed in part by the postulate, *then* the further relations stated in the proposition will of necessity be predicable of it. But the proof does not show, and does not claim to show, that water is in fact such a liquid.

Now Archimedes may have believed that it was "self-evident" that the postulate was true for all fluids. If he did, he was assuredly mistaken. As we shall have occasion to see repeatedly, and as we have already pointed out, the apparent self-evidence of a proposition is not conclusive evidence for its truth. But whether he did so believe or not, the truth or falsity of the postulate plays no part in the *demonstration.* The foregoing proof, to repeat, does *not* demonstrate the material truth of the proposition. What sort of evidence is required for the material truth of the proposition is considered in Chapters VIII, XI, XIII and XIV. Here we must make it perfectly plain that the *only* sense in which the proof, if conclusive, can be said to demonstrate is that it reveals a necessary connection between the defining properties of fluids and solids and their other properties. *The demonstration uncovers relations of implication between propositions,* and does nothing else. Whether any actual fluid does embody the properties stated by the postulate is not decided by the proof.

The reader will also note that the volume of the liquid and the size of the solid immersed in it play no part in the proof, because the proposition follows from the assumption concerning liquids *as such,* not from assumptions concerning liquids and solids of a certain volume. A proposition is demonstrated, therefore, if the premises *imply* the proposition, or, in other words, if the proposition is a necessary consequence of the premises.

2. It is time to proceed to the second question: Is the proof conclusive? Before the reader commits himself, let us remind him that the proof is conclusive only if the proposition is a necessary consequence of the premises. A proof is *not* conclusive if *other* premises are required besides those explicitly stated. How, then, can we be sure that no premises other than those stated are needed to imply the proposition? There is only one way to find out. We must break up the demonstration above into a series of implications, each of

which is to require no assumptions other than those explicitly granted. Let us therefore analyze the proof in some detail.

The first part of the proof may be put as follows:

1. The part of the liquid which suffers greater pressure than other parts moves out of the way the parts which are under lower pressure.	*Postulate*
The part of the fluid immediately beneath the solid suffers greater pressure than the parts of the fluid under this part.	*The solid is by hypothesis denser than the liquid.*
∴ The part of the fluid immediately beneath the solid moves out of the way the parts underneath itself.	

The second part of the proof can be stated as follows. We shall letter each step as a convenience for reference.

2. a. The weight of A is equal to $(G+H)$.	*Hypothesis*
The weight of B is equal to G.	*Hypothesis*
∴ The weight of $(A+B)$ is equal to $(G+H)+G$.	*The weight of a body is equal to the arithmetical sum of the weights of its parts, that is, the property of weight is assumed to be additive.*
b. The volume of A is equal to the volume of fluid weighing G.	*Hypothesis*
The volume of B is equal to the volume of fluid weighing $(G+H)$.	*Hypothesis*
∴ The volume of $(A+B)$ is equal to the volume of fluid weighing $(G+H)+G$.	*The volume of a body is equal to the arithmetical sum of the volumes of its parts, that is, the property of volume is assumed to be additive. Furthermore, the density of the fluid must be assumed to be constant.*
c. $(G+H)+G$ is the weight of the fluid equal in volume to the volume of $(A+B)$.	*Conclusion of 2b*
The weight of $(A+B)$ is equal to $(G+H)+G$.	*Conclusion of 2a*
∴ The weight of $(A+B)$ is equal to the weight of the fluid whose volume is $(A+B)$ or $(A+B)$ has the same density as the fluid.	
d. Solids which have the same density as a fluid will, if let down into it, be immersed so that they do not project above the surface but do not sink lower.	*Proposition 3*
$(A+B)$ has the same density as the fluid, and is let down into it.	*Conclusion of 2c*

$\therefore (A + B)$ will not project above the surface nor sink lower or $(A + B)$ will remain stationary in the fluid.	
e. If a solid lighter than a fluid be forcibly immersed in it, then the solid will be driven upwards by a force equal to the difference between its weight and the weight of the fluid displaced.	*Proposition 6*
B is a solid lighter than the fluid and forcibly immersed in it.	*Hypothesis*
$\therefore B$ is driven upwards by a force equal to the difference between its weight and the weight of the displaced fluid.	
f. G is the weight of B.	*Hypothesis*
$(G + H)$ is the weight of the fluid displaced by B.	*Hypothesis*
$\therefore$ The difference between the weight of B and the weight of the displaced fluid, is the difference between G and $(G + H)$, or H.	
g. The difference between the weight of B and the weight of the displaced fluid is H.	*Conclusion of 2f*
B is driven upwards by a force equal to the difference between its weight and the weight of the displaced fluid.	*Conclusion of 2e*
$\therefore B$ is driven upwards by a force H.	
h. $(A + B)$ remains stationary in the fluid.	*Conclusion of 2d*
B is driven upwards by a force H.	*Conclusion of 2g*
$\therefore A$ is depressed by a force H or The weight of A in the fluid is equal to H.	*The forces acting on a body in the same straight line can be added algebraically. This is an assumption concerning the additive nature of forces.*

The entire demonstration, we now see, can be analyzed into a number of distinct steps. The demonstration is conclusive, therefore, if each distinct step is conclusive. In this way we discover that the proposition cannot be demonstrated if we assume *only* the postulate. We require in addition four other assumptions concerning the additive nature of weights, volumes, and forces, and the constancy of density in a fluid. Archimedes did not state them explicitly, and in so far as he did not do so the proof is *not* conclusive. These assumptions, however, are of so general a nature that in almost all physical inquiries they are silently taken for granted. Nevertheless, it is very important that we note them explicitly, for without them or their equivalents we cannot prove the hydrostatic principle of Archimedes. Moreover, in some branches of modern physics evidence has been found for doubting the universal truth of several of them. The careful enumeration of all the premises or assumptions of an argument is of extraordinary value in the development of the sciences.

3. We are prepared now to answer the third question: Upon what factors or aspects of the subject matter does the conclusiveness of the demonstration depend? We have seen that the demonstration is conclusive if each step is conclusive. Why, then, is each step conclusive? We have already discussed the answer in the introductory chapter. Each step is conclusive because if the premises in that step are true the conclusion of the step *must also* be true: the relations between premises and conclusion are such that it is not possible to find a universe in which the premises of *this form* are true and the conclusion false.

§ 2. SOME FALLACIOUS DEMONSTRATIONS

We can see more clearly the need for the careful analysis of inferences if we examine two further examples of historically famous inferences.

1. The first is an attempt to improve upon Euclid. Euclid began his great work *The Elements* (of geometry) with twenty-three definitions, five axioms (which were unproved assumptions common to all the sciences), and five postulates (which were unproved propositions relating solely to geometry). The fifth postulate (Book I) is a proposition about parallel lines, but Euclid did not find it necessary to employ it until he came to the twenty-ninth proposition of his first book. Now while the other axioms and postulates struck Euclid's successors as "self-evident," the fifth postulate seemed to them to require proof. As Proclus, a mathematician of the fifth century, remarked, ". . . the fact that, when the right angles are lessened, the straight lines converge is true and necessary; but the statement that, since they converge more and more as they are produced, they will sometimes meet is plausible but not necessary, in the absence of some argument showing that this is true in the case of straight lines." [2] That the fifth postulate was stated without proof has been for centuries regarded as a blemish in the *Elements,* and many attempts have been made to demonstrate it.

We shall examine a proof by Ptolemy, as reported by Proclus. But we must first state the relevant definitions and postulates of Euclid. Parallel straight lines, according to him (Definition 23), are

[2] *The Thirteen Books of Euclid's Elements,* tr. by Sir T. L. Heath, 1926, 3 vols., Vol. I, p. 203.

"straight lines which, being in the same plane and being produced indefinitely in both directions, do not meet one another in either direction." The following are the five postulates:

Postulate 1. "To draw a straight line from any point to any point."

Postulate 2. "To produce a finite straight line continuously in a straight line."

Postulate 3. "To describe a circle with any center and distance."

Postulate 4. "All right angles are equal to one another."

Postulate 5. "If a straight line falling on two straight lines make the interior angles on the same side less than two right angles, the two straight lines, if produced indefinitely, meet on that side on which are the angles less than the two right angles."

Euclid introduced this last postulate in order to demonstrate Proposition 29: "A straight line falling on parallel straight lines makes the alternate angles equal to one another, the exterior angle equal to the interior and opposite angle, and the interior angles on the same side equal to two right angles." Ptolemy tried to prove the parallel postulate by first proving Proposition 29 without its aid, and then showing that the postulate was a consequence of this theorem. We reproduce his attempted proof for the theorem:

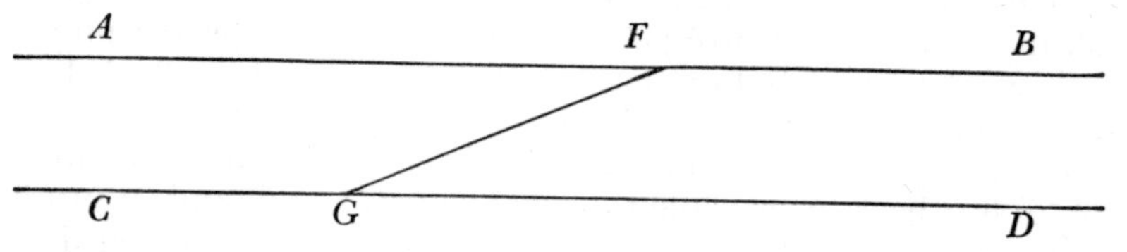

"The straight line which cuts the parallels must make the sum of the interior angles on the same side equal to, greater than, or less than two right angles.

"Let *AB, CD* be parallel, and let *FG* meet them. I say (1) that *FG* does not make the interior angles on the same side greater than two right angles.

"For, if the angles *AFG, CGF* are greater than two right angles the remaining angles *BFG, DGF* are less than two right angles.

"But the same two angles are also greater than two right angles; *for* AF, CG *are no more parallel than* FB, GD, *so that, if the straight line falling on* AF, CG *makes the intertor angles greater than two right angles, the straight line falling on* FB, GD *will also make the interior angles greater than two right angles.*

"But the same angles are also less than two right angles; for the four angles *AFG, CGF, BFG, DGF* are equal to four right angles: which is impossible.

"Similarly (2) we can show that the straight line falling on the parallels does not make the interior angles on the same side less than two right angles.

"But (3), if it makes them neither greater nor less than two right angles, it can only make the interior angles on the same side *equal* to two right angles."[3]

Is Ptolemy's proof valid? Does Proposition 29 follow necessarily from the axioms and postulates, omitting Postulate 5? Let us examine carefully the reasoning we have italicized above. Ptolemy argues that if we suppose angles *AFG, CGF,* are greater than two right angles, we must also suppose angles *BFG, DGF,* to be greater (as well as less) than two right angles, because *whatever is true of the interior angles on one side of the transversal* FG *to the parallel lines is necessarily true at the same time of the interior angles on the other side*. But this assumption is *not* included among the postulates. Ptolemy defends it by asserting that *AF, CG,* are no more parallel in one direction than *FB, GD,* are in the other. However, this simply amounts to saying that *through the point* F *only one parallel can be drawn to the line* CD. And this assumption is precisely *equivalent* to Postulate 5 which he is trying to prove.[4]

Ptolemy's proof, therefore, is unsuccessful, and a more careful analysis of his reasoning could have shown him that this was so. As a matter of fact, we know that Postulate 5 cannot be shown to be a necessary consequence of the remaining postulates. For it can be *demonstrated* that Postulate 5 is *independent* of the others. We discuss the method of proving independence in Chapter VIII. The reader should note at present that a rigorous analysis of an argument into a series of steps makes possible the discovery of all the assumptions required to validate the proof. A recognition of what assumptions we are making, and a readiness to consider all possible alternatives to them, is the one outstanding trait of the method of science. It is the only safeguard we can erect against intellectual dogmatism and arrogance.

2. The second example of a historically significant "proof" is the attempt to demonstrate an important proposition in elementary algebra. The reader is doubtless familiar with the rule that the

[3] *The Thirteen Books of Euclid's Elements,* tr. by Sir T. L. Heath, 1926, 3 vols., Vol. I, p. 205.

[4] *Ibid.,* p. 206.

product of two negative integers is always positive, thus: $(-3) \times (-4) = (+12)$. Can this proposition be demonstrated? Of course a demonstration is possible only if other propositions are assumed as premises. It turns out that algebra too may be developed systematically on the basis of axioms concerning the addition and multiplication of quantities. Our question must therefore be put as follows: Can the proposition that the product of two negatives is positive be shown to be a logical consequence of the assumptions concerning the addition and multiplication of *positive* numbers alone?

Unfortunately, the systematic exposition of algebra is highly abstract, requiring considerable intellectual maturity for its com-

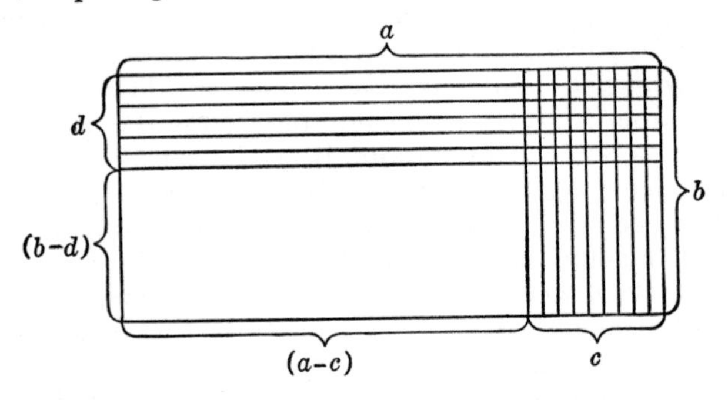

prehension, so that beginners are introduced to the subject as if it were simply a collection of rules. However, proofs are sometimes attempted of the important rules, and the following argument is frequently offered in support of the proposition concerning the product of two negative numbers. The argument tries to show that the rule for multiplying negative numbers is a necessary consequence of the rules for multiplying and adding positive ones.

The accompanying rectangle has sides equal to a and b respectively. Its area, in accordance with a theorem in plane geometry, is ab. The area of the smaller, unshaded rectangle, with sides equal to $(a-c)$ and $(b-d)$ respectively, is equal to $(a-c)\ (b-d)$. Let us now express this last area in terms of the large rectangle and the smaller, shaded ones. An examination of the figure shows that the area of the unshaded figure may be obtained by first *subtracting* from the large rectangle the vertically shaded rectangle (its area is bc) as well as the horizontally shaded rectangle (its area is ad), and then *adding* the rectangle shaded both ways (its area is cd). Hence we may write equation 1: $(a-c)\ (b-d) = ab - bc - ad + cd$.

Let us now give to both a and b the value zero. Then we get equation 2: $(0-c)\ (0-d) = 0.0 - 0.c - 0.d + cd$; or equation 3: $(-c)\ (-d) = (+cd)$. In general, the proof concludes, the product of two negatives is positive.

But is the proof valid? The reader will see easily that it is not. For equation 1 was developed on the assumption that a and b are *not* equal to zero. We cannot obtain equation 3 from equation 1 unless we make the *additional* assumption that equation 1 will be true for *all possible values* of a and b. But this addition is equivalent to the assumption that *all the laws which hold for the addition and multiplication of positive numbers are also true for negative ones.* And it is just this assumption which was being demonstrated.

We know, in fact, that the rules of operation upon negative numbers are *independent* of the rules for positive ones. Once more the value of analyzing an argument into a series of steps becomes evident. Just as the study of the assumptions required to prove Euclid's Postulate 5 led Lobatchevsky and Bolyai to the discovery of non-Euclidean geometries, so an examination of the fundamental rules of algebra led to the discovery by Sir William R. Hamilton and H. G. Grassmann of different algebraic systems. Without non-Euclidean geometries and the more general algebras, the advances of modern physics would hardly have been possible. It is well to note the far-reaching consequences of the method which insists on exhibiting all the assumptions required in demonstrations, and studies impartially the alternatives to such assumptions. The significance for civilization of logical method cannot be made clearer than by a contemplation of its rôle in the history of science.

EXERCISES

CHAPTER I: THE SUBJECT MATTER OF LOGIC

1. Is it on the basis of immediate knowledge or on evidence that you know
   a. that there is a center to the earth?
   b. that there is a king of Italy?
   c. that you have lungs with which you breathe?
   d. that there is a Belgian Congo?
   e. that there is a country called Tibet between India and China?

2. What evidence have you for the belief that a remote ancestor of yours must have lived in 2000 B.C.? or that Washington, Napoleon, King Arthur, King David, Homer, or Moses lived on this earth?

3. Prove that $5 + 2 = 7$. Formulate all the assumptions involved in this proof. (See Poincaré, *The Foundations of Science*, pp. 34-42.)

4. What is meant by saying that logic is formal?

5. Examine the proof that:

$$a^2 - b^2 = (a + b)\ (a - b)$$

   What fundamental assumptions are involved?

6. State the propositions necessary to prove
   a. that the earth is round,
   b. that it revolves around its axis and also around the sun.

7. What evidence is there that under the present Constitution a President of the United States might be elected by a minority vote?

8. Are the following arguments valid?
   a. San Francisco is west of New York.
      Peking is west of San Francisco.
      Berlin is west of Peking.
      ∴ Berlin is west of San Francisco.
   b. *A* is to the right of *B*.
      B is to the right of *C*.
      ∴ *A* is to the right of *C*.
      Is this true of people sitting in a circle?

9. Is the following a valid argument simply because premises and conclusion are true?
   a. Fish live in water.
      Monkeys are not fish.
      ∴ Monkeys do not live in the water.

b. Compare with:
Fish live in water.
Whales are not fish.
∴ Whales do not live in the water.

10. Examine the following as to their validity:

a. All lions are fierce.
Some lions do not drink coffee.
∴ Some creatures that drink coffee are not fierce.

b. Some pillows are soft.
No pokers are soft.
∴ Some pokers are not pillows.

c. No bankrupts are rich.
Some merchants are not bankrupts.
∴ Some merchants are rich.

d. No emperors are dentists.
All dentists are dreaded by children.
∴ No emperors are dreaded by children.

e. All the clever boys got prizes.
All the hard-working boys got prizes.
∴ All the clever boys were hard-working.

f. All liberals hold these opinions.
He holds these opinions.
∴ He is a liberal.

11. Discuss the following as to their validity:

a. If war is declared, the enemy country will be invaded.
But war is not declared.
∴ The enemy country will not be invaded.

b. If a number is divisible by 4, it is even.
It is not even.
∴ It is not divisible by 4.

c. If the street is sprinkled, there is no dust in the air.
The street is not sprinkled.
∴ There is dust in the air.

d. Either the child is ill or it is spoiled.
But the child is ill.
∴ It is not spoiled.

e. If he fulfills his election promises, he will be a popular mayor.
He will be a popular mayor.
∴ He will fulfill his election promises.

f. If he was born in Paris, he cannot become a cabinet officer.
He was born in Paris.
∴ He cannot become a cabinet officer.

g. Either he is indifferent or he is forgetful.
He is not forgetful.
∴ He is indifferent.

12. Examine the following arguments:

a. Ninety-nine Cretans in a hundred are liars.
Epimenides is a Cretan.
∴ Epimenides is a liar.

b. Minos, Sarpedon, Rhadamantus, Deucalion, and Epimenides are all the Cretans I know.
They were all atrocious liars.
∴ Pretty much all Cretans must have been liars.

c. All the beans taken from bag A are white.
These beans are white.
∴ These beans are from bag A.

13. Distinguish between inference and implication.

14. What is the difference between logic and grammar, logic and psychology, logic and physics?

15. In Book IV, Chapter XVII of his Essay Concerning Human Understanding, Locke comments as follows upon the value of the study of formal logic:

"If syllogisms must be taken for the only proper instrument of reason and means of knowledge, it will follow that, before Aristotle, there was not one man that did or could know anything by reason; and that, since the invention of syllogisms, there is not one of ten thousand that doth. But God has not been so sparing to men to make them barely two-legged creatures, and left it to Aristotle to make them rational."

Examine the force of this jibe against Aristotle.

For further study: L. S. Stebbing, *Modern Introduction to Logic,* 1930, Chap. I; John Dewey, *How We Think,* new ed., 1933, Pt. II; John Venn, *Principles of Empirical or Inductive Logic,* 2d ed., Chap. 5, Sec. 1.

CHAPTER II: THE ANALYSIS OF PROPOSITIONS

1. Classify the following propositions:

a. A bird in the hand is worth two in the bush.
b. Truth crushed to earth will rise again.
c. Six is the first perfect number.
d. All the king's horsemen could not put Humpty Dumpty together again.
e. Never will the dead speak.
f. There is nothing in the contract which prevents you from going.
g. Many men labor in vain.
h. Most adults are married.
i. Eight out of every thousand soldiers were killed.
j. Youth is always full of hope.
k. Few trees bear fruit.
l. My coat is on the chair.
m. A few eminent men had distinguished sons.
n. Few eminent men had distinguished sons.
o. Not a few eminent men had distinguished sons.
p. A few facts are better than much rhetoric.
q. One or the other of the members of the committee must have divulged the secret.
r. Literacy is a necessary but not a sufficient condition for voting.
s. Being without funds is a necessary but not a sufficient condition for obtaining help.
t. The virtuous alone are happy.
u. Only the unemployed are lazy.

v. None but foreigners are ill-treated.
w. All students except freshmen are welcome.
x. No one is admitted unless on business.
y. All but the wounded went hungry

2. Exhibit the logical structure of the following arguments:

a. "For we assumed that the world had no beginning in time, then an eternity must have elapsed up to every given point of time, and therefore an infinite series of successive states of things must have passed in the world. The infinity of a series, however, consists in this, that it never can be completed by means of a successive synthesis. Hence an infinite past series of worlds is impossible, and the beginning of the world is a necessary condition of its existence." (This is part of Kant's proof (in his first antinomy) of the thesis that the world had a beginning in time.) [1]

b. "Every compound substance in the world consists of simple parts, and nothing exists anywhere but the simple, or what is composed of it.

"For let us assume that compound substances did not consist of simple parts, then if all composition is removed in thought, there would be no compound part, and (as no simple parts are admitted) no simple parts either, that is, there would remain nothing, and there would therefore be no substance at all. Either, therefore, it is impossible to remove all composition in thought, or, after its removal, there must remain something that exists without composition, that is, the simple. In the former case the compound could not itself consist of substances (because with them composition is only an accidental relation of substances, which substances, as permanent beings, must subsist without it). As this contradicts the supposition, there remains only the second view, namely, that the substantial compounds in the world consist of simple parts.

"It follows as an immediate consequence that all the things in the world are simple beings, that their composition is only an external condition, and that, though we are unable to remove these elementary substances from their state of composition and isolate them, reason must conceive them as the first subjects of all composition, and therefore, antecedently to it, as simple beings." (This is part of the proof in Kant's second antinomy.) [2]

c. Proof of God's existence: "Whatever is in motion is moved by another: and it is clear to the sense that something, the sun for instance, is in motion. Therefore it is set in motion by something else moving it. Now that which moves it is itself either moved or not. If it be not moved, then the point is proved that we must needs postulate an immovable mover: and this we call God. If, however, it be moved, it is moved by another mover. Either, therefore, we must proceed to infinity, or we must come to an immovable mover. But it is not possible to proceed to infinity. Therefore it is necessary to postulate an immovable mover." [3]

3. Discuss the existential import of these propositions:

a. The present King of France is bald.
b. A square circle is a contradiction in terms.
c. The fifth root of a quadratic equation is an integer.

[1] *Critique of Pure Reason,* tr. by Max Müller, p. 344.
[2] *Ibid.*, pp. 353-54.
[3] St. Thomas Aquinas, *Summa contra Gentiles,* tr. by the English Dominican Fathers.

4. What is the distinction between a proposition and a sentence; between a proposition and a judgment?

5. What assumptions, statable as propositions, are involved in the following:

   a. Beware of the dog!
   b. Take your hats off!
   c. May the Lord have mercy on your soul!
   d. Will you join the dance?

6. Arrange the following terms in order of increasing intension: animal, organic being, vertebrate, man, body, featherless biped. What can you say about their extensions?

7. Distinguish between the subjective intension, the connotation, and the comprehension of: circle, man.

8. Which terms are distributed in the propositions in example 1?

9. Classify the following propositions:

   a. Unless he comes soon, we shall not wait.
   b. Either he is drunk or sober.
   c. He sold his horse in town.
   d. Either he answers all the questions, or he will fail to pass.

10. Distinguish between the following with respect to form:

    a. Newton is a physicist.
    b. All Englishmen are good-natured.
    c. Smith is older than Brown.
    d. This leaf is green.
    e. He gave a book to his daughter for her birthday.

11. Express the following as general propositions:

    a. Only the brave deserve the fair.
    b. Some men are teetotalers.
    c. All but children are admitted.
    d. No mathematician is musical.
    e. Some soldiers are not wounded.

For further study on existential import of propositions:

J. N. Keynes, *Formal Logic,* 4 Ed., p. 210 ff.
W. E. Johnson, *Logic,* Part I, Chap. IX.

On the analysis of propositions by contemporary logicians, see:

B. Russell, *Introduction to Mathematical Philosophy,* Chaps. XV, XVI, XVII; also, "Philosophy of Logical Atomism" in *Monist,* Vols. XXVIII and XXIX.

On the significance of negation, see

J. Royce, in Hasting's *Encyclopaedia of Religion and Ethics,* Article on Negation.
R. Demos, *Journal of Philosophy,* 1932. Article "Non-Being."
F. H. Bradley, *Principles of Logic,* Bk. I, Chap. 3.

On disjunction, see

F. H. Bradley, *Principles of Logic,* Bk. I, Chap. 4.

CHAPTER III. THE RELATIONS BETWEEN PROPOSITIONS

1. If *All giraffes have long necks* is true, what may be inferred concerning the following?

    a. No giraffes have short necks.
    b. No giraffes have long necks.
    c. Most giraffes have long necks.
    d. Thirty per cent of the giraffes have not long necks.
    e. All long-necked animals are giraffes.
    f. No short-necked animals are giraffes.
    g. All animals which are not giraffes have short necks.

2. If *No mammals weigh less than ten pounds* is false, what may be inferred concerning the following?

    a. All mammals weigh not less than ten pounds.
    b. All mammals weigh less than ten pounds.
    c. Few mammals weigh less than ten pounds.
    d. No animals weighing less than ten pounds are mammals.
    e. Some mammals do not weigh less than ten pounds.
    f. Some mammals weigh not less than ten pounds.
    g. All animals weighing not less than ten pounds are mammals.
    h. No nonmammalian animals weigh less than ten pounds.

3. If *Most men die young* is true, what may be inferred concerning the following?

    a. Some men do not die young.
    b. Some men die old.
    c. All men die young.
    d. All men die old.
    e. Most of those dying young are men.
    f. Some of those dying old are not men.
    g. Few creatures which are not men die old.

4. If *Some of the American population have no savings* is false, what may be inferred concerning the following?

    a. Forty per cent of the American population have savings.
    b. Some of the American population have savings.
    c. All of the American population have savings.
    d. No part of the American population has any savings.
    e. Some people who have no savings belong to the American population.
    f. Some of the non-American population have savings.
    g. Some people who have savings do not belong to the American population.

5. What information about "mortals" may be obtained from *All Struldbrugs are immortal?*

6. What information about "noncapitalists" may be obtained from *No capitalists are far-sighted?*

7. What information about "a solid body" can be derived from *No bodies which are not solids are crystals?* (Jevons.)

8. Examine the following:

a. If *All* $=$ *lateral* $\triangle$*s are* $=$ *angular,* then *All* $\triangle$*s with* $=$ ***angles have*** $=$ ***sides.***
b. If *All pacifists are social radicals,* then *All social radicals are pacifists.*
c. If *All wealthy men are generous,* then *All poor men are niggardly.*
d. If *Few old men do not desire immortality,* then *Few of those who desire immortality are not old men.*
e. If *Heat expands bodies,* then *Cold contracts bodies.* (Jevons.)
f. If *A false balance is abominative to the Lord,* then *A just weight is His delight.* (Jevons.)

9. State equivalent propositions for the following:

a. If the treasury was not full, the taxgatherers were to blame.
b. Through any three points not in a straight line a circle may be described.
c. It is false to say that only the notorious prosper in life.

10. What can be inferred from *The* $\angle$*s at the base of an isosceles* $\triangle$ *are equal* by obversion, conversion, and contraposition? (Jevons.)

11. Are the following equivalent?

a. All who were there talked sense.
b. All who talked nonsense were away. (De Morgan.)

12. Is the obverse of *No numbers are both odd and even, All numbers are either odd or even?*

13. "When I was a boy I have seen plenty of people puzzled by the following: An elderly nun was often visited by a young gentleman, and the worthy superior thought it necessary to ask who it was. 'A near relation,' said the nun. 'But what relation?' said the superior. 'Oh, madam,' said the nun, 'very near, indeed; for his mother was my mother's only child.' The superior saw that this was very close, and did not trouble herself to disentangle it. And a good many people on whom it was proposed used to study and bother to find out what the name of the connexion was." (De Morgan.) What was it?

14. What is the opposition between each pair of the following?

a. None but Democrats voted against the proposal.
b. Amongst those who voted against the proposal were some Democrats.
c. It is untrue that those who voted against the motion were all Democrats.

15. What is the contradictory of each of the following?

a. Few men are wealthy.
b. Denmark was not the only nation to remain neutral.
c. Forty per cent of the population is in dire want.
d. "You can fool some of the people all of the time, or all of the people some of the time; but you can't fool all of the people all of the time." (Abraham Lincoln.)
e. The King of Utopia did not die on Tuesday last.

16. Give a contrary of each of the following:

a. He was the tallest man in the German army.
b. Descartes was certain of his own existence.

17. Prove that if two propositions are equivalent their contradictories are equivalent.

18. If p, q, r, s, are *any* propositions such that p and q are *contradictories*

of one another, r and s are *contradictories* of one another, and p and r are *contraries* of one another, what is the relation of s to p, q to r, and s to q?

19. Give the contrapositive and inverse of each of the following:

   a. Whom the gods wish to destroy they first make mad.
   b. All are not saints that go to church.

20. *All that love virtue love angling.* Arrange the following propositions in the three following groups: (a) Those which can be inferred from the proposition above; (b) those which are consistent with it but which cannot be inferred from it; and (c) those which are inconsistent with it.

   a. None that love not virtue love angling.
   b. All that love angling love virtue.
   c. All that love not angling love virtue.
   d. None that love not angling love virtue.
   e. Some that love not virtue love angling.
   f. Some that love not virtue love not angling.
   g. Some that love not angling love virtue.
   h. Some that love not angling love not virtue.

21. A denies that *None but the native-born are citizens;* B denies that *None but citizens are native-born.* Which of the five class relations between "native-born" and "citizens" must they agree in *rejecting* and which may they agree in *accepting?*

22. Give the contrapositive and contradictory of each of the following:

   a. If any nation prospers under a protective tariff, its citizens reject all arguments in favor of free trade.
   b. If all men are mortal, then none of us will live to know his great-grandchildren.

23. Are the following contradictories?

   a. When there is thunder there is lightning and it either rains or hails.
   b. Sometimes when it thunders either there is no lightning or no rain or no hail.

24. Examine: If anyone is punished, he should be responsible for his actions. Hence, if some lunatics are not responsible for their actions, they should not be punished.

25. What is the logical relation between each of the propositions in the following pairs:

   a. Ten is greater than five.
      Five is less than ten.
   b. All angels fear to tread there.
      Some who fear to tread there are angels.
   c. Seven plus five equals twelve.
      The base angles of an isosceles triangle are equal.
   d. Smith is older than Brown.
      Smith is younger than Brown.
   e. This book is not in English.
      This book is not in French.
   f. No crocodiles shed tears.
      Some crocodiles shed tears.

g. The base angles of an isosceles triangle are equal.
The base angles of an equilateral triangle are equal.

For further study on the relations between propositions, see:

W. E. Johnson, *Logic,* Part I, Chaps. III, IX.
E. J. Nelson, *Monist,* 1932, "The Square of Opposition."
J. N. Keynes, *Formal Logic,* 4 ed. Appendix C, Chaps. II, IV, and V.

CHAPTER IV: THE CATEGORICAL SYLLOGISM

1. The first four axioms of the categorical syllogism are not independent of one another. Prove the second, third, and fourth by assuming the first axiom, together with the general principle of contraposition as well as the processes of conversion and obversion.

Example: Proof of Axiom 2. Suppose *p* and *q* are the premises and *r* the conclusion of a syllogism in which a term *N* is distributed in *r* but not in *p.* Now by the general principle of contraposition, the *contradictory* of *r,* together with *p,* must imply the *contradictory* of *q.* The term *N* will now be the middle term of the new syllogism. But since *N* is, by hypothesis, distributed in *r,* it will not be distributed in the contradictory of *r;* since if a term is distributed in one proposition, it is not distributed in its contradictory. Hence the middle term of the new syllogism is not distributed. It follows that a syllogism violating Axiom 2 implicitly violates Axiom 1.

2. Given that the major term is distributed in the premises and undistributed in the conclusion of a valid syllogism. Determine the syllogism.

3. Prove that, if three propositions involving three terms, each of which occurs in two of the propositions, are together incompatible, (a) each term is distributed at least once, and (b) one and only one of the propositions is negative.

4. Examine the validity of the following argument which sometimes appears in mathematical texts: "A magnitude required for the solution of a problem must satisfy a particular equation; since the magnitude *x* satisfies this equation, it is the magnitude required."

5. Is the following argument valid?

No *P* is *M.*
No *S* is *M.*
∴ Some *non-S* is *P.*

6. Is it possible to construct an *invalid* syllogism in which the major premise is *universal negative,* the minor premise *affirmative,* and the conclusion *particular negative?*

7. How many more distributed terms may there be in the premises of a valid syllogism than in the conclusion?

8. What can be determined respecting a valid syllogism under each of the following conditions?

a. Only one term is distributed, and that one only once.
b. Only one term is distributed, and that one twice.
c. Two terms only are distributed, each only once.
d. Two terms only are distributed, each twice.

9. In what cases will contradictory *major* premises both yield conclusions when

combined with the same *minor?* How are the conclusions related? Are there any cases in which contradictory *minor* premises both yield conclusions when combined with the same *major?*

10. Do c and d of the following propositions follow from a and b?

a. All just actions are praiseworthy.
b. No unjust actions are expedient.
c. Some inexpedient actions are not praiseworthy.
d. Not all praiseworthy actions are inexpedient.

11. Prove that any mood valid in both the second figure and the third figure is valid also in the first figure and the fourth figure.

12. Show that any given mood may be *directly* reduced to any other mood, provided that (a) the latter contains neither a strengthened premise nor a weakened conclusion; and (b) if the conclusion of the former is universal, the conclusion of the latter is also universal.

13. Construct a valid sorites consisting of five propositions and having *Some acorns do not grow into oaks* as its first premise. What is the mood and figure of each of the distinct syllogisms into which the sorites may be resolved?

14. Whatever *P* and *Q* may stand for, we can show *a priori* that: Some *P* is *Q*, for everything that is both *P* and *Q* is *Q*, and everything that is both *P* and *Q* is *P*, therefore some *P* is *Q*. Examine this paradox.

15. In what figures is the mood *AEE* valid? (Jevons.)

16. If the major term of a syllogism be the predicate of the major premise, what do we know about the minor premise? (Jevons.)

17. Prove that *O* cannot be a premise in the first or fourth figure; and that it cannot be the major in the second, or the minor in the third. (Jevons.)

18. If the middle term is distributed in both premises, what can we infer as to the conclusion? (Jevons.)

19. Show that if the conclusion of a syllogism be a universal proposition, the middle term can be distributed but once in the premises. (Jevons.)

20. Prove that when the minor term is predicate in the premise, the conclusion cannot be an *A* proposition. (Jevons.)

21. Prove that the major premise of a syllogism whose conclusion is negative can never be an *I* proposition. (Jevons.)

22. If the major term be universal in the premise and particular in the conclusion (which is not weakened), determine the mood and figure.

23. If the minor premise of a syllogism be *O*, what is the figure and mood?

24. Examine the validity of the following syllogisms:

a. None but whites are civilized; the ancient Germans were whites; therefore they were civilized. (Whately.)

b. None but civilized people are whites; the Gauls were whites; therefore they were civilized. (Whately.)

c. All books of literature are subject to error; and they are all of man's invention; hence all things of man's invention are subject to error. (Jevons.)

25. Show that if either of two given propositions will suffice to expand a given enthymeme of the first or second order, the two propositions are equivalent provided neither is a strengthened premise.

26. Prove special rules for the sorites:

$$\begin{array}{ll} S_1 & M_1 \\ M_1 & M_2 \\ \vdots & \vdots \\ M_n & P \\ \hline \therefore S & P \end{array}$$

a. Only one negative premise, and that one must be the last.
b. Only one particular premise and that one must be the first.

27. Construct a valid sorites of five propositions in which: Some *A* is not *B*: is the first premise.

28. Complete: Opponents of Robespierre declared in the Convention that he had identified his own enemies with those of the state. This he denied, adding, "And the proof is that you still live."

29. Examine the following as to their validity:

a. No Frenchmen like plum pudding.
All Englishmen like plum pudding.
∴ Englishmen are not Frenchmen.

b. No pigs can fly.
All pigs are greedy.
∴ Some greedy creatures cannot fly.

c. Nothing intelligible ever puzzles me.
Logic puzzles me.
∴ Logic is unintelligible.

d. No bald creatures need a hairbrush.
No lizards have hair.
∴ No lizards need a hairbrush.

e. All lions are fierce.
Some lions do not drink coffee.
∴ Some creatures that drink coffee are not fierce.

f. No fossil can be crossed in love.
An oyster can be crossed in love.
∴ Oysters are not fossils.

g. Some pillows are soft.
No pokers are soft.
∴ Some pokers are not pillows.

h. No fish suckles its young.
The whale suckles its young.
∴ The whale is no fish.

i. No ducks waltz.
No officers ever decline to waltz.
All my poultry are ducks.
∴ My poultry are not officers.

j. All babies are illogical.
No one is despised who can manage a crocodile.

Illogical persons are despised.
∴ No babies can manage crocodiles.

For further study:

A. De Morgan, *Transactions of the Cambridge Philosophical Society*, Vols. 8, 9, 10, series of papers "On the Syllogism."
Wm. P. Montague, *Ways of Knowing*, pp. 93 to 99.

CHAPTER V: HYPOTHETICAL, ALTERNATIVE, AND DISJUNCTIVE SYLLOGISMS

1. Examine the following:

a. If all men were capable of perfection, some would have attained it.
But no men have attained perfection.
∴ No men are capable of perfection.

b. If you needed food, I would give you money.
But since you do not care to work, you cannot need food.
∴ I will give you no money.

2. What may be inferred from the following?

a. He always stays in when it rains, but he often goes out when it is cold.
b. Either luck is with us or the robbers have been frightened away.
c. Either the robbers have not been frightened away or our jewels are gone.

3. Examine:

a. If I am fated to be drowned now, there is no use in my struggling; if not, there is no need of it.
But either I am fated to drown now or I am not.
∴ It is either useless or needless for me to struggle against it.

b. Patriotism and humanitarianism must be either incompatible or inseparable.
But although family affection and humanitarianism are compatible, either may exist without the other.
∴ Family affection may exist without patriotism.

4. If *P*, then *Q*; and if *R*, then *S*.
But *Q* and *S* cannot both be true.
Prove that *P* and *R* cannot both be true. (De Morgan.)

5. Examine:

a. "Either our soul, said they [the ancient philosophers], perishes with the body, and thus, having no feeling, we shall be incapable of any evil; or if the soul survives the body, it will be more happy than it was in the body; therefore death is not to be feared." (*The Port Royal Logic*, tr. by T. S. Baynes.)

b. If Abraham was justified, it must have been either by faith or by works. Now he was not justified by faith (according to James) nor by works (according to Paul).
∴ Abraham was not justified.

c. If the Mosaic account of the cosmogony is strictly correct, the sun was not created till the fourth day. And if the sun was not created till the fourth day, it could not have been the cause of the alternation of day and night for the first three days. But either the word "day" is used in Scripture in a different sense to that in which it is commonly accepted now, or else the sun must have been the cause of the alternation of day and night for the first three days. Hence either the Mosaic account of the cosmogony is not strictly correct, or else the word "day" is used in Scripture in a different sense to that in which it is commonly accepted now. (Keynes.)

d. Everyone is either well informed of the facts or already convinced on the subject.

No one can be at the same time both already convinced on the subject and amenable to argument.

∴ Only those who are well informed of the facts are amenable to argument.

e. Poetry must be either true or false: if the latter, it is misleading; if the former, it is disguised history and savors of imposture as trying to pass itself off for more than it is.

Some philosophers have, therefore, wisely excluded poetry from the ideal commonwealth. (Keynes.)

f. You say that there is no rule without an exception. I answer that, in that case, what you have just said must have an exception, and so prove that you have contradicted yourself. (Keynes.)

g. If all the accused were innocent, some at least would have been acquitted; but since none have been acquitted, none were innocent.

h. Given that *A* is *B*, to prove that *B* is *A*. Now *B* (like everything else) is either *A* or *not-A*. If *B* is *not-A*, then by our first premise we have the syllogism, *A* is *B*, *B* is *not-A*, hence *A* is *not-A*, which is absurd. Therefore *B* is *A*. (Keynes, after Jastrow.)

i. If all the soldiers had been English, they would not all have run away; but some did run away; hence some of them at least were not English. (Keynes.)

6. Examine:

a. Logic is indeed worthy of being cultivated, if Aristotle is to be regarded as infallible; but he is not. Logic, therefore, is not worthy of being cultivated (Whately.)

b. We are bound to set apart one day in seven for religious duties, if the fourth commandment is obligatory on us: but we are bound to set apart one day in seven for religious duties; and hence, it appears that the fourth commandment is obligatory on us. (Whately.)

c. "He [Robert Simson] used to sit at his open window on the ground floor, . . . deep in geometry. . . . Here he would be accosted by beggars, to whom he generally gave a trifle, he roused himself to hear a few words of the story, made his donation, and instantly dropped down into his depths. Some wags one day stopped a mendicant who was on his way to the window with 'Now, my man, do as we tell you, and you will get something from that gentleman, and a shilling from us besides. You will go and say you are in distress, he will ask you who you are, and you will say you are Robert Simson, son of John Simson of Kirktonhill.' The man did as he was told; Simson quietly gave him a coin, and dropped off. The wags watched a little, and saw him rouse himself again, and exclaim 'Robert Simson, son of John Simson of Kirktonhill! why, that is myself. That man must be an impostor.' " (De Morgan.)

7. Analyze the argument in the following:

Which is better, a clock that is right only once a year, or a clock that is right twice every day? "The latter," you reply, "unquestionably." Very good, now attend.

I have two clocks: one doesn't go *at all,* and the other loses a minute every day: which would you prefer? "The losing one," you answer, "without a doubt." Now observe: the one which loses a minute a day has to lose 12 hours or 720 minutes, before it is right, whereas the other is evidently right as often as the time it points to comes round, which happens twice a day.

So you've contradicted yourself *once.*

"Ah, but," you say, "what's the use of its being right twice a day, if I can't tell when the time comes?"

Why, suppose the clock points to 8 o'clock, don't you see that the clock is right *at* eight o'clock? Consequently, when 8 o'clock comes round your clock is right.

"Yes, I see *that,*" you reply.

Very good, then you've contradicted yourself *twice:* now get out of the difficulty as best you can, and don't contradict yourself again if you can help it.

You *might* go on to ask, "How am I to know when 8 o'clock *does* come? My clock will not tell." Be patient. You know that when 8 o'clock comes your clock is right; very good; then your rule is this: Keep your eye fixed on your clock, and the *very moment it is right* it will be 8 o'clock. "But—," you say. There, that'll do; the more you argue, the farther you get from the point, so it will be as well to stop. (Lewis Carroll.)

For further study, see:

J. Venn, *Empirical Logic,* Chap. X.

J. N. Keynes, *Formal Logic,* 4 ed. Appendix C, Chap. V, for a discussion of more complicated forms of agreement.

CHAPTER VI: GENERALIZED OR MATHEMATICAL LOGIC

1. State whether the relation in each of the following is transitive, intransitive, symmetrical, asymmetrical, one-one, one-many, or many-many.

a. He is the shortest man in the army.
b. Joseph had the same parents as Benjamin.
c. Adam is the ancestor of all of us.
d. Impatience is not the characteristic of a good teacher.
e. Smith is a next-door neighbor of Jones.
f. Russia was defeated by Japan.
g. Romeo is the lover of Juliet.
h. The ticket agent is on speaking-terms with many notables.
i. Brown is an employee of Jackson.

2. Discuss the following:

"It is a profoundly erroneous truism, repeated by all copybooks and by eminent people when they are making speeches, that we should cultivate the habit of thinking of what we are doing. The precise opposite is the case. Civilization advances by extending the number of important operations which we can perform without thinking about them. Operations of thought are like cavalry charges in battle—they are strictly limited in number, they require fresh horses, and must only be made at decisive moments."[4]

3. State the following expressions in words and simplify:

   a. $ad + \bar{a}\bar{c}d$.
   b. $a\bar{d} + a\bar{e} + \bar{a}b + \bar{a}c + \bar{a}e + \bar{b}c + \bar{b}\bar{d} + \bar{b}\bar{e} + \bar{b}\bar{e} + \bar{c}\bar{d} + \bar{c}\bar{e}$. (Keynes.)

4. Show that:

   a. $bc + \bar{b}d + cd$ is equivalent to $bc + \bar{b}d$. (Keynes.)
   b. $ab + ac + bc + \bar{a}b + \bar{a}\bar{b}\bar{c} + c = \bar{a} + b + c$.

5. Give the contradictories of the following terms:

   a. $ab + bc + cd$.
   b. $ab + \bar{b}c + \bar{c}d$. (Keynes.)

6. Give the contradictory of each of the following:

   a. Flowering plants are either endogens or exogens, but not both.
   b. Flowering plants are vascular, and either endogens or exogens but not both. (Keynes.)

7. What are some of the difficulties in employing everyday language for special scientific purposes?

8. Consult a large dictionary and discover what changes have occurred in the meaning of the following words: kind, manuscript, watch, genus, doctor.

9. What is meant by a class?

10. What is meant by the logical sum of two classes; by the logical product of two classes?

11. What is meant by saying that one of two classes is included in the other?

12. What is the null-class, and what relations hold between it and every other class?

13. State the following symbolically:

a. Only the perseverant succeed.
b. Some professors are not gray-haired.
c. None but the young are capable of heroism.
d. All logic books contain misprints.
e. No athletes live long.

14. Prove symbolically the following:

a. All a's are b's; therefore all non-b's are non-a's.
b. No a's are b's; therefore all a's are non-b's.
c. Some a's are b's; therefore some b's are a's.
d. Some a's are not b's; therefore some non-b's are not non-a's.

15. State the following propositions symbolically:

a. If p implies q, and q implies r, then p implies r.
b. If p and q imply r, then p and *not-r* imply *not-q*.
c. If either p or q implies r, then either p implies r or q implies r.

For further study:

On the logic of relations:

B. Russell, *Principles of Mathematics,* Chaps. II and IX.

On the algebra of logic:

L. Couturat, *Algebra of Logic.*

[4] A. N. Whitehead, *Introduction to Mathematics,* 1911, p. 61.

A. N. Whitehead, *Universal Algebra,* Bk. I, Chap. I and Book II.
C. I. Lewis, and Langford, *Symbolic Logic.*
R. M. Eaton, *Symbolism and Truth,* Chap. VII.

On the nature of symbols:

W. E. Johnson, *Logic,* Part II, Chap. III.
A. N. Whitehead, *An Introduction to Mathematics,* Chap. V.
R. M. Eaton, *Symbolism and Truth,* Chap. II.

CHAPTER VII: THE NATURE OF A LOGICAL OR MATHEMATICAL SYSTEM

1. How would you establish that the following postulates are *consistent?* Show also that they are *independent* of one another.

Postulate 1. If a and b are distinct elements of a class K, and $<$ an otherwise unspecified relation, then either $a < b$ or $b < a$.

Postulate 2. If $a < b$, then a and b are distinct.

Postulate 3. If $a < b$, then $b < c$, then $a < c$.[5]

2. Show that the set of operations (addition, multiplication, division, and subtraction) upon *integers* is isomorphic with the set of operations upon *fractions.*

3. Discuss the relation of algebra to analytic geometry with respect to isomorphism.

4. Prove that Aristotle's *dictum de omni et nullo* for the categorical syllogism is *equivalent* to the five axioms of validity stated in § 3 of Chapter IV.

5. Show that Axiom 1 for categorical syllogism (the middle term must be distributed at least once) is equivalent to Axiom 2 (no term can be distributed in the conclusion if it is not distributed in the premises).

6. "Write any odd number, say 35, on one card, and any even number, say 46, on another. Ask someone to give one of the cards to A and another to B, but not to tell you who has which. You undertake to tell A which number he was given. Ask A to multiply the number on his card by any even number, and B to multiply his by any odd number. Ask A and B to add their results and tell you what the sum is. If the sum is even, A was given the odd number; if the sum is odd, A was given the even number."[6] Demonstrate that this result will always hold.

7. Show that it is *impossible* to factor an integer into its *prime* factors in more than one way.

8. What is an axiom?

9. What is meant by saying that a proposition is self-evident?

10. What kinds of questions is logic in the position to ask and settle concerning a proposition?

11. Prove by mathematical induction:

$$1 + 2 + 3 + \cdots + n = \frac{n\ (n + 1)}{2}$$

12. Read Chapters 6 and 7 in Whitehead's *Introduction to Mathematics.* Discuss the ways in which integral, rational, real, and imaginary numbers differ from one another.

[5] E. V. Huntington, *The Continuum,* 1917, p. 10.
[6] E. T. Bell, *Numerology,* 1933, p. 174.

For further study:

J. W. Young, *Fundamental Concepts of Algebra and Geometry,* Chaps. II, III, IV, and V.
R. D. Carmichael, *The Logic of Discovery,* Chaps. II, III, IV, V, VI.
A. N. Whitehead, *Introduction to Mathematics*
P. E. B. Jourdain, *The Nature of Mathematics.*
B. Russell, *Mysticism and Logic,* Chaps. IV and V.

CHAPTER VIII: PROBABLE INFERENCE

1. What is the probability that a coin will fall tails up 3 times in succession if the coin can fall with either face uppermost, each of which events is as probable as the other?

2. What is the probability of getting heads 3 times in 5 throws of a fair coin?

3. A purse contains 2 quarters, 3 dimes, and 6 nickels. Two coins are taken out at random. What is the probability that the 2 coins are:

a. The 2 quarters.
b. One dime and 1 nickel.

4. The probability that A will die within ten years is $\frac{9}{100}$; that B will, is $\frac{7}{100}$, and that C will, is $\frac{11}{100}$.

a. What is the probability that A will live for another ten years?
b. What is the probability that both A and B will live another ten years?
c. What is the probability that at least one of the three will live another ten years?

5. If 7 people seat themselves at random at a round table, what is the probability that 2 designated men will be neighbors? What is the probability if 3 people so seat themselves?

6. The probability of getting a head with a coin is $\frac{1}{2}$; of getting a six with a die is $\frac{1}{6}$; and of drawing a white ball from an urn is $\frac{3}{5}$. What is the probability, if these are independent events, that:

a. One of these events, and only 1, will happen?
b. Two, but not more than 2, of these events will happen?
c. That at least 1 will happen?
d. That at least 2 will happen?
e. That at most 1 will happen?
f. That at most 2 will happen?

7. One of two externally similar boxes contains 4 pens and 2 pencils; the other contains 5 pens and 3 pencils. A box is opened at random and an item taken out. What is the probability that it is a pen? If the contents of two boxes had been in one bag, what is the probability that an item taken at random from it would be a pen?

8. Examine the following argument:

"The world that we know contains a quantity of good which, though limited, is still far in excess of what could be expected in a purely mechanistic system.

"If the Universe were composed entirely of a vast number of elementary entities, particles of matter or electricity, or pulses of radiant energy, which preserved themselves and pushed and pulled one another about according to merely physical laws, we should expect that they would occasionally agglutinate into unified structures, which in turn, though far less frequently, might combine to form structures still more complex, and so on. But that any considerable number of these higher aggregates would come about by mere chance would itself be a chance almost infinitely small. Moreover, there would be a steady tendency for such aggregates, as soon as they were formed, to break down and dissipate the matter and energy that had been concentrated in them. . . .

"Now the serious atheist must take his world seriously and ask: What is the chance that all this ascent is, in a universe of descent, the result of chance? And of course by chance, as here used, we mean not absence of any causality, but absence of any causality except that recognized in physics. Thus it would be 'chance' if a bunch of little cards, each with a letter printed on it, when thrown up into the breeze, should fall so as to make a meaningful sentence like 'See the cat.' Each movement of each letter would be mechanically caused, but it would be a chance and a real chance, though a small one, that they would so fall. . . . Let the atheist lay the wager and name the odds that he will demand of us. Given the number of corpuscles, waves, or what not, that compose the universe, he is to bet that with only the types of mechanistic causality . . . that are recognized in physics, there would result, I will not say the cosmos that we actually have, but any cosmos with an equal quantity of significant structures and processes. He certainly will not bet with us on even terms, and I am afraid that the odds that he will feel bound to ask of us will be so heavy that they will make him sheepish, because it is, after all, the truth of his own theory on which he is betting.

"But what is the alternative to all this? Nothing so very terrible; merely the hypothesis that the kind of causality that we know best, the kind that we find in the only part of matter that we can experience directly and from within, the causality, in short, that operates in our lives and minds, is not an alien accident but an essential ingredient of the world that spawns us." [7]

9. Analyze the following arguments to exhibit their probable nature:

"Our visitor bore every mark of being an average commonplace British tradesman, obese, pompous, and slow. He wore rather baggy gray shepherd's check trousers, a not over-clean black frock-coat, unbuttoned in the front, and a drab waistcoat with a heavy brassy Albert chain, and a square pierced bit of metal dangling down as an ornament. A frayed top hat and a faded brown overcoat with a wrinkled velvet collar lay upon a chair besides him. Altogether, look as I would, there was nothing remarkable about the man save his blazing red head and the expression of extreme chagrin and discontent upon his features.

"Sherlock Holmes's quick eye took in my occupation, and he shook his head with a smile as he noticed my questioning glances. 'Beyond the obvious facts that he has at some time done manual labor, that he takes snuff, that he is a Freemason, that he has been in China, and that he has done a considerable amount of writing lately, I can deduce nothing else.'

"Mr. Jabez Wilson started up in his chair, with his forefinger upon the paper, but his eyes upon my companion.

" 'How, in the name of good-fortune, did you know all that, Mr. Holmes?' he asked. 'How did you know, for example, that I did manual labor. It's as true as gospel, for I began as a ship's carpenter.'

[7] W. P. Montague, *Belief Unbound,* 1930, pp. 70-73.

" 'Your hands, my dear sir. Your right hand is quite a size larger than your left. You have worked with it, and the muscles are more developed.'

" 'Well, the snuff, then, and the Freemasonry?'

" 'I won't insult your intelligence by telling you how I read that, especially as, rather against the strict rules of your order, you use an arc-and-compass breast-pin.'

" 'Ah, of course, I forgot that. But the writing?'

" 'What else can be indicated by that right cuff so very shiny for five inches, and the left one with the smooth patch near the elbow where you rest it upon the desk.'

" 'Well, but China?'

" 'The fish which you have tattooed immediately above your right wrist could only have been done in China. I have made a small study of tattoo marks, and have even contributed to the literature of the subject. That trick of staining the fishes' scales of a delicate pink is quite peculiar to China. When, in addition, I see a Chinese coin hanging from your watch-chain, the matter becomes even more simple.' " [8]

10. Examine the following:

a. "In the controversies raised on the subject of Phrenology, the opponents of the system have considered that they disproved it by instancing decided exceptions to the phrenological allocations of faculties—cases of mathematicians with a small organ of number, or musicians with a small organ of tune. The facts supposed, however, are not conclusive against the system. For, in the first place, the disproof of the coincidence alleged, in respect of one or two faculties, or any number, would not disprove all the rest. But, in the second place, a few exceptions would not thoroughly disprove the alleged connexion; they would only disprove its unfailing uniformity. . . . For, if the coincidences of a certain distinguished mental aptitude—as number, music, colour—with the unusual size of a certain region of the head, were more frequent than it would be on mere chance, or in the absence of all connexion, he [the phrenologist] would be entitled to infer a relationship between the two." [9]

b. "The prevalence of the different forms of Christianity after the Reformation shows a coincidence with Race that chance would not account for. The Greek church was propagated principally in the Slavonic race; the Roman Catholic church coincides largely with the Celtic race; and the Protestant church has found very little footing out of the Teutonic races. From this coincidence must be presumed a positive affinity between the several forms and the mental peculiarities of the races." [10] (Bain.)

11. Point out the error in the following alleged "correction," written to the editor of a New York newspaper.

Sir:

Some little time ago a very interesting problem was propounded to your readers—a head being thrown in two flips of a coin, and an ace being cast in six throws of a die, which has the greatest probability of happening?

[8] Sir A. Conan Doyle, "The Red-Headed League," in *Adventures of Sherlock Holmes*.

[9] Alexander Bain, *Logic,* 2d ed., 1895, 2 vols., Vol. II, pp. 87-88.

[10] *Ibid.*, p. 88.

You have had several mathematical answers printed, but, strange to say, these answers are based on fallacies and are incorrect.

The chances with these two propositions are equal—that is, you have the same chance of casting an ace in six throws of a die as you have in turning up a head in two flips of a coin.

The chances of flipping a head in two throws of a coin are not one to two, or one to three, as given in two of your correspondents' answers, but it amounts to no chance at all. There are but two faces to a coin, therefore mathematically in every two throws a head must come up once. With the die the odds are: ace in one throw, five to nine; in two throws, four to two; in three throws, three to three; in four throws, two to four; in five throws, one to five, and in six throws no odds at all, because the ace, mathematically, is certain to come up.

12. What is probable inference?

13. Examine the following arguments:

a. From a bag of coffee a handful is taken out and found to have nine-tenths of the beans perfect. It is inferred that about nine-tenths of all the beans in the bag are probably perfect.

b. A man once landed at a seaport in a Turkish province; and as he walked up to the house he was to visit, he met a man on horseback surrounded by four horsemen holding a canopy over his head. Since the governor of the province was the only person the stranger could think of who would be so greatly honored, he inferred that the person he saw was the governor.

For further study:

C. S. Peirce, *Chance, Love, and Logic.* Part I, Chaps. 3, 4, 5, and 6. Also *Collected Papers.* Vol. II, Chaps. 5, 6, 7, and 8.

M. R. Cohen, *Reason and Nature,* Bk. I, Chap. III, § 3 and § 4.

J. M. Keynes, *Treatise on Probability,* Part I.

J. Laird, *Knowledge, Belief and Opinion,* Chap. XVII.

CHAPTER IX: SOME PROBLEMS OF LOGIC

1. How would you distinguish between principles of logic and principles of physics?

2. How would you solve the following difficulties?

a. A barber is defined as one who shaves *all* those and *only* those who do *not* shave themselves. Does the barber shave himself?

b. A word standing for an adjective will be said to be *autologous* if it itself possesses the property denoted by it; it will be said to be *heterologous* otherwise. Thus the word "short" is itself short and is therefore autologous; the word "long" is not itself long and is heterologous.

Consider the word "heterologous." If it is autologous it has the property denoted by itself, and so is heterologous; if it is heterologous, it has not the property denoted by itself (that is, is *not* heterologous) and so is autologous. It seems then, that if it is, it isn't; and if it isn't, it is.

c. "The number of syllables in the English names of finite integers tends to increase as the integers grow larger, and must gradually increase indefinitely, since only a finite number of names can be made with a given finite number of syllables. Hence the names of some integers must consist of at least nineteen syllables, and among these there must be a least. Hence 'the least integer not nameable in fewer than nineteen syllables' must denote a definite integer; in fact, it denotes 111,777. But 'the least integer not nameable in fewer than nineteen syllables' is itself a name consisting of eighteen syllables; hence the least integer not nameable in fewer than nineteen syllables can be named in eighteen syllables, which is a contradiction.[11]

3. Analyze the following discussion of the "laws of thought":

"If they are principles of thinking, we have to consider how they may be applied to reality and with what success. If they are principles of being, we want exhaustive evidence that all reality obeys these laws, and must at least grapple with the paradox of Change. For the reality of Change seems flatly to defy them all. A thing that changes neither remains itself, nor is it incapable of assuming contrary attributes in time or even simultaneously. It both is and is not, and cannot strictly be said to 'be' either one thing or another. If the moving arrow ever 'were' at the points it passes through, if we were ever right in *saying* that it was, Zeno's inference would be inevitable that motion is impossible. . . ."[12]

4. Examine the following argument: The principle of contradiction does not hold, since according to it *animal* cannot both be vertebrate and invertebrate, although in fact some animals are vertebrate and some are not.

5. What is the alleged paradox of inference?

For further study:

J. Venn, *Empirical Logic*, Chaps. I, II, III.

M. R. Cohen, "Subject Matter of Formal Logic," in *Journal of Philosophy*, Vol. XV, 1918.

B. Bosanquet, *Essentials of Logic*, Chaps. I, II, III.

J. Dewey, "Notes on Logical Theory" in *Journal of Philosophy*, Vol. I, 1904.

C. S. Peirce, *Collected Papers*, Vol. II, Bk. I, Chap. I.

CHAPTER X: LOGIC AND THE METHOD OF SCIENCE

1. What is the distinction between formal logic and scientific method?

2. Read the first essay in William James *The Will to Believe*. Discuss the issues raised in connection with the scope of scientific method.

In connection with this chapter, students are advised to read:

Osler, *Evolution of Medicine*, Chap. I.

H. Gomperz, *Greek Thinkers*, Vol. I, Bk. III.

A. D. White, *Warfare of Science and Theology*, Chap. I.

C. S. Peirce, *Chance, Love, and Logic*. Part I, Chaps. I and II.

CHAPTER XI: HYPOTHESES AND SCIENTIFIC METHOD

1. Discuss the following statement:

". . . Science, though it starts from observation of the particular, is not con-

[11] A. N. Whitehead and Bertrand Russell, *Principia Mathematica*, 2d ed., 1925, 3 vols., Vol. I, p. 61.

[12] F. C. S. Schiller, *Formal Logic*, 1912, p. 117.

cerned essentially with the particular, but with the general. A fact, in science, is not a mere fact, but an instance." [13]

2. Has the Ptolemaic hypothesis of planetary motion been disproved?

3. Analyze the following argument:

". . . What remains to be said upon the quantity and source of the blood which thus passes, is of so novel and unheard-of character, that I not only fear injury to myself from the envy of a few, but I tremble lest I have mankind at large for my enemies. . . . When I surveyed my mass of evidence, whether derived from vivisections, and my various reflections on them, or from the ventricles of the heart and the vessels that enter into and issue from them, the symmetry and size of these conduits,—for nature doing nothing in vain, would never have given them so large a relative size without a purpose,—or from the arrangement and intimate structure of the valves in particular, and of the other parts of the heart in general, with many things besides, I frequently and seriously bethought me, and long revolved in my mind, what might be the quantity of blood which was transmitted, in how short a time its passage might be effected, and the like; and not finding it possible that this could be supplied by the pieces of the ingested aliment without the veins on the one hand becoming drained, and the arteries on the other getting ruptured through the excessive charge of blood, unless the blood should somehow find its way from the arteries into the veins, and so return to the right side of the heart: I began to think whether there might not be *a motion, as it were, in a circle*. . . .

"But lest anyone should say that we give them words only, and make mere specious assertions without any foundation, and desire to innovate without sufficient cause, three points present themselves for confirmation, which being stated, I conceive that the truth I contend for will follow necessarily, and appear as a thing obvious to all. First,—the blood is incessantly transmitted by the action of the heart from the vena cava to the arteries in such quantity, that it cannot be supplied from the ingesta, and in such wise that the whole mass must very quickly pass through the organ. . . .

"Let us assume either arbitrarily or from experiment, the quantity of blood which the left ventricle of the heart will contain when distended to be, say two ounces, three ounces, one ounce and a half—in the dead body I have found it to hold upwards of two ounces. Let us assume further, how much less the heart will hold in the contracted than in the dilated state; and how much blood it will project into the aorta upon each contraction;—and all the world allows that with the systole something is always projected . . . and let us suppose as approaching the truth that the fourth, or fifth, or sixth, or even but the eighth part of its charge is thrown into the artery at each contraction; this would give either half an ounce, or three drachms, or one drachm of blood as propelled by the heart at each pulse into the aorta; which quantity, by reason of the valves at the root of the vessel, can by no means return into the ventricle. Now, in the course of half an hour, the heart will have made more than one thousand beats, in some as many as two, three, and even four thousand. Multiplying the number of drachms propelled by the number of pulses, we shall have either one thousand half ounces, or one thousand times three drachms, or a like proportional quantity of blood, according to the amount which we assume as propelled with each stroke of the heart, sent from this organ into the artery; a larger quantity in every case than is contained in the whole body. . . .

[13] Bertrand Russell, *The Scientific Outlook*, pp. 57-58.

"Upon this supposition, therefore, assumed merely as a ground for reasoning, we see the whole mass of blood passing through the heart, from the veins to the arteries, and in like manner through the lungs."[14]

4. Compare the use of the word "hypothesis" in the present chapter with its use in mathematics, where it denotes the conditions under which a theorem holds.

5. "It is the immediate and in a sense the most important task of our conscious knowledge of nature to enable us to anticipate future experiences, so that we may order our present activity in the light of such anticipations. In all cases we employ as the means of carrying out this task previous knowledge obtained either through chance observation or planned experiment. The process used to deduce the future from the past and so to accomplish the desired anticipation, is as follows: We make ourselves subjective pictures or symbols of external objects. We construct these in such a manner that the logically necessary sequences of these pictures are always symbols for the physical series of represented objects. In order that this be possible, there must be certain concordances between nature and our intellect. Experience teaches us that this process is indeed possible, and so that such concordances do in fact exist. If we are fortunate enough to construct pictures of the desired kind from our accumulated past experience, we can, using them as models, deduce the series of events which will happen in external nature at some remote time or in consequence of our own purposive activity. In this way we can anticipate the facts and so adjust our present decisions. The pictures of which we speak are our representations or ideas of things; they must share with external objects the single property already mentioned, but need have no other property in common with physical things in order that they should satisfy the purpose for which they had been constructed. In fact, we do not know, and indeed have not the means to find out, whether our representations of things coincide with the things in any other manner besides the aforesaid one.

"But the pictures we construct of things are not uniquely determined by the condition that the series of pictures represent the series of external events. Different pictures of the same objects are possible, and these pictures may differ among themselves in several ways. We shall designate as *not permissible* such pictures which contradict the laws of thought, and we shall postulate that all our representations be logically *permissible*. Furthermore, we shall say that pictures are *incorrect* if their essential relations to one another contradict the relations between the represented external objects if they do not satisfy the fundamental condition we require of them. We thus demand, in the second place, that our pictures be *correct*. But two permissible and correct pictures of the same external objects may still differ in *appropriateness*. We shall designate one of two pictures of the same object as *more appropriate*, if it mirrors more essential relations between the objects; or in other words if it is clearer or more distinct. If two pictures are equally clear, that will be the more appropriate which contains the fewest number of superfluous or irrelevant relations; or in other words, which is the *simpler*. Superfluous features cannot be completely eliminated from representations, precisely because they are representations made by

[14] William Harvey, *An Anatomical Disquisition on the Motion of the Heart and Blood in Animals,* first published in 1628, Chaps. VIII-IX.

our own specific intellect and thus must be characterized by the traits of its ways of symbolization."[15]

Compare this discussion of the conditions for satisfactory hypotheses with the treatment in the text.

6. Mill declared that Kepler, in ascertaining the nature of the planetary orbits, was simply "describing" a complex fact apprehended in direct observation. According to Mill, no inference was required to do this, no use of hypotheses.[16]

Peirce commented as follows on these remarks of Mill:

"What Kepler had given was a large collection of observations of the apparent places of Mars at different times. He also knew that, in a general way, the Ptolemaic theory agrees with the appearances, although there were various difficulties in making it fit exactly. He was furthermore convinced that the hypothesis of Copernicus ought to be accepted. Now this hypothesis, as Copernicus himself understood its first outline, merely modifies the theory of Ptolemy so far as [to] impart to all the bodies of the solar system one common motion, just what is required to annul the mean motion of the sun. It would seem, therefore, at first sight, that it ought not to affect the appearances at all. If Mill had called the work of Copernicus mere description he would not have been so *very far* from the truth as he was. But Kepler did not understand the matter quite as Copernicus did. Because the sun was so near the centre of the system, and was of vast size (even Kepler knew its diameter must be at least fifteen times that of the earth) Kepler, looking at the matter dynamically, thought it must have something to do with causing the planets to move in their orbits. This retroduction, vague as it was, cost great intellectual labor, and was most important in its bearings upon all Kepler's work. Now Kepler remarked that the lines of apsides of the orbits of Mars and of the earth are not parallel; and he utilized various observations most ingeniously to infer that they probably intersected in the sun. Consequently, it must be supposed that a general description of the motion would be simpler when referred to the sun as a fixed point of reference than when referred to any other point. Thence it followed that the proper times at which to take the observations of Mars for determining its orbit were when it appeared just opposite the sun—the true sun—instead of when it was opposite the *mean* sun, as had been the practice. Carrying out his idea, he obtained a theory of Mars which satisfied the longitudes at all the oppositions observed by Tycho and himself, thirteen in number, to perfection. But unfortunately, it did not satisfy the latitudes at all and was totally irreconcilable with observations of Mars when far from opposition.

"At each stage of his long investigation, Kepler has a theory which is approximately true, since it approximately satisfies the observations (that is, within 8′, which is less than any but Tycho's observations could decisively pronounce an error), and he proceeds to modify this theory, after the most careful and judicious reflection, in such a way as to render it more rational or closer to the observed fact. Thus, having found that the centre of the orbit bisects the eccentricity, he finds in this an indication of the falsity of the theory of the equant and substitutes, for this artificial device, the principle of the equable description of areas. Subsequently, finding that the planet moves faster

[15] Heinrich Hertz, *Principles of Mechanics,* Introduction.

[16] *A System of Logic,* Bk. III, Chap. II, § 3.

at ninety degrees from its apsides than it ought to do, the question is whether this is owing to an error in the law of areas or to a compression of the orbit. He ingeniously proves that the latter is the case.

"Thus, never modifying his theory capriciously, but always with a sound and rational motive for just the modification he makes, it follows that when he finally reaches a modification . . . which exactly satisfies the observations, it stands upon a totally different logical footing from what it would if it had been struck out at random . . . and had been found to satisfy the observation. Kepler shows his keen logical sense in detailing the whole process by which he finally arrived at the true orbit. This is the greatest piece of Retroductive reasoning ever performed."[17]

Analyze Kepler's procedure as described by Peirce, and make explicit the hypotheses he was employing.

7. In washing glasses in hot soapsuds and placing them mouth downward on a plate, it was found that bubbles appeared on the *outside* of the mouth of the glass and then went *inside*. The hypothesis was made that this happens because air leaves the glass. Show that this hypothesis explains the observed fact, if it is further assumed that:

a. The soapy water on the plate prevents the escape of air except as it is caught in bubbles.

b. The air in the glass expands because of an increase of either heat, or pressure, or both.

c. The air could not have been heated *after* the glass was taken from the hot suds.

d. Cold air enters the glass in transferring it from the suds to the plate.

e. The air in the glass contracts when cooled.

8. Hiero, the ruler of Syracuse, ordered Archimedes to discover without destroying the crown, whether a gold crown contained silver alloy. Archimedes noticed one day while taking a bath that his body seemed lighter, and it occurred to him that any body immersed in a liquid loses a weight equal to the weight of the displaced liquid.

Show that this suggestion is sufficient to solve the problem put to Archimedes.

9. Before the eighteenth century, heat was regarded as an "imponderable fluid" or caloric, which lodged in the pores of substances. According to this, when an object gets colder the caloric fluid flows out, and conversely it flows in when the object gets warmer. This theory accounted for all known facts about heat. But an alternative theory of heat was suggested, according to which heat is a form of motion. This theory also explained the known facts. At the beginning of the nineteenth century, however, Sir Humphry Davy performed an experiment which was allegedly crucial between the two theories. The experiment consisted in rubbing together two pieces of ice which were isolated from all sources of heat. The ice melted, and according to the caloric theory it must have combined with the caloric fluid to produce water. The caloric theory, however, could not explain the source of this caloric. On the other hand, the melting of the ice was easily explained on the kinetic theory of heat. Hence Davy's experiment is regarded as a crucial one.

In what sense does this claim hold?

[17] C. S. Peirce, *Collected Papers*, 1931, Vol. I, pp. 30-31.

10. Show that an important condition for hypotheses is not fulfilled by the part of Freud's theory discussed in the following:

"[Freud declares that] 'the *libido is regularly and lawfully of a masculine nature, be it in the man or in the woman; and if we consider its object, this may be either the man or the woman.*' . . . Those individuals whose sex life seeks an object he calls the anaclitic type, and this is essentially a masculine type, since it is originally the woman who tends the infant. . . . *Later on he states that where woman is anaclitic or object-loving in her makeup, in that degree is she masculine.* This is a perfect example of the unassailable position, and has its analogs in much of male estimation of woman. Woman is primarily unintelligent, many men from Plato's time have said. But if they are shown a woman who is intelligent, their answer is, well, in that respect she is masculine!" [18]

For further study see:

A. D. Ritchie, *Scientific Method,* Chaps. III, IV, VI.

N. R. Campbell, *What is Science?,* Chaps. III, IV, V.

F. C. S. Schiller, "Hypothesis," in Chas. Singer's *Studies in the History and Methods of Science,* Vol. II.

CHAPTER XII: CLASSIFICATION AND DEFINITION

1. Examine the rôle played in modern astronomy by the classification of stars into constellations.

2. Attempts have been made to define several of the ethical concepts in terms of others taken as undefined. One such attempt consists in taking "better" as undefined. The following definitions are then offered:

A is worse than B	=	B is better than A.	*Df.*
A is good	=	A is better than the nonexistence of A.	*Df.*
A is bad	=	A is worse than the nonexistence of A.	*Df.*
A is as good as B	=	A is not better than B, and B is not better than A.	*Df.*
A is ethically indifferent	=	A is not better than the nonexistence of A, and the nonexistence of A is not better than A.	*Df.*

Discuss these definitions from the point of view of (a) the psychological objective of definition, and (b) the logical objective.

3. What is the difference between natural and artificial classification?

4. Discuss the statement: "All description is classification."

5. What is the difference between a real and a nominal definition?

6. In what sense is it correct to say that the genus is part of the species, and in what sense that the species is part of the genus?

7. State the definition, a property, and an accident for each of the following: triangle, circle, star, animal, professor.

8. Point out the ambiguities in each of the following: bill, law, bolt, star, end, interest.

For further study:

J. Venn, *Empirical Logic,* Chaps. XI, XII, and XIII.

[18] Abraham Myerson, "Freud's Theory of Sex," in *Sex in Civilization,* ed. by V. F. Calverton and S. D. Schmalhausen, 1929, pp. 519, 520.

J. S. Mill, *System of Logic,* Bk. I, Chaps. VII and VIII.
P. W. Bridgman, *Logic of Modern Physics,* Chap. I.
W. E. Johnson, *Logic,* Part I, Chap. VII.

CHAPTER XIII: THE METHODS OF EXPERIMENTAL INQUIRY

1. Examine the following:

". . . To find the solution of a definite problem requires a greater effort of genius than to resolve one not specified; for in the latter case hazard, chance, may play the greater part, while in the former all is the work of the reasoning and intelligent mind. Thus, we are certain that the Dutchman, the first inventor of the telescope, was a simple spectacle-maker, who, handling by chance different forms of glasses, looked, also by chance, through two of them, one convex and the other concave, held at different distances from the eye; saw and noted the unexpected result; and thus found the instrument. On the other hand, I, on the simple information of the effect obtained, discovered the same instrument, not by chance, but by the way of pure reasoning. Here are the steps: the artifice of the instrument depends on one glass or on several. It cannot depend on one, for that must be either convex, or concave, or plain. The last form neither augments nor diminishes visible objects; the concave diminishes them, the convex increases them, but both show them blurred and indistinct. Passing then to the combination of two glasses, and knowing that glasses with plain surfaces change nothing, I concluded that the effect could not be produced by combining a plain glass with a convex or a concave one; I was thus left with the two other kinds of glasses, and after a few experiments I saw how the effect sought could be produced. Such was the march of my discovery, in which I was not assisted in any way by the knowledge that the conclusion at which I aimed was a verity." [19]

2. Examine the following:

"If Sarsi insists that I must believe, on Suidas's credit, that the Babylonians cooked eggs by swiftly whirling them in a sling, I will believe it; but I must say, that the cause of such an effect is very remote from that to which it is attributed, and to find the true cause I shall reason thus. If an effect does not follow with us which followed with others at another time, it is because, in our experiment, something is wanting which was the cause of the former success; and if only one thing is wanting to us, that one thing is the true cause. Now we have eggs, and slings, and strong men to whirl them, and yet they will not become cooked; nay, if they were hot at first they more quickly became cold; and since nothing is wanting to us but to be Babylonians, it follows that being Babylonians is the true cause why the eggs became cooked, and not the friction of the air, which is what I wish to prove." [20]

3. Examine the following investigation, pointing out the assumptions and types of arguments used:

"The *North-East wind* is known to be specially *injurious* to a great many persons. . . . What circumstance or quality is this owing to? . . . We can distinguish various qualities in winds;—the degree of violence, the temperature, the humidity or dryness, the electricity, and the ozone. We then refer to the actual instances to see if some one mode of any one of these qualities uniformly accom-

[19] Galileo, *Il Saggiatore,* quoted by J. J. Fahie, *Galileo,* 1903, pp. 80-81.
[20] *Ibid.,* pp. 187-88.

panies this particular wind. Now we find, that as regards *violence,* easterly winds are generally feeble and steady, but on particular occasions, they are stormy; hence, we cannot attribute their noxiousness to the intensity of the current. Again, while often *cold,* they are sometimes comparatively warm; and although they are more disagreeable when cold, yet they do not lose their character by being raised in temperature; so that the bad feature is not coldness. Neither is there one uniform degree of *moisture;* they are sometimes wet and sometimes dry. Again, as to *electricity,* there is no constant electric charge connected with them, either positive or negative, feeble or intense. . . . Farther, as respects *ozone,* they have undoubtedly less of this element than the South-West winds; yet an easterly wind at the sea shore has more ozone than a westerly wind in the heart of a town. It would thus appear that the depressing effect cannot be assigned to any one of these five circumstances. When, however, we investigate closely the conditions of the north-easterly current, we find that it blows from the pole towards the equator, and is for several thousand miles *close upon the surface of the ground;* whereas the south-west wind coming from the equator descends upon us from a height. Now, in the course of this long contact with the ground, a great number of impure elements—gaseous effluvia, fine dust, microscopic germs—may be caught up and may remain suspended in the lower stratum breathed by us. On this point alone, so far as we can at present discover, the agreement is constant and uniform." [21] (Bain.)

4. Discuss the issues raised in the following:

"Bodies are put in motion by . . . different agencies . . . animal strength, wind, water, steam, combustion . . . etc. Finding a body in motion, therefore, we cannot ascribe it to any special agent, merely from the fact that it is in motion: we see a wheel turning and doing work, but we may not be able to attribute its motion to one agent rather than another." [22] (Bain.)

5. How would you proceed to establish the relation between the weather and the barometer?

6. If the theory were advanced that poverty was due to failure to attend church regularly, how could this be tested? What bearing would the disproof of the theory have on the question as to the advisability of attending church regularly?

7. Examine the evidence for the conclusion drawn in the following argument:

"Last week I got into trouble through imbibing too much brandy and gin. The other day it was ale and gin. And two months ago I spent a sorry day after an evening with beer and gin. I see, accordingly, that it is the gin that must be responsible."

8. Discuss the value of the inference in the following:

"I notice that when my children are spoken to in a quiet tone of voice, they pay no attention, but that when I address them harshly they obey at once. I must therefore form the habit of always speaking to them sternly the first time."

9. What is meant by: necessary condition; sufficient condition; necessary and sufficient condition. Illustrate each.

For further study:

J. S. Mill, *System of Logic,* Bk. III, Chaps. VIII, IX, and X.

F. H. Bradley, *Principles of Logic,* Bk. II, Part II, Chaps. 1, 2, and 3.

[21] *Loc. cit.,* p. 53.

[22] *Ibid.,* p. 77.

On the uniformity of nature:

J. Venn, *Empirical Logic,* Chap. IV.
C. S. Peirce, *Chance, Love, and Logic,* Part I, Chap. 5. *Collected Papers,* Vol II, Chap. VIII, § 8 and § 9.
J. M. Keynes, *Treatise on Probability,* Chap. XXII.

CHAPTER XIV: PROBABILITY AND INDUCTION

1. Examine the following argument from analogy used by Besian Array, doctor of the Sorbonne, in 1671:

"Theology teaches that the sun has been created in order to illuminate the earth. But one moves the torch in order to illuminate the house, and not the house in order to be illuminated by the torch. Hence it is the sun which revolves around the earth, and not the earth which revolves around the sun."

2. Examine the merits of the following argument:

"Look round the world [said Cleanthes], contemplate the whole and every part of it: you will find it to be nothing but one great machine, subdivided into an infinite number of lesser machines, which again admit of subdivisions to a degree beyond what human senses and faculties can trace and explain. All these various machines . . are adjusted to each other with an accuracy, which ravishes into admiration all men who have ever contemplated them. The curious adapting of means to ends, throughout all nature, resembles exactly, though it much exceeds, the productions of human contrivance; of human design, thought, wisdom, and intelligence. Since therefore the effects resemble each other, we are led to infer, by all the rules of analogy, that the causes also resemble; and that the Author of Nature is somewhat similar to the mind of man; though possessed of much larger faculties, proportioned to the grandeur of the work which he has executed. By this argument . . . alone, do we prove at once the existence of a Deity, and his similarity to human mind and intelligence. . . .

"If we see a house, Cleanthes [said Philo], we conclude, with the greatest certainty, that it had an architect or builder; because this is precisely that species of effect which we have experienced to proceed from that species of cause. But surely you will not affirm, that the universe bears such a resemblance to a house, that we can with the same certainty infer a similar cause, or that the analogy is here entire and perfect. The dissimilitude is so striking that the utmost you can here pretend to is a guess, a conjecture, a presumption concerning a similar cause." [23]

3. Why is a single instance sometimes sufficient to establish a universal conclusion, while in other cases the greatest possible number of instances which verify a theory without exception are not sufficient?

4. What is perfect induction?

5. Examine the value of the following argument: A man forgot whether, in the ritualistic churches, a bell is tinkled at the elevation of the Host or not. But knowing that the services resembled somewhat decidedly those of the Roman Mass, he concluded that it is not unlikely that the bell is used in the ritualistic, as in the Roman, churches.

For further study:

W. E. Johnson, *Logic,* Part II, Chaps. VIII, IX, X, XI; Part III, Chaps. II, IV.

[23] Hume, *Dialogues concerning Natural Religion,* Pt II.

C. A. Mace, *Principles of Logic*, Chaps. XII, XIII, XIV, XV, XVI, XVII, XVIII.
J. Royce, "Principles of Logic" in *Encyclopaedia of Philosophical Sciences*, Vol. I.
H. W. B. Joseph, *Introduction to Logic*, 2d ed., Chaps. XVIII, XIX.

CHAPTER XV: MEASUREMENT

1. Examine the following for the assumptions it makes concerning the measurement of values:

"To take an exact account . . . of the general tendency of any act, by which the interests of a community are affected, proceed as follows. Begin with any one person of those whose interests seem most immediately to be affected by it: and take an account,

"1. Of the value of each distinguishable *pleasure* which appears to be produced by it in the *first* instance.

"2. Of the value of each *pain* which appears to be produced by it in the *first* instance.

"3. Of the value of each pleasure which appears to be produced by it *after* the first. This constitutes the *fecundity* of the first *pleasure* and the *impurity* of the first *pain.*

"4. Of the value of each *pain* which appears to be produced by it after the first. This constitutes the *fecundity* of the first *pain,* and the *impurity* of the first pleasure.

"5. Sum up all the values of all the *pleasures* on the one side, and those of all the pains on the other. The balance, if it be on the side of pleasure, will give the *good* tendency of the act upon the whole, with respect to the interests of that *individual* person; if on the side of pain, the *bad* tendency of it upon the whole.

"6. Take an account of the *number* of persons whose interests appear to be concerned; and repeat the above process with respect to each. *Sum up* the numbers expressive of the degrees of *good* tendency, which the act has, with respect to each individual, in regard to whom the tendency of it is *good* upon the whole: do this again with respect to each individual, in regard to whom the tendency of it is *bad* upon the whole. Take the *balance;* which, if on the side of *pleasure,* will give the general *good tendency* of the act, with respect to the total number or community of individuals concerned; if on the side of pain, the general *evil tendency,* with respect to the same community." [24]

2. If pressure, temperature, and volume all vary in "ideal" gases, the following relation holds:

$$\frac{P_1 V_1}{P_2 V_2} = \frac{T_1}{T_2}$$

Discuss the types of measurements required to establish this law.

3. Discuss the assumptions and types of measurement required to measure the thickness of gold leaves, as described in the following:

"Gold is reduced by the gold-beater to leaves so thin, that the most powerful microscope would not detect any measurable thickness. If we laid several hun-

[24] Jeremy Bentham, *An Introduction to the Principles of Morals and Legislation,* Chap. IV.

dred leaves upon each other to multiply the thickness, we should still have no more than $\frac{1}{100}$th of an inch at the most to measure, and the errors arising in the superposition and measurement would be considerable. But we can readily obtain an exact result through the connected amount of weight. Faraday weighed 2,000 leaves of gold, each $3\frac{3}{8}$ inches square and found them equal to 384 grains. From the known density of gold it was easy to calculate that the average thickness of the leaves was $\frac{1}{282000}$ of an inch." [25]

4. It is stated in most books on astronomy that the earth in rotating on its axis slows down at the rate of 22 seconds per century.

a. How is this retardation measured?

b. How is it possible to maintain that the earth is slowing down if the period of the earth's rotation is taken as the standard unit of time? (Read Jevons, *Principles of Science,* 2d ed., Chap. XIV, section on "Standard Unit of Time.")

5. The standard length in the metric system is the *meter.* It is the distance between two lines ruled on a bar of platinum-iridium, which is kept in the Internal Bureau of Weights and Measures at Paris. What meaning, if any, do you assign to the expression, "The length of the standard meter has changed"?

6. a. Is probability a magnitude obtainable by *fundamental* or by *derived* measurement?

b. Is probability an *extensive* magnitude?

7. a. What is meant by the statement, "All measurements of lengths can be only *approximately* true"?

b. Consequently, what is meant by an "error" in measurement and by methods of "eliminating" such "errors"?

8. Which of the following are extensive and which intensive qualities? State roughly the procedure, if any, required to measure each.

number
mass
length
period of time
angle
area
volume
force
energy
brightness

temperature
heat
pressure
velocity
acceleration
color
shape
electrical resistance
hardness
kindness

intelligence
humidity
beauty
pleasure
humor
durability
noisiness
piety
comfort
vivacity

scholarship
desire
poverty
patience
fragrance
cleanliness
permeability
viscosity
smoothness

For further study:

N. R. Campbell, *What is Science?* Chaps. VI and VII.

N. R. Campbell, *Measurement and Calculation.*

W. S. Jevons, *Principles of Science,* Chaps. 13, 14, and 15.

W. E. Johnson, *Logic,* Part II, Chap. VII.

CHAPTER XVI: STATISTICAL METHODS

1. Examination of death rates for urban and rural areas in the United States during 1901-11 shows that the death rate for urban parts was higher than for rural parts.

[25] Jevons, *Principles of Science,* 2d ed., p. 296.

a. Does this mean that cities were less healthful places in which to live?

b. Does it mean there were larger proportions of infants or of old persons in the population of cities than in that of rural places?

2. Comparison of average age at time of death for any occupation with average age of death for all occupations discloses whether persons in that occupation are dying unusually young or unusually old. Does it follow that the occupation is the cause of the shortening or lengthening of lives of persons engaged in it?

3. An administrative officer of a hospital finds that 1 out of 10 typhoid cases in his care dies of the disease. Is it safe to infer that the fatality of typhoid is 10 per cent?

4. A physician concluded that because he had inoculated two children with the nasal washings and blood from another child sick with measles, and because the two children did not come down with measles, measles were not contained in the washings and blood.

Examine this argument.

5. Discuss the issues involved in the following:

In 1888, a doctor claimed that summer diarrheas of infancy are due to poisonous milk. An opponent claimed that, on the other hand, the high infant mortality was due to the growing use of the baby perambulator, since the death rate among children had increased since the baby cab had come into fashion. The first doctor replied that he withdrew his claim, but thought he could claim, with a right equal to that of his opponent, that the high infant mortality was due to the growing use of umbrellas.

6. The following is the table of death rates from tuberculosis in Richmond, Virginia, and in New York City in 1910:

	Population		*Deaths*		*Death Rate per 100,000*	
	New York	*Richmond*	*New York*	*Richmond*	*New York*	*Richmond*
White	4,675,174	80,895	8,365	131	179	162
Colored	91,709	46,733	513	155	560	332
Total	4,766,883	127,628	8,881	286	187	226

a. Does it follow that tuberculosis caused a greater mortality in Richmond than in New York?

b. Notice that the death rate for whites and that for Negroes were *lower* in Richmond than in New York, although the *total* death rate was *higher*. Are the two populations compared really *comparable*, that is, homogeneous?

7. A leading British statistician reported that the curve of seasonal distribution of typhoid was similar to that of the temperature of the water supply. He concluded that the warmer the drinking water, the more rapidly bacteria multiplied, and hence the larger the number of typhoid cases.

Examine this conclusion in the light of the following: It is a fact that, up to a certain point, the higher the temperature, the more rapidly do bacteria grow.

But it is also a fact that in drinking-water bacteria do not grow, for lack of food, and that they die *faster* in it, the warmer the water.

8. The *registered* fatalities from cancer were 2½ times as great among males and ¾ times as great among females in 1896 as those registered in 1867-70. Does this prove that fatalities from cancer have increased?

9. Illegitimacy is sometimes estimated as the ratio of the number of illegitimate births to total births, and sometimes as the ratio of the number of illegitimate births to the number of unmarried women. Which do you think is a more reliable measure? (Compare the two measures in the case when the marriage rate is low and the number of illegitimate births is low.)

10. The influence of birth rate on death rate may be seen from the following account:

". . . If, owing to a high birth-rate, there is a larger proportion of children in one community than in another, and the relative hygienic conditions of the two are equal, there will be more deaths of children in the former; and inasmuch as the rate of mortality among young children is higher than that of all others except the aged, the general death-rate will be raised [in that community]. But if the high birth-rate be *continued*, there will not only be a large proportion of children, but of others between 10 and 40 years of age, at which ages a low rate of mortality holds; and this factor counterbalances the other, and causes a continued high birth-rate to be associated with a low death-rate. Speaking generally, the mortality of a population in which there is an excess of births over deaths should be lower than that of a stationary population . . . because in the latter case there is a larger proportion of old people than in the former."[26] Construct a set of figures to illustrate the points made.

11. In the British Army in 1860-64, 32,324 examinations of recruits were made, of which 371.67 per thousand were rejected. During 1882-86, 132,563 were examined, of whom 415.58 per thousand were rejected. May we conclude that the masses from whom the army recruits are taken were inferior in quality in the later years to those of the earlier ones?

12. The percentage of the population in receipt of poor-relief during a certain year in the different registration districts in England is given by the following table:

[26] Arthur Newsholme, *Elements of Vital Statistics,* 3d ed., 1899, p. 96.

Percentage of Population Receiving Relief	*Number of Districts with Given Percentage in Receipt of Relief*
0.75 – 1.25	18
1.25 – 1.75	48
1.75 – 2.25	72
2.25 – 2.75	89
2.75 – 3.25	100
3.25 – 3.75	90
3.75 – 4.25	75
4.25 – 4.75	60
4.75 – 5.25	40
5.25 – 5.75	21
5.75 – 6.25	11
6.25 – 6.75	5
6.75 – 7.25	1
7.25 – 7.75	1
7.75 – 8.25	0
8.25 – 8.75	1
	632

a. Calculate the arithmetic mean, the mode, and the median.

b. Calculate the range, the quartile deviation, the mean deviation, and the standard deviation.

13. Find the coefficient of correlation for the following values of X and Y.

X	*Y*
1	3
2	5
3	2
4	8
5	7
6	10
7	11
8	6
9	9
10	12

14. Examine critically the following investigation reported in a New York newspaper:

Students living in New York are more radical than those from other parts of the country, according to a survey conducted by Dr. Clara Eliot of the Social Science Department at Barnard College. The full report of her work, which appeared in the *Barnard Bulletin* yesterday, shows that the students at Barnard are more radical than their professors and that the freshman class is the most conservative group in college.

Tests were given to 341 girls, including students in history, government, elementary psychology, statistics, and sociology, to get a cross-section of the entire school. Taking the tests were 86 freshmen, 111 sophomores, 81 juniors and 63 seniors. It had been originally designed by Manly H. Harper as a social study among 3,000 teachers. Seventy-one questions were asked, to which a negative or affirmative answer would indicate a radical or a conservative viewpoint.

The survey revealed that students majoring in mathematics and the natural science departments were the most conservative, while those majoring in the humanities were less so. Students of the social science departments were found to be the most radical. Students from private schools had a higher medium for conservatism than those who had attended public schools, but this difference was only noticeable during the freshman and sophomore years, the comparison by types of schools showing very little difference thereafter.

Classification by fathers' occupations showed that daughters of professional men were more liberal than daughters of men engaged in scientific, technical, or commercial occupations, but the influence of the parent's occupation had entirely disappeared by the time the student had reached the senior class.

The liberal students were found to be more consistent in their work and in their replies to the various questions, according to the bulletin report, the greatest number of inconsistencies being found among the most conservative, as shown by the Harper tests.

"This may serve to indicate that students of a radical tendency have 'thought things through,' whereas the others have been content to accept opinion," says the report. "The most pronounced variation in opinion is to be found in the classification by years. Freshmen are decidedly more conservative than sophomores and the remainder of the college. A year of collegiate work seems to upset the precise beliefs with which new students come. This truth has been found not only in this social survey but through the observations of those who have worked with first-year classes."

15. The death-rate in a certain town being unusually low, its citizens advertise it as being unusually healthful. Is this claim well founded? What would your answer be if you knew that it was a college town?

16. Suppose that life insurance companies all report that their incomes have for the last twenty years exceeded their expenses. Does this prove that their business is in sound condition? Does the assumption that the total insurable population will keep on growing indefinitely influence your answer?

17. Suppose that Montana has a lower birth rate (that is, the ratio of the number of children born to the number of the total population) than Massachusetts. Does this mean that the character of the climate of Montana or the greater fertility of the women of Massachusetts is the cause of this difference? If not, what cause can you suggest?

18. Suppose that the number of births in the United States exceeds the number of deaths by 900,000 every year, and this holds true for the next fifty years. Will the prospects of increasing population be greater at the end of fifty years than now? (See *The Nation*, New York, November 4, 1931.)

19. An airplane-travel company advertises that travelling by plane is not dangerous, since the number of fatalities and accidents per passenger-mile (the number of passengers times the distance that each travels) is less for travel by plane than for travel by railroad. Assuming the facts as stated, is this a good index of safety? Would you consider taking the number of persons using each type of service as a basis of comparison?

20. It has been argued that immigrants to this country do not engage in the so-called productive activities (that is, the production of raw materials, such as agricultural products), since only 2 per cent of those engaged in productive activities are immigrants.

Assuming the facts to be as stated, examine the cogency of the argument.

For further study:

N. S. Jevons, *Principles of Science,* Chaps. 16 and 17.
J. M. Keynes, *Treatise on Probability,* Part V.
G. U. Yule, *Introduction to Statistics.*
T. Merz, *History of European Thought in the Nineteenth Century,* Vol. II, Chap. 12.

CHAPTER XVII: PROBABLE INFERENCE IN HISTORY AND ALLIED INQUIRIES

1. Read the History of Susanna, in the Apocrypha. Examine the logic of Daniel's procedure, and state the argument to exhibit formally its probable nature.

2. "If the French alphabet is treated like the Hebrew system of enumeration, by which the first ten letters represent the units, and the next the tens, and so on, the letters have the following value:

a	b	c	d	e	f	g	h	i	k	l	m	n	o	p	q	r	s	t	u	v	w	x	y	z
1	2	3	4	5	6	7	8	9	10	20	30	40	50	60	70	80	90	100	110	120	130	140	150	160

Turning the words *l'Empereur Napoléon* into ciphers on this system, it happens that the sum of these numbers equals 666, and Napoleon is thereby seen to be the beast prophesied in the Apocalypse ["Here is wisdom; let him that hath understanding, count the number of the beast; for it is the number of a man, and his number is six hundred three score and six. And power was given unto him to continue forty and two months."] Moreover, working out the same way the words *quarante-deux,* that is the term for which the beast was permitted to continue, the sum of these numbers again equals 666, from which it is deduced that the term of Napoleon's power had come in 1812, when the French Emperor reached his forty-second year." [27]

What logical difficulties do you find in this attempt to interpret a text?

3. a. What do the theory of special creation and the Darwinian theory of evolution aim to explain?

b. Does the Darwinian theory offer a causal account of the origin of species?

c. Is a description of the historical sequence of forms of life an "explanation" of such forms? Your answer should indicate in what sense you understand "explanation."

4. Solve the following cryptograms:

a. BO VOFYBNJOFE MJGF JT OPU XPSUI MJWJOH
b. TVG GSVV GL Z MFMMVIB
c. VHYHG SOXV ILYH HTXDOV WZHOYH

5. Read the following cases cited in J. H. Wigmore, *The Nature of Judicial Proof,* 1931:

[27] Tolstoy, *War and Peace,* Vol. III, Pt. I, Chap. XIX.

a. No. 17, p. 72.
b. No. 72, p. 164.

Analyze each, state what is to be proved and the evidence advanced. State the argument formally.

For further study, see:

J. G. Droysen, *Principles of History* (1897).
H. Sidgwick, *Philosophy and Its Problems,* Chaps. 6, 7, 8, 9.
Allen Johnson, *History and the Historian.*
J. M. Vincent, *Historical Research.*

CHAPTER XVIII: LOGIC AND CRITICAL EVALUATION

1. It has been said that history is philosophy teaching by example. Can all kinds of history or historians do this? Or must certain conditions be met before this is possible?

2. What rôle would you assign to history in constructing a theory of economic value?

3. In what sense are those nations or animal species which survive in a given struggle the fittest? In this connection read Huxley's essay *On Evolution* and *Ethics.*

4. It has been urged that the highest music is pure music free from subordination to other arts. Do you think that the music of a song suffers because of its association with words?

5. Do you think that a program describing the various movements of a symphony helps or hinders the enjoyment of the music itself?

6. In what sense can a poem be translated? In what sense is it true that a poem cannot be translated?

7. Do you think it is necessary to know the biography of Sophocles or Shakespeare to understand the *Antigone* or *Hamlet?*

8. What distinction, if any, would you make between legal and moral imperatives? What distinction, if any, between courtesy and kindness?

9. How would you determine the question as to whether our duties to our country may lead to conduct generally regarded as unpatriotic?

10. Does the question whether the end justifies the means assume the doctrine of plurality of causes? If the means are necessary to the end, could we not say rather that an end is bad if it involves bad means?

11. In what sense, if any, is the unattainable to be condemned as a goal of conduct? Would you regard holiness as attainable?

12. A French legal philosopher has argued that legal science must study only *what is,* and not *what should be.* He has also contended that all laws grow out of the principle of solidarity, and that those legislative enactments which contravene that principle should be declared void. Assuming that the French courts do not follow this conclusion, what merits, if any, has his argument?

13. Does a legal rule which is generally flouted in a given community cease to be a law? Should a good citizen continue to obey it? Should a court refuse to enforce it?

14. In what sense was the eighteenth amendment to the United States Consti-

tution an expression of the will of the people? Were the Minimum Wage Laws, passed by Congress and some of the States but declared unconstitutional by the courts, the expression of the will of the people?

15. What is the difference between a hyperbole and a lie. Give an example of a synecdoche? Have you ever been mislead by one?

16. When Socrates said that he did not know anything, was he guilty of a lie?

17. On what grounds could you prove to a friend that he ought to go to a piano recital rather than go to a prize-fight?

18. "A criterion of taste is nothing but taste itself in its more deliberate and circumspect form. Reflection refines particular sentiments by bringing them into sympathy with all rational life." Explain this quotation from Santayana.

19. Aristotle declared that "we deliberate not about ends but about means to end." Discuss this statement, showing in what sense it could be true and in what sense not.

20. To what extent does the following quotation from the *Oedipus at Colonos* of Sophocles represent an arguable position:

> Far best were ne'er to be;
> But, having seen the day,
> Next best by far for each to flee
> As swiftly as each may,
> Yonder from whence he came.

21. In his famous "Ten O'Clock" lecture, Whistler declared that only a painter was competent to judge a picture. Examine the implications of this statement. To what extent do you think it is true?

22. The following problem is sometimes discussed in classes in ethics: "A man returning from his day's work was crossing a railroad track near his home when he discovered a switch left open by a careless switchman. This he saw at once would mean death or injury to the several hundred people on a rapidly approaching train. At the same moment he saw his own child playing upon the track in front of the engine. He had time only to turn the switch and save the train, or else save the child. What was it his duty to do?"

Examine how questions of evidence may enter in the discussion of this problem.

23. What would you think of a proposal to establish the relative merits of two poems by submitting questionnaires to several hundred people, and regarding that poem as the more excellent which receives the larger number of votes?

24. What significance do you attach to the following everyday expressions:

a. How do you do?
b. Good evening!
c. It is a fine day!
d. Pleased to meet you.
e. You must call me up sometime.

25. Examine the following passage from Bradley's *Principles of Logic* for the use of metaphors:

"The notion that existence could be the same as understanding strikes as cold and ghost-like as the dreariest materialism. That the glory of this world in the end is appearance leaves the world more glorious, if we feel it is a show of some fuller splendor; but the sensuous curtain is a deception and a cheat, if it hides some colourless movement of atoms, some spectral woof of impalpable abstractions, or unearthly ballet of bloodless categories."

26. What difference is there between a fiction and a hypothesis?

For further study, see:

M. R. Cohen, *Reason and Nature,* Bk. III, Chaps. II, IV.

M. R. Cohen, "Impressionism and Authority" in *New Republic,* 1921, Vol. 28, p. 252.

M. R. Cohen, *Law and the Social Order,* pp. 229-247.

G. Santayana, *Life and Reason,* Vol. IV, *Reason in Art,* Chaps. I, II, IX, X, and XI.

F. S. Cohen, *Ethical Systems and Legal Ideals,* Chap. III.

J. Dewey, *Experience and Nature,* Chaps. IX and X, also *The Philosophy of John Dewey;* ed. by J. Ratner, Chap. XVII.

G. E. Moore, *Principia Ethica,* Chap. I.

H. Vaihinger, *Philosophy of As If,* Part I, Section A.

CHAPTER XIX: FALLACIES

Examine the following arguments for their validity:

1. The more I struggle to improve this book, the less does it satisfy me. It would be better, therefore, if I erase all my revisions.

2. Murderers should be put to death; therefore, it is vicious to pardon anyone convicted of murder.

3. The inhabitants of the town consist of men, women, and children of all ages; those who voted at the town meeting were inhabitants of the town; therefore, the voters consisted of men, women, and children of all ages.

4. A spoonful of this medicine cured a light cold that I had last month. Half a cupful, therefore, ought to rid me of this severe one.

5. The end of the thing is its perfection; death is the end of life; hence death is the perfection of life.

6. Mr. A asked Mr. B if he knew that all meat is nourishing. The latter replied that this was a fact of which he was most certain. Mr. A then asked if Mr. B also knew whether what he had in a covered dish was nourishing; to this Mr. B. replied that he did not. Uncovering the dish, Mr. A displayed some roast meat, and charged that Mr. B had contradicted himself.

7. The use of brandy does people much harm. It is therefore a mistake to use it to revive a man who has just escaped drowning.

8. When we buy abroad, the domestic consumer will obtain the goods beyond doubt, but the foreign producer obtains the money. On the other hand, when we sell abroad the producer at home, while he gains the money, loses his goods. It will be better then to buy and sell at home, for in that case we retain both the goods and the money. Hence, a high protective tariff should be enacted.

9. The present restrictions on sex relations supported by custom, grew up in different social conditions, when people did not know how to control the results of such relations. These restrictions, therefore, have pertinence at the present time.

10. "A servant who was roasting a stork for his master was prevailed by his sweetheart to cut off a leg for her to eat. When the bird came upon the table, the master desired to know what had become of the other leg. The man an-

swered that storks had never more than one leg. The master, very angry, but determined to strike his servant dumb before he punished him, took him next day into the fields where they saw storks, standing each on one leg, as storks do. The servant turned triumphantly to his master, on which the latter shouted and the birds put down their other legs and flew away. 'Ah, sir,' said the servant, 'you did not shout to the stork at dinner yesterday; if you had done so, he would have shown his other leg too.' " (De Morgan)

11. This booklet tells of many people who have been cured of various diseases by taking A's patent medicine, and also of the unfortunate people who have died through failure to take it. Hence, it is just the medicine I need to cure me.

12. In an oration on Renan, the French philosopher Boutroux declared: "The best men in the nation are those it crucifies. Hence martyrdom is the ransom of superiority."

13. Who thrusts a knife into another person should be punished; a surgeon in operating does so; hence he should be punished.

14. Charity is always a good; therefore giving alms to beggars is always a good.

15. Milk is a wholesome food; therefore it may be taken in combination with all acid foods.

16. Explain the equivocation which may arise in the following:

a. He went to Washington and then to Chicago by the express train.

b. Did you make a long speech at the meeting?

17. What do you think of the argument that because the explanation of some phenomena in terms of mechanism breaks down, therefore vitalism or spiritualism is true.

18. If it were true that no individual can relieve his economic distress by increased expenditure, would it therefore be true that no nation can do so?

19. Under what conditions are propositions which are true of every citizen of a nation, also true of the nation, and under what conditions are they not?

20. It has been urged that Latin should be taught in the secondary schools because of its value in illuminating the etymology of many English words. Assuming the value established, is the value sufficient to justify making Latin compulsory? If not, what would be sufficient evidence?

21. In the course of a debate A asserts that some of his opponent's facts are not true and challenges him to prove them. The opponent (B) replies that facts are facts and cannot be false and that since A himself has called them facts, it is absurd to challenge their truth. Comment on the logical force of B's argument.

22. What do you think of the motto *Fiat justitia, pereat mundus* (Let justice prevail, even though the heavens fall)?

23. What fallacy is illustrated by those who argue that since the rules of courtesy make life pleasant, they should under no condition be violated?

24. Evidence was produced that a certain statesman, although he denied that he had done so, had read a certain secret treaty. This evidence was met with a rebuttal: "How could a man of his character have told an untruth?" Comment on the rebuttal.

25. In Bradley's *Logic* there occurs the statement: "If reasoning is from an axiom, how did people reason before axioms were invented?" Assuming that by *inventing* Mr. Bradley means *discovering,* what do you think of his implied rejection of the need for axioms?

26. In his *Formal Logic* Mr. Schiller declares: "A formally valid thought may

be actually false, and a formally invalid thought may be actually true. Hence, the presence of a formal fallacy is no disproof of the real worth of an argument." Assuming that by a formally valid thought Mr. Schiller means a validly drawn conclusion, comment on the character of his argument.

27. Examine the following argument used by Herbert Spencer to prove that great men do not make history but are the products of society: "Whence comes the great man? The question has two conceivable answers: his origin is supernatural, or it is natural. Is his origin supernatural? Then he is a deputy god, and we have theocracy once removed. Is this an unacceptable solution? Then the origin of the great man is natural; and immediately this is recognized, he must be classed with all other phenomena in the society that gave him birth as a product of its antecedents. Along with the whole generation of which he forms a minute part, along with its institutions, language, knowledge, manners, and its multitudinous arts and appliances, he is a *resultant*. Before he can remake his society, his society must make him. All those changes of which he is the proximate initiator have their chief causes in the generations he descended from."

28. Berkeley's argument for the proposition *To be is to be perceived* is in part as follows: "Can there be a nicer strain of abstraction than to distinguish the *existence* of sensible objects from their *being perceived,* so as to conceive them existing unperceived? Light and colors, heat and cold, extension and figures—in a word, the things we see and feel—what are they but so many sensations, notions, ideas, or impressions on the sense? And is it possible to separate, even in thought, any of these from perception? I will not deny, I can abstract—if that may properly be called *abstraction* which extends only to the conceiving separately such objects as it is possible may really exist or be actually perceived asunder. But my conceiving or imagining power does not extend beyond the possibility of real existence or perception. Hence, as it is impossible for me to see or feel anything without an actual sensation of that thing, so is it impossible for me to conceive in my thoughts any sensible thing or object distinct from the sensation or perception of it."

Comment on this argument.

29. In his book *Historical Materialism* Bukharin argues (p. 167) "that the class structure of society impressed its class stamp on mathematics" on the ground that mathematical study has historically been closely connected with religion, surveying, commerce, and architecture. Comment on this argument.

30. The following argument has been used to maintain that individuals play no rôle in determining social change:

"The differences between one nation and another ultimately depend simply and solely upon the physical circumstances to which they are exposed. If the people who went to Hamburg had gone to Timbuctoo, they would now be indistinguishable from the semi-barbarian tribes which inhabit Central Africa; and if the people who went to Timbuctoo had gone to Hamburg, they would now have been white-skinned merchants driving a roaring trade in imitation sherry and indigestible port. The differentiating agency must be sought in the great permanent geographical features of land and sea; these have moulded the characters and histories of every nation upon the earth. To suppose otherwise is to suppose that the mind of man is exempt from the universal law of causation. There is no caprice, no spontaneous impulse, in human endeavors. Even tastes and inclinations *must* themselves be the result of surrounding causes."

What abuses of scientific method, if any, does this argument contain?

For further study:

A. Sidgwick, *Fallacies.*
J. Bentham, *Book of Fallacies,* Bowring ed. of Collected Works, Vol II.
Sydney Smith, *Works,* Vol. 2, pp. 387-415.
A. De Morgan, *Formal Logic,* Chap. XIII.
H. W. B. Joseph, *An Introduction to Logic,* Chap. XXVII.

CHAPTER XX: CONCLUSION

1. a. What is meant by "explanation" in the sciences?

b. What happens when boiling water is poured into two glasses one of which is very thin and the other thick? Examine the explanation of this phenomenon as stated in most elementary texts in physics.

2. "There are two tables before me as I sit down to write. One of them has been familiar to me from earliest years. It is a commonplace object of that environment which I call the world. It has extension; it is comparatively permanent; it is coloured; above all it is *substantial,* it is a *thing:* not like space nor like time. Table No. 2 is my scientific table. It does not belong to that world which spontaneously appears around me when I open my eyes. It is part of a world which in more devious ways has forced itself on my attention. My scientific table is mostly emptiness. Sparsely scattered in that emptiness are numerous electric charges rushing about with great speed; but their combined bulk amounts to less than a billionth of the bulk of the table itself. There is nothing *substantial* about my second table. It is nearly all empty space—space pervaded, it is true, by fields of force, but these are assigned to the category of "influences," not of "things." It makes all the difference in the world whether the paper before me is poised as it were on a swarm of flies and sustained in shuttlecock fashion by a series of tiny blows from the swarm underneath, or whether it is supported because there is substance below it, it being the intrinsic nature of substance to occupy space to the exclusion of other substance; all the difference in conception at least, but no difference to my practical task of writing on the paper. I need not tell you that modern physics has by delicate test and remorseless logic assured me that my second scientific table is the only one which is really there—wherever 'there' may be." [28]

a. Examine the distinction between the two tables Eddington describes.

b. It is sometimes said that scientific explanation consists in "reducing" the first table to a table of the second kind. Just what does this "reduction" amount to?

3. Socrates was dissatisfied with the kind of explanations of phenomena he found among the scientists (in particular, Anaxagoras) and he stated his discontent as follows, while he was awaiting the execution of the death-sentence imposed by the Athenians:

". . . I found my philosopher altogether forsaking mind or any other principle of order, and having recourse to air, and ether, and water, and other eccentricities. I might compare him to a person who began by maintaining generally that mind is the cause of the actions of Socrates, but who, when he

[28] Abridged and slightly altered from the Introduction to Sir A. S. Eddington's *The Nature of the Physical World,* 1929, pp. xi-xiv.

endeavoured to explain the causes of my several actions in detail, went on to show that I sit here because my body is made up of bones and muscles; and the bones, as he would say, are hard and have joints which divide them, and the muscles are elastic, and they cover the bones, which have also a covering or environment of flesh and skin which contains them; and as the bones are lifted at their joints by the contraction or relaxation of the muscles, I am able to bend my limbs, and this is why I am sitting here in a curved posture—that is what he would say; and he would have a similar explanation of my talking to you, which he would attribute to sound, and air, and hearing, and he would assign ten thousand other causes of the same sort, forgetting to mention the true cause, which is, that the Athenians have thought fit to condemn me, and accordingly I have thought it better and more right to remain here and undergo my sentence. . . . There is surely a strange confusion of causes and conditions in all this. It may be said, indeed, that without bones and muscles and the other parts of the body I cannot execute my purposes. But to say that I do as I do because of them, and that this is the way in which mind acts, and not from the choice of the best, is a very careless and idle mode of speaking." [29]

a. State clearly the point of Socrates' criticism.

b. What light does this criticism throw on the possibility of different kinds of explanations?

c. In what sense, if any, is one kind of "explanation" more "fundamental" or "truer" than another?

4. Examine critically the arguments advanced for the view that explanation of phenomena in terms of matter in motion is the only possible or genuine one.

For further study:

G. Santayana, *Life of Reason*, Vol. V, *Reason in Science*, Chaps. I, III, IX, X, and XI.

L. T. Hobhouse, *Theory of Knowledge*, Part II, Chap. XIX; Part III, Chap. VIII.

T. Merz, *History of European Thought in the 19th Century*, Vol. I, Chaps. I, II, III.

A. Aliotta, *Idealistic Reaction Against Science*, Part I.

A. D. Ritchie, *Scientific Method*, Chap. VII.

[29] Plato, *Phaedo*, in *op. cit.*, Vol. II, p. 244.

INDEX

D09307740